Anatomical Adaptations of Halophytes

Marius-Nicuşor Grigore • Constantin Toma

Anatomical Adaptations of Halophytes

A Review of Classic Literature and Recent Findings

Springer

Marius-Nicuşor Grigore
Faculty of Biology
Alexandru Ioan Cuza University
Iasi, Romania

Constantin Toma
Faculty of Biology
Plant Anatomy & Ecology Laboratory
Alexandru Ioan Cuza University
Iasi, Romania

ISBN 978-3-319-66479-8 ISBN 978-3-319-66480-4 (eBook)
DOI 10.1007/978-3-319-66480-4

Library of Congress Control Number: 2017951218

Printed on acid-free paper

This Springer imprint is published by Springer Nature
The registered company is Springer International Publishing AG
The registered company address is: Gewerbestrasse 11, 6330 Cham, Switzerland

Foreword

When my colleague and good friend, Marius-Nicuşor Grigore, asked me to write the Foreword to this book, "Anatomical adaptations of halophytes," I was flattered… but I thought I was not the most appropriate person to do so. I have worked for over 30 years on various aspects of plant development and plant responses to stress and the last 15 more specifically on salt tolerance mechanisms in various plant species, including several halophytes. However, I am a plant biochemist and molecular biologist and I do not consider myself an expert in plant anatomy … well, let's be honest: my knowledge of plant anatomy is practically zero. So, I felt I had nothing to say to the potential readers of the book, who I assumed should be mostly botanists and plant anatomists. While looking for a polite way to decline the invitation, I glanced at these pages that are now in your hands, dear reader—or, nowadays, more probably on your computer screen—and I completely changed my mind. The book fascinated me, and I realized that it could be just as interesting for any scientist working with this amazing group of salt-tolerant plants, regardless of their particular field of research. What follows are, therefore, some brief comments from a nonspecialist, addressed to those researchers with various backgrounds and expertise, about a much specialized—and very special—book, which I believe should be present on the shelves (or hard disks) of any laboratory involved in the study of halophytes.

I am not going in detail into the contents of the book, its scope, or the authors' reasons to write it; I will not even mention the (multiple) definitions of "halophytes," apart from the fact that they are plants able to survive and reproduce under salinity conditions that will kill the vast majority of plant species. The readers can look themselves at the preface, the table of contents, or the introduction of the book. I would like, instead, to frame the study of halophytes in a wider context: as the most appropriate models to investigate the mechanisms of salt tolerance in plants—which should be obvious, despite the opinion of some hardcore plant molecular biologists (no, this book *does not* deal with *Arabidopsis thaliana*). The study of these mechanisms is at present a very active field in plant biology research, because of its unquestionable academic interest but also due to its practical implications in agriculture. Soil salinity is—together with drought—the most important environmental stress factor responsible for the reduction of crop yields worldwide. In the context of global climate change, increasing scarcity of good-quality water for irrigation, and population growth, the biotechnological improvement of the salt tolerance of our crops has become an urgent need for the future of agriculture and food production, especially in arid and semiarid regions; this, in turn, requires a deep knowledge of the mechanisms used by plants to counteract the deleterious effects of high soil salinity. Halophytes can not only contribute basic knowledge but also provide biotechnological tools—salt tolerance genes and salt-induced promoters—which could be eventually used for the genetic improvement of salt tolerance of our crops. Some of them could even be domesticated to generate new crops for food, feed, fiber, biofuel production, or other industrial uses. They would be the basis of a "saline agriculture," could be grown in salinised soils and irrigated with brackish or sea water, avoiding in this way competition with our conventional crops for scarce resources: fertile farmland and freshwater for irrigation.

Coming back to the book itself, there are several aspects that, in my opinion, make it special and should be highlighted. First, the book should appeal to a wide readership; due to their interest in basic research and possible practical applications—as mentioned above—halophytes are being studied from many different points of view by researchers with quite different scientific backgrounds: ecologists, botanists, agronomists, plant physiologists, biochemists, and molecular biologists, many of them (especially the latter) with little knowledge, if any, of plant anatomy. The information contained in this book is organized basically around specific anatomic adaptations, which represent structural strategies of halophytes to withstand high soil salinity (succulence, salt glands, Kranz anatomy, etc.), and not based on taxonomic classifications, as one could expect for a book of this kind. I find this approach very convenient and accessible, particularly for nonspecialists. Those scientists investigating the physiological, biochemical, and molecular mechanisms activated in plants, in general, and in halophytes, in particular, in response to salt stress—I know it from my own experience—often neglect constitutive mechanisms of defense based on structural adaptations of their investigated species. This book can help all of us to look at our own research with a wider perspective. I also find very interesting and intellectually rewarding (this is obviously a personal opinion) the historical approach underlying the entire book's contents and organization, with numerous references to the work of classical botanists—which in other scientific fields would be considered "ancient" but in plant anatomy will never be outdated. I would like to point out, especially, the high quality of the (also "classical") ink drawings in many of the figures; probably no "modern" micrograph can show anatomic structures with such clarity and detail.

In a time of "omics" technologies, full genome sequences, DNA "barcodes," and Apps that can supposedly identify any plant species from a picture taken with your mobile phone, it is good to look back and acknowledge the enormous contribution of classical botanists to our present knowledge in this field. It is as well refreshing to look at the beauty of nature, also reflected in the anatomic adaptations to salinity shown here. This is, basically, what this book is about.

Institute of Plant Molecular and Cellular Biology (IBMCP, UPV-CSIC)
Universitat Politècnica de València
Valencia, Spain

Oscar Vicente

Preface

Succulence, tracheoidioblasts (spiral cells), salt secretion, Kranz anatomy, successive cambia, and bulliform cells represent major anatomical adaptations found in halophytes; they are treated in this monograph with an emphasis on literature from the end of the nineteenth century and the beginning of the twentieth century. Of course, the list of adaptations is not exhaustive; other anatomical features in halophytes may be associated with saline environments. Naturally, the reviewed literature has no pretension to be complete. Actually, part of recent literature might be regarded—at first glance—as being repetitive, at least when it refers to purely anatomical descriptive approaches in species that have also been investigated in the past. This is an inherent risk in plant anatomy research, and the real challenge would be to find the right frame in which to integrate this kind of approach.

Another important observation in relation to this book is that we tried—as much as possible—to remain close to anatomy and not to go into details of associated issues emerging from many of the discussed adaptations. For instance, Kranz anatomy is related to C_4 photosynthetic pathway, and this requires a very sophisticated approach in terms of physiology, biochemistry, and paleo-ecology of halophytes. In this respect, a luxuriant literature has been produced in the past 20 years, including edited books and published papers—original and reviews. Of course, our intention did not interfere with these research directions. The same is true for salt glands, where many results derived from ultramicroscopic observations or are focused on the ecophysiology of salt secretion. Again, we tried to remain at the border of this approach.

However, these two examples clearly suggest that the anatomy of halophytes is a refreshing field that opened many advanced research directions, such as plant ecophysiology, salt stress physiology, and molecular approaches.

The chapter dedicated to saline environments is intended only to familiarize the reader with several operational terms related to salt areas. This should not be regarded a soil science chapter, since we are aware that the terminology and taxonomy of salt-affected soils is very problematic and the strongest evidence is the multitude of classification systems existing worldwide. Therefore, in this chapter several concepts are discussed, since the chosen terms are largely used in the botanical literature, especially in that from the past decades.

Because of the specificity of this book, mentioned species were kept in the genuine nomenclatural state, with no intention to find actual synonyms. This would have been tricky, especially for species from *Chenopodiaceae* and *Plumbaginaceae*; in addition, consistent with the vision of this book, readers interested in the history and evolution of plant taxonomy should be invited to discover by themselves the exciting journey toward—likely—the most romantic and fruitful period in the history of botany.

Iasi, Romania

Marius-Nicuşor Grigore
Constantin Toma

Acknowledgements

We have to thank several persons:

Oscar Vicente (Instituto de Biología Molecular y Celular de Plantas, UPV-CSIC) was willing to write a Foreword for our book. In the past, he helped me very much to integrate in his research group from UPV, giving me the possibility to work and extend my research interests in halophytes. Our collaboration was and still is very fruitful, and I'm deeply indebted to him for all the support during the last 7 years. Together with Monica Boscaiu (Instituto Agroforestal Mediterráneo, UPV), we managed to create a strong research interest network between Romania and Spain.

Universidad Politécnica de Valencia (UPV) hosted me three times at Instituto de Biología Molecular y Celular de Plantas, for developing scientific missions related to the study of halophytes. Two stays were funded by COST Action FA0901: "Putting Halophytes to Work—From Genes to Ecosystems," in a frame of Short-Term Scientific Missions. One stay was funded by Universidad Politécnica de Valencia that supported me in the anatomical study of Spanish halophytes. During these stays, I had the excellent opportunity to collect a large number of halophytes, which were anatomically investigated. Part of the obtained results is included in this book.

COST Action FA0901: "Putting Halophytes to Work—From Genes to Ecosystems" also funded me for attending a meeting in Torun (Poland), when I collected two halophyte species from salt-affected areas, during a field trip. On this occasion, I thank Agnieszka Piernik (Nicolaus Copernicus University, Torun) and Alexander Fehér (Slovak University of Agriculture, Nitra, Slovakia) for helping me identifying several taxa in the field.

Survey and analysis of old botanical literature represent a pivotal aspect for this book. Not all the bibliographic references were available to us. A part of old papers—that was not on our disposal—has been provided by few persons. Roberta Gasparri from the Department of Agricultural, Food and Environmental Sciences, Marche Polytechnic University (Ancona, Italy), sent us Licopoli's papers.

Valentina Voicu from the National Institute of Research-Development for Pedology, Agrochemistry and Environmental Protection (Bucharest, Romania) helped us to obtain several papers in soil science.

Somayeh Safiallah from Islamic Azad University (Tehran, Iran), a promising researcher in halophytes study, provided us color micrographs with several halophytic species of *Bassia* from Iran; with her kind permission, some of them have been included in this book.

Ana Cojocariu from "Anastasie Fătu' Botanical Garden" (Iasi, Romania) provided us an original photo with *Nitraria schoberi*, a rare species vegetating on saline soils from Vulcanii Noroioşi (Pâclele Mici, Buzău county).

Marius-Nicuşor Grigore

Introduction

Halophytes are intriguing plants able to survive in a multitude of saline environments subjected to physiological drought conditions. For this reason, our vision about halophytes is based on the idea that they should be actually regarded as a special case of xerophytes.

Despite unprecedented advances in the study of halophytes, yet dominated by molecular and proteomics approaches, rediscovering basic concepts and theories can act as an exciting and revitalizing odyssey in order to get a clear and accurate picture about this ecological group of plants.

The interest in the study of halophytes is manifold: in the current context of human condition, salinization and aridization continue to be a threat to the condition of actual agriculture. Salinity has affected agriculture for millennia, having a deep negative impact on agriculture and, most likely, being involved in the fall of some ancient flourishing civilizations. Halophytes can provide salt-tolerant genes for prospective plants to be used successfully in future agriculture. Not in the least, strengthening the core knowledge about biology of halophytes is a *sine qua non* condition in order to avoid complex approaches without appropriate background documentation.

Anatomical study of halophytes represents perhaps the first modern research direction in the frame of biology of these species; starting from the end of the nineteenth century, Danish, French, and German botanists commenced to investigate the structure of plants vegetating in "extreme" environments, such as salt marshes and salt deserts and mangrove ecosystems. Actually, perhaps this period of time was the most fruitful from the entire epoch of plant anatomy study. It is worth mentioning Danish plant ecologist Eugenius Warming (1841–1924), who extensively dealt with the structure and ecology of halophytes. German botanist Georg Volkens (1855–1917) investigated the structure of many halophytic species from Egyptian and Arabian Desert and described, *inter alia*, the structure of interesting halophytic chenopods. French botanist Henri Chermezon (1885–1939) had a great contribution in the structure of littoral plants, many of them being halophytes. German botanist Andreas Franz Wilhelm Schimper (1856–1901) investigated the structure of leaves of many mangrove species; he contributed, together with Warming, to theorize about the physiological drought of saline environments. Italian botanist Gaetano Licopoli (1833–1897) focused on the disputed structure of salt glands in *Plumbaginaceae* species.

In parallel, many other botanists worldwide sectioned organs of halophytes and published their results, either as entire study on halophytes or as included contributions in extensive studies on plant anatomy. A large part of these valuable contributions are still unknown to readers even today, because of their limited availability; perhaps the publishing journals were too "local" and had a limited frequency or the foreign languages acted as a barrier to the circulation of knowledge. Fortunately, many of these papers have been greatly revealed in the past 10 years through digital collections; today, due to their complex, fine, and elegant details and drawings, some of these data are also suitable for a documentary or text analysis—at least for a plant anatomist dealing also with the history of this field.

The high taxonomical diversity of halophytes makes their anatomical study quite difficult. Halophytes' adaptations represent very interesting structural strategies that help plants to cope with harsh environments; for many other species (glycophytes, plants living in freshwater conditions), these habitats would not be suitable for surviving and reproduction.

Any reader should have in mind that many of the advanced, modern approaches on halophytes have as starting point rigorous anatomical observations.

Contents

About the Authors

Marius-Nicuşor Grigore received his BSc and PhD in Biology from Alexandru Ioan Cuza University in Iaşi, Romania. He was subsequently a postdoctoral fellow in the field of halophyte ecophysiology at the same university, as well as an invited researcher at the Polytechnic University of Valencia, Spain. His primary research interests are in the ecology, anatomy, and physiology of halophytes. He has published 6 books and over 60 peer-reviewed papers on halophytes. In 2016, he was awarded the Romanian Academy's "Emil Racoviţă" prize for the book—*Halophytes: An Integrative Anatomical Study*, published with Springer in 2014. He is currently an editor at Alexandru Ioan Cuza University Press.

Constantin Toma received his BSc and PhD in Biology from Alexandru Ioan Cuza University in Iaşi, Romania. He is full member of the Romanian Academy and an honorary member of the Sciences Academy of the Republic of Moldova. He has extensive expertise in plant morphology and anatomy and has published 30 books and more than 500 papers. Four of his books were awarded by the Romanian Academy, the last in 2016—*Halophytes: An Integrative Anatomical Study* (Springer). He is currently a professor emeritus at Alexandru Ioan Cuza University and honorary president of the National Society of Biological Sciences.

About the Authors

Part I
Halophytes and Saline Environments

Chapter 1
Definition and Classification of Halophytes

1.1 General Considerations

At this moment, there is a plethora of definitions that are attributed to halophytes; due to their taxonomical and ecological complexity, there is no consensus on a unique definition of a halophyte. This issue has been a matter of debate in several papers focused on halophytes' biology and inherent difficulties (Grigore 2008a, b, 2012; Grigore and Toma 2010; Grigore et al. 2010, 2012, 2014).

As shown, this heterogeneity is suggested, first of all, by the large number of interpretations made by various authors over time. Thus, Chapman (1960) describes halophytes as "salt-tolerant plants," Fernald (1950) regards them as plants "growing in saline soils," and Dansereau (1957) defines them as "plants that grow exclusively on salt soil, e.g., *Salicornia* species." Other definitions include plants of salty or alkaline soils (Correll and Johnston 1970); plants that can tolerate salt concentrations found in salty soils (Oosting 1956); and plants tolerant of various mineral salts in the soil solution, usually sodium chloride (Lawrence 1951). Waisel (1972) defines halophytes as plants that grow and complete their entire life cycle in habitats where the salt content is high. Usually, this term is restricted only to plants that appear constantly on salt areas. Waisel also uses another term—*pseudohalophytes* ("false" halophyte) to refer to plants that occupy only local nonsaline ecological niches in an overall saline environment, or those that occur in such environments only for short periods, i.e., during the rainy season. Duncan (1974), making a list of halophytes which vegetate on the Atlantic and Gulf coasts of North America and Mexico, regards them as species that can tolerate seawater, "pure or diluted."

Grigore et al. (2010) chronologically reviewed more than 40 definitions of halophytes, stating that the huge variability in approaching such an ecological group derives from historical period, from authors' background, and from reinterpretations of previously stated definitions.

Closely related to the richness of halophyte definitions, the multitude of saline habitats, whose terminology is still problematic, can explain the lack of a consensus

M.-N. Grigore, C. Toma, *Anatomical Adaptations of Halophytes*,
DOI 10.1007/978-3-319-66480-4_1

about definition of halophytes. From an ecological point of view, it is natural to link halophytes (as definition, classification, taxonomy, and adaptations) to saline environments that directly influenced them during time (Grigore 2008a, b; Grigore and Toma 2010). Nevertheless, the precise effects and how do salts act on plants' biology are far to be fully understood. Among others, this is because there are differences between the concentration and composition of salts from one habitat to another one; in addition, the "salinity" term seems to refer rather to the concentration than to composition of the soil solution. Salinity itself is a term applied in plant biology with multiple and sometimes confusing meanings (Grigore 2010, 2012; Grigore et al. 2014). The term *salinity* is not, per se, a biological one; thus, the scenario could become complicated, when adopted by other natural science.

An ecological definition of halophytes seems the more appropriate solution, and consequently, they must be considered all species that vegetate in saline habitats (Grigore 2008a, b; Grigore and Toma 2010). This definition seems simple and accessible but only at first, because saline habitats are again imprecisely defined.

From a historical point of view, and from analysis of the reviewed definitions, Grigore (2012) suggested that many existing definitions are in fact terms composed by an adjective and a noun (see Table 1.6).

The typical example is that of "*maritime plants*," a term that has been used since decades with the meaning of "*halophytes*." Of course, not all the halophytes are maritime plants, but the use of this expression is of great importance, since an ecological group of plants has been clearly linked to (maritime) salinity. Perhaps, *maritime plants* have been used as an expression from the beginning of modern plant ecology; Willdenow (1792, 1799) uses it in German and it has also been translated in English edition (Willdenow 1805). More likely, the term *halophyte* as we know, nowadays, has been created and introduced in botanical language after the middle of the nineteenth century, as in the last decade it is well established (see Grigore 2012).

Some definitions could be considered as "ecological," every time the plants are correlated with saline habitats. It seems very logical when taking into consideration that we deal with an ecological group of plants. Sometimes, the authors talk about the condition of "completing life cycle" characterizing halophytes. Here, some additional comments are required. Complete life cycle means, of course, that the plant needs to flower, in order to produce fruits with seeds. These will germinate and thus will ensure the plant survival and its stability in a given habitat. Germination in a saline environment is a very delicate and sensitive issue regarding halophytes' biology (see Ungar 1991 and references therein). Perhaps, every definition of halophytes suggesting the absolute necessity of completing the entire life cycle must be discussed with caution. It is well known that the success of halophyte populations, especially for *annuals* that have only one opportunity in their life history to reproduce, is greatly dependent on seed germination responses (Ungar 1991). Seed germination for most halophytes occurs during periods of the year when soil salinity levels are reduced (Ungar 1978). In addition, laboratory investigations with halophytes suggest that optimal germination percentages are usually found in nonsaline conditions (Grigore et al. 2012). Anyway, it must be emphasized that, generally, the seeds of halophytes can tolerate higher salinity concentrations than those of glycophytes. In a salt marsh, the halophytes must adopt therefore different survival strategies. It was shown that the majority of salt marsh species are *perennial*, and in fact, relatively few species of annuals have become adapted to the true salt marsh habitat (Ranwell 1972). This would imply that perennial halophytes, having rhizomes, for instance, would be able to assure the persistence at a location on the salt marsh for several decades. So, they would be able to survive in a saline habitat, without "completing the entire life cycle" (hypothetically, without flowering, producing seeds which will germinate generating seedlings).

Other definitions induce a subtle nuance: halophytes are those species growing in saline habitats *only* (or in conditions of an excess of salts, high levels of salt or plants that need a high concentration of salts in their media for an optimal growth). This is, in background, a definition of euhalophytes (obligatory halophytes). There are still many discussions regarding the "absolute" requirement of these species for a high salt content and thus remaining types of halophytes would be eliminated.

Some other definitions could be regarded as "physiological." Establishing a numerical boundary between halophytes and glycophytes could be useful for standardization, but perhaps many of such definitions are the result of experimental approaches, when the natural situation is completely different. Nevertheless, the value of these definitions should not be denied, especially when we need to compare different species in terms of their salinity tolerance.

Therefore, in time, the emphasis has moved from halophytes' perception in the broadest sense to a narrower sense of terms that finally overlapped with the definition of obligatory (true) halophytes (*euhalophytes*).

Therefore, we believe that a broader definition of halophytes, in an ecological way (especially supported by field data), would be more appropriate in an operational sense. Plants that grow in saline environments should be regarded as halophytes, since there is always a close dependence on soil salinity, despite it varying within very large values. The common view is inclined to consider as halophytes only those euhalophytes ("true" halophytes), but those designated as preferential, supporting, and accidental halophytes deserve their place next to the "classic" category of obligatory halophytes, since they occupy distinct niches within a saline environment. Their relation with environmental factors has an adaptive value that cannot be denied to the detriment of ecological relationships between true halophytes and soil salinity.

Nevertheless, soil salinity, and—of course, excessive salinity—that many plants, the so-called glycophytes, cannot face, is the main factor that influences halophyte distribution. This explains the universal taxonomical occurrence of halophytes in very different ecological and climatic areas of the Globe. For instance, several genera and even species (*Atriplex halimus, Salicornia herbacea*, and *Juncus maritimus*) occur in salt marshes from temperate to tropical and subtropical zone (Waisel 1972). It is amazing how so many plants, different in taxonomical, evolutionary, and geographically ways, adapted to extreme soil conditions, expressed by a significant high salinity.

1.2 Halophytes and Glycophytes

Since decades, a clear distinction between halophytes and glycophytes has been maintained and propagated; traditionally, glycophytes were assimilated to crop plants, which are, to a large extent, sensitive to high salinity conditions (especially NaCl salt). In the common language, these two terms are regarded as antonyms; etymologically, *halophytes* means "salt-loving plants" (*hals—salt; phyton—plant; philein—to love*), while *glycophytes* refer to plants that prefer "sweet" substrates (*glykos—sweet*). In fact, there is a very broad spectrum of salt tolerance in plants, ranging from the most sensitive plants that are severely affected even by a lower concentration of 50 mM NaCl (about a tenth of the concentration of seawater) to those able to complete their life cycle in terms of a concentration of 500 mM NaCl (close to seawater concentration) (Sharma and Gupta 1986). Often, in usual language, the term "salt plant" is literally used instead of the halophyte term, especially for plants that can grow in the presence of high concentrations of sodium salts (Table 1.5). Jennings (1976) provides an ecological definition of halophytes: they form the "native flora of saline environments."

Another ecological definition of halophytes refers to them as those plants that can grow satisfactorily and can compete with other species in the same habitat and thus completing their life cycle (Waisel 1972). The final goal of the plant would be in this case to ensure the survival and perpetuation of the species; however, halophytes possess adaptive mechanisms that allow them to avoid, in certain circumstances, the action of harmful salinity. As long as the exact and appropriate definition of saline environments will not be a stringent issue, perhaps the definition of halophytes as plants that occur in these saline environments would be the most appropriate. In other words, Frey and Basan (1985) regard the presence of halophytes a necessary part of salt marsh definition. Interestingly, perhaps it would be more correct, from an ecological and causal point of view, to consider as halophytes plants that grow in saline environments instead of arbitrarily listing species whose habitats are to be found. This broader and ecological approach (and more flexible) seems to be the most acceptable when talking about this heterogeneous ecological group of plants. From this point of view, halophytes are those species for which saline habitats are the major and, in the most cases, the only habitat. Glycophytes from maritime flora of salt marshes would include those species regarded as occurring in nonsaline inland habitats (this approach proposes an external, nonbiological definition of maritime salt marshes, sometimes contradicting the suggestion offered by Frey and Basan 1985).

Adam (1990) suggests that the term "halophyte" to be applied to species that are more or less confined to maritime or other saline habitats; the use of this term should be made in the broadest sense for all species occupying saline environments.

The term "glycophyte" is being traditionally used for the species more widely distributed in nonsaline than in saline habitats. Of course, this not necessarily implies that glycophytes are not, to some extent, tolerant to a certain degree of salinity; moreover, frequently populations of glycophytes from salt marshes are genetically adapted to saline habitats (Adam 1990).

Almost all crop plants are glycophytes. They have a selective advantage over halophytes on nonsaline soils, because they have a higher growth rate (Sharma and Gupta 1986).

Halophytes including salt marshes and mangroves species are well-developed and specialized organisms, with morphological and physiological traits that allow them to reproduce on soils with high concentrations of salt (Khan and Duke 2001).

Poljakoff-Mayber and Gale (1975) define halophytes as plants that normally grow on salt marshes, in seawater, or in saline soils.

From a physiological point of view, halophytes are recognized as plants that can survive at high concentrations of electrolytes in their substrates (Flowers et al. 1977); these substrates are typically dominated by NaCl, but they can also contain other various salts, such as Na_2SO_4, Mg SO_4, $CaSO_4$, $MgCl_2$, KCl, and Na_2CO_3.

Flowers et al. (1986) defined halophytes as plants found growing under naturally saline conditions. For terrestrial plants, this means a minimum salt concentration of about 100 mM in the soil solution. According to them, the most useful criterion for separating the halophytes from glycophytes remains the ability of the halophyte to complete its life cycle at salt concentrations in excess of 100–200 mM NaC1.

Nevertheless, the distinction between glycophytes and halophytes is functional especially for plant physiology and agronomy, and it is based on the value of electrical conductivity (EC) of the saturation extract. Thus, according to the definition of US Salinity Laboratory Staff (Richards 1954), the saturation extract of a saline soil has an electrical conductivity greater than 4 dS m^{-1} and an exchangeable sodium percentage (ESP) of less than 15. In evaluations of the suitability of saline soils for crop production, the measurement of EC offers a rapid and accessible method for characterizing the salt content (Marschner 1995). To have a complete picture of these values, one must say that the electrical conductivity of seawater corresponds to about 44 dS m^{-1}; the salt concentration of seawater (3%: 480 mM Na^+, 50 mM Mg^{2+}, and 560 mM Cl^-) corresponds to an osmotic potential of −2.7 MPa (Schulze et al. 2005).

Using EC, the osmotic potential of the saturation extract can also be calculated: osmotic potential (MPa) $=$ EC $\times -0.36$. According to Richards (1954), the salt tolerance of a crop (glycophyte) may be appraised according to three criteria:

a. The ability of the crop to survive on saline soils
b. The yield of the crop on saline soils
c. The relative yield of the crop on a saline soil as compared with its yield on a nonsaline soil under similar growing conditions

Accordingly, several ways in which glycophytes and halophytes respond in relation to soil salinity (EC) have been proposed (Figs. 1.1, 1.2, and 1.3).

Bucur et al. (1957) intensively studied interrelationships between salt-tolerant plants and soil salinity; his advanced results (commented by Grigore 2013) show, among other interesting conclusions that in a group of halophytes, the increasing plant biomass occurs at high salinity levels (Fig. 1.2), while, in other, plant biomass

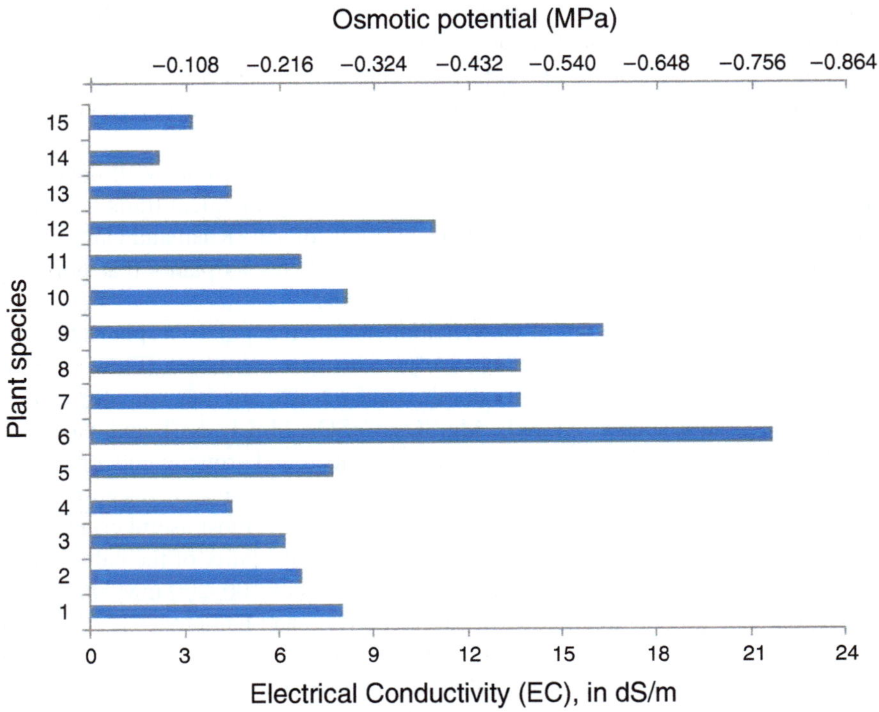

Fig. 1.1 Several Romanian crop species and their tolerance to soil salinity (*1* barley; *2* wheat; *3* oat; *4* rye; *5* maize; *6* castor bean; *7* sun flower; *8* sugar beet; *9* beet; *10* Sorghum Sudan grass; *11* common sainfoin; *12* alfalfa; *13* spring vetch; *14* common garden pea; *15* been) (based on Măianu et al. 1965)

increases only under lowered salinity (usually below 100 mg soluble salts %) (Fig. 1.3).

Lambers et al. (2008) consider halophytes those species that typically grow in soils with high levels on NaCl and, therefore, on a low water potential. They accumulate NaCl in their vacuoles. By contrast, glycophytes have a limited capacity to transport NaCl into their vacuoles and are unable to tolerate high salinity levels. However, cytoplasmic enzymes of both glycophytes and halophytes are very similar with respect to their sensitivity to high concentrations of inorganic solutes.

1.3 Halophytes' Classification

This issue is very complex and even controversial, due to special aspects related to halophyte definitions and diversity of saline environments. Over time, several classifications have been proposed, but none seems to be entirely satisfactory, because there is no unique criterion of classification; consequently, attempts to

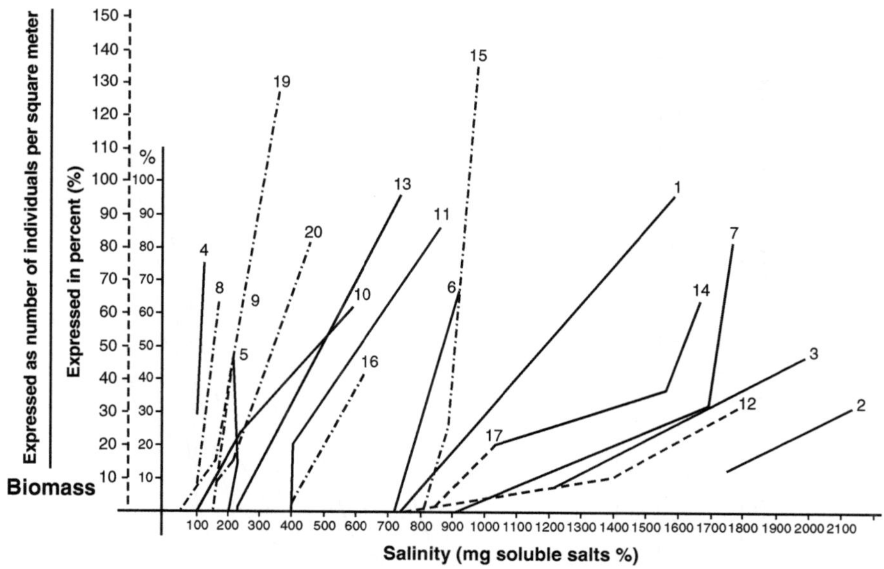

Fig. 1.2 Relationship between plant biomass and increasing soil salinity in the rhizosphere of several halophytes (euhalophytes): 1—*Atriplex hastata*; 2—*Scorzonera austriaca*; 3—*Atriplex littoralis*; 4—*Heleochloa schoenoides*; 5—*Podospermum canum*; 6—*Aster tripolium*; 7—*Camphorosma ovata*; 8—*Silaum flavescens*; 9—*Aster cinereus*; 10—*Artemisia maritima*; 11—*Taraxacum bessarabicum*; 12—*Kochia prostrata*; 13—*Puccinellia distans*; 14—*Limonium gmelini*; 15—*Salicornia europaea*; 16—*Plantago schwarzenbergiana*; 17—*Lepidium cartilagineum*; 18—*Peucedanum latifolium*; 19—*Leuzea salina*; 20—*Plantago tenuiflora* (modified after Bucur et al. 1957)

harmonize different systems are debatable. Attempts of classification can be divided mainly into three major criteria (Waisel 1972). Thus, some authors took into account the salt content of the native habitats. Others have tended to emphasize the importance of origin of salts and classified halophytes accordingly. Finally, some authors have focused on plant responses to salinity (Waisel 1972).

In addition to several classifications well known (Stocker 1928; Steiner 1935; Iversen 1936; Van Eijk 1939; Chapman 1942; Waisel 1972) and discussed several times (Grigore 2008a, b, 2012; Grigore and Toma 2010; Grigore et al. 2014), in the following paragraphs other classifications will be mentioned, emphasizing Romanian contribution especially to this part and to biology of halophytes in general.

Contejean (1881) (then discussed in Vilbouchevitch 1892) refers to the three major classes of maritime (halophyte) species:

1. Exclusive (or almost exclusive) maritime species, which may be found only accidentally outside littoral zones and which cannot propagate in nonsaline soils;
2. Less exclusive maritime species that often occur in soils less saline or completely nonsaline;
3. Almost indifferent maritime species that are found more often in inland areas than in littoral regions; most of them seem to be established next to the sea rather

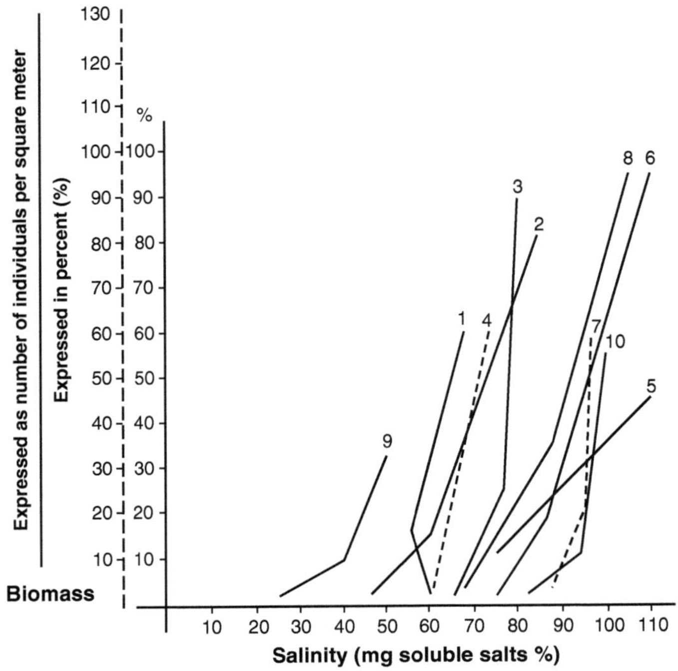

Fig. 1.3 Relationship between plant biomass and decreasing soil salinity in the rhizosphere of several halophytes (neohalophytes): 1—*Trifolium hybridum*; 2—*Lythrum virgatum*; 3—*Gratiola officinalis*; 4—*Medicago lupulina*; 5—*Festuca pratensis*; 6—*Phalaris arundinacea*; 7—*Glyceria aquatica*; 8—*Beckmannia eruciformis*; 9—*Plantago lanceolata*; 10—*Taraxacum officinale* (modified after Bucur et al. 1957)

because of the influence of climatic conditions and local factors than to a real need of salt derived from the sea.

As it can be seen, this classification is based on the influence of sea (marine) salinity; accordingly, plants far located from the sea are less tolerant than those located close to the seashore, where salt concentration is higher.

Chermezon (1910), referring to littoral flora (divided by him in xerophilous and halophilous), divides plants into three categories:

1. Flora of the beaches (sea shores), occupying a very small area, where there are few or no continental species; many species can move to some extent from here to the dunes and vice versa; the most characteristic species are *Cakile maritima, Eryngium maritimum, Statice bellidifolia, Polygonum maritimum*, etc. Many species of beaches appear to be less demanding in terms of salinity, which is less significant, at least in the more distant area from the sea. It is, however, sufficient to eliminate continental species. These plants make the transition between psammophytes and other halophyte species;

2. Flora of the cliffs occupies a restricted area consisting of rocks exposed to salt spray in the vicinity of the sea. Action of seawater is stronger here than on beaches and characters of plants belong obviously to halophilous plants: *Crambe maritima, Silene maritima, Plantago macrorhiza, State dodartii*, etc.;
3. Flora of salt marshes consists of halophyte vegetations that occupy considerable areas; it forms a band on lower shores and along estuaries whose width varies with terrain profile and has a large number of characteristic species: *Frankenia pulverulenta, Spergularia marginata, Aster tripolium, Plantago maritima, Atriplex littoralis, Salsola soda, Triglochin maritimum*, etc. This is the well-individualized flora of the coastal region and the most halophilous. Salinization is here quite considerable, which explains the rarity of continental species. The features of plants are of halophilous nature, reaching its maximum in this area.

The Romanian botanist Prodan (1922, 1923) suggested an interesting classification, based on a cause–effect relationship, meaning that "*natural saline environments which, according to their degree of humidity and partly, to characteristic plants can be divided into: dry salt areas, salt marshes and salt lakes*."

Therefore, he indirectly divided halophytes according to characteristics of their environments; it is an ecological classification, which takes into account environmental factors. Correspondingly, he distinguished halophytes from dry salt areas, from salt marshes, and from salt lakes. In his important monograph related to ecology of halophytes (1922, 1923), species belonging to these three categories of halophytes (Figs. 1.4, 1.5, and 1.6) are described in detail; sometimes elements of morphological and anatomical value are also given.

Prodan (1922) also proposed another classification of halophytic flora: spring flora, including plants that bloom in spring and last until summer, and autumn flora, including plants that bloom in summer and last until late autumn.

In a book chapter from 1939, Prodan offers a classification of halophytes, based on "*the way in which plants withstand salt*." He delineated three categories of halophytes:

1. "First category" includes species that "*grow* ***exclusively*** *in salt areas and only exceptionally in other places*": *Zostera marina, Ruppia rostellata, Juncus gerardii, Atriplex hastata, Aster tripolium*, and *Artemisia salina;*
2. "The second category" represents "*the species which besides salty areas can also vegetate in several habitats (waters, marshes, sands)*": *Najas minor, Beckmannia eruciformis, Carex distans, Spergularia marginata*, etc;
3. "Third category" comprises species that "*grow in other environments and can pass only rarely or exceptionally in saline areas*." Some of them suggest an incipient halophilous affinity: *Triglochin palustris, Andropogon ischaemum, Polygonum aviculare*, and *Tamarix pallasii*.

Several classifications are based on other criteria—the contact between salts and different organs and tissues of plants exposed to salinity. Thus, in terrestrial habitats, the contact occurs between roots and soil salinity (terrestrial halophytes).

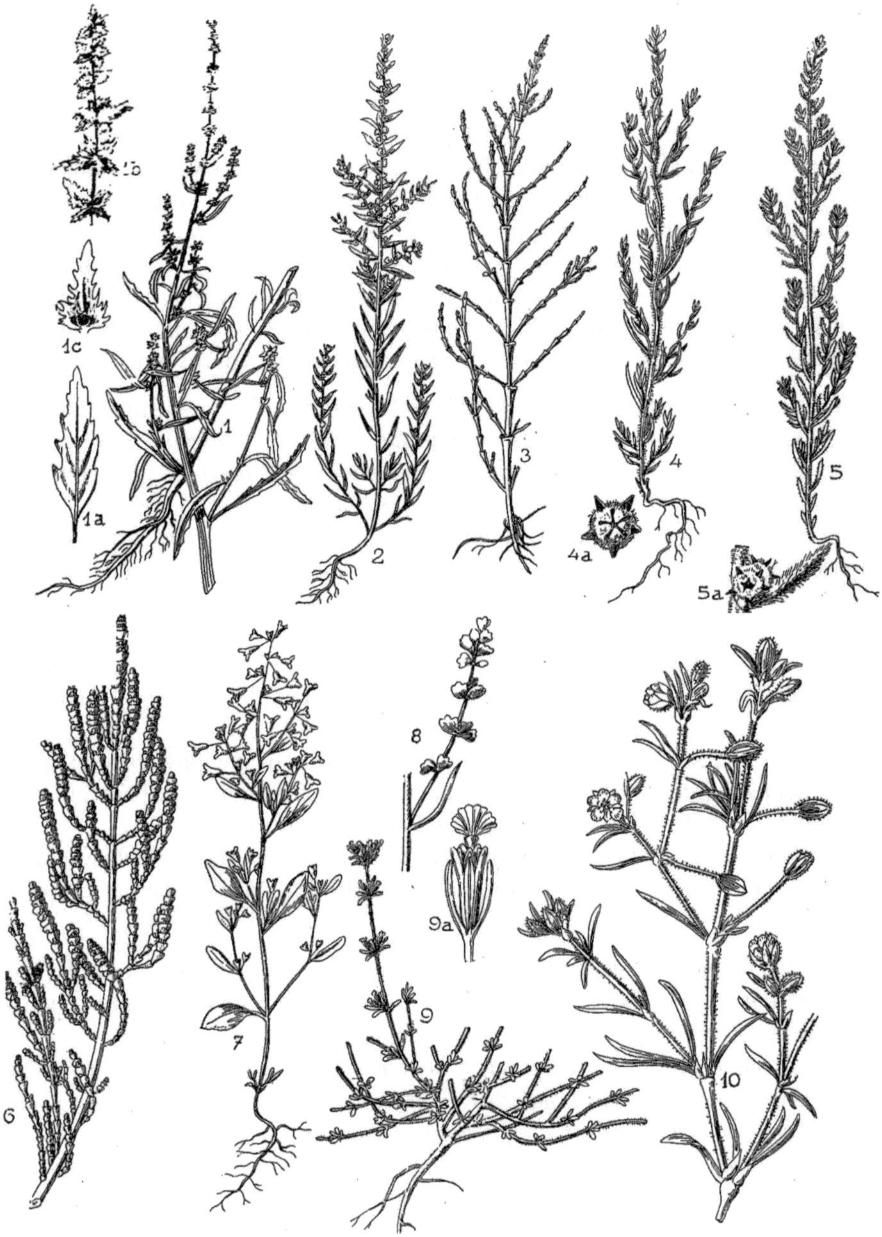

Fig. 1.4 Pictures of Romanian halophytes: *Atriplex littoralis* (1; 1a—basal leaf; 1b—fruiting shoot; 1c—fruit), *Suaeda maritima* (2), *Salicornia herbacea* (3), *Bassia hirsuta* (4; 4a—fruiting perianth), *B. sedoides* (5; 5a—fruiting perianth), *Arthrocnemum glaucum* (6), *Obione pedunculata* (7), *O. portulacoides* (8), *Frankenia hispida* (9), *Spergularia rubra* (10) (Prodan 1922)

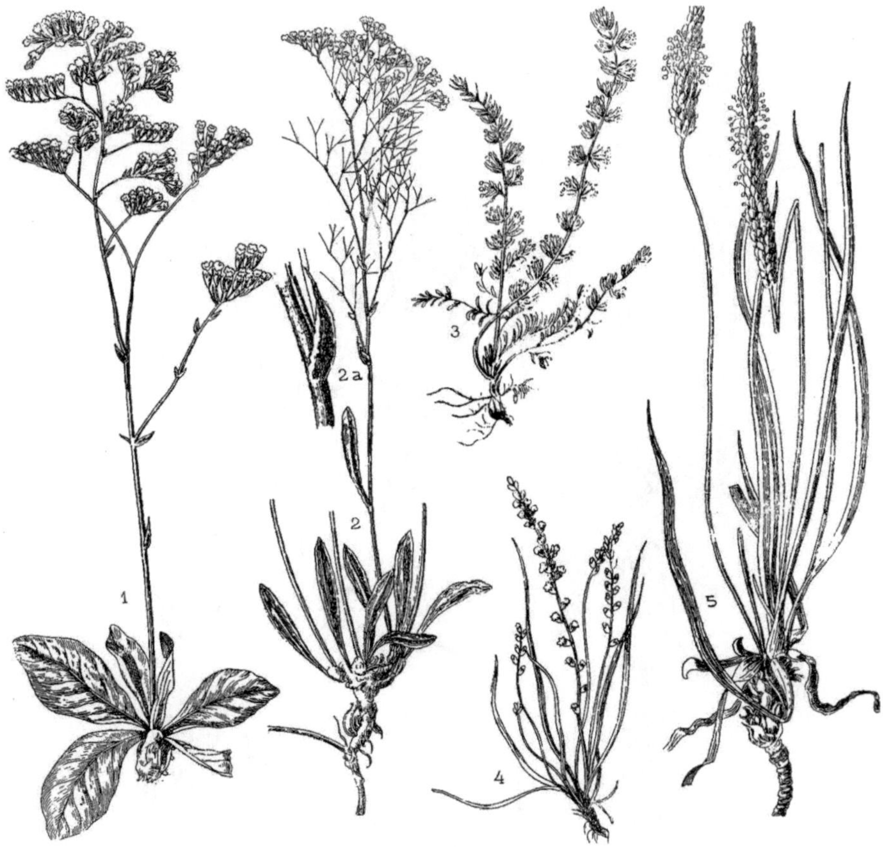

Fig. 1.5 Pictures of Romanian halophytes: *Statice gmelini* (1), *S. caspia* (2; 2a—a bract from flowering shoot), *Camphorosma ovata* (3), *Plantago tenuiflora* (4), *P. maritima* (5) (Prodan 1922)

In marine habitats, or salt marshes, the salts come in contact either with plant roots (emersed halophyte or hygro-halophytes) or with the entire organs of the plant (submerged halophyte or hydro-halophytes). Coastal or desert plants, those whose organs are affected by salt particles or salt spray, are called aero-halophytes (Waisel 1972). The total content of salts and soil humidity should also be taken into consideration. In this respect, there are halophytes that prefer undrained constantly moist places (hygro-halophytes and hydro-halophytes), while others resist only in saline environments, but relatively dry (xero-halophytes).

The Romanian botanist Guşuleac (1933), studying the Northern Romanian halophilous vegetation and natural colonization of saline environments, establishes three groups of plants:

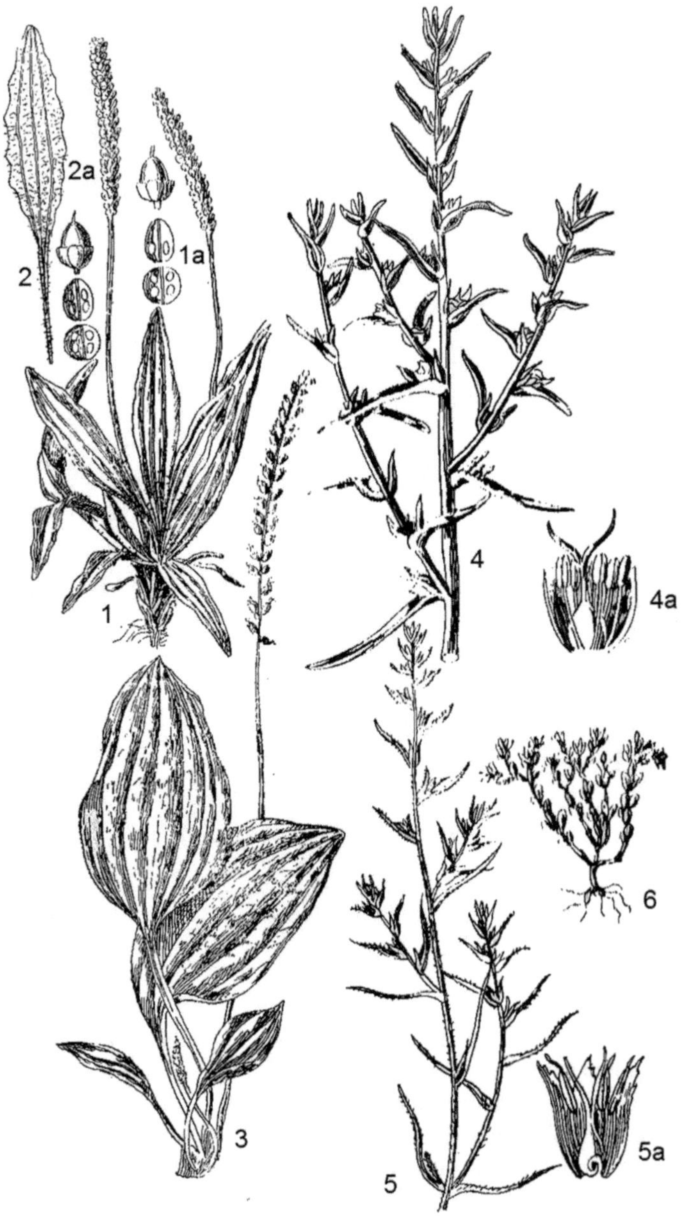

Fig. 1.6 Pictures of Romanian halophytes: *Plantago schwarzenbergiana* (1; 1a—covered and uncovered fruits), *P. sibirica* (2; 2a—covered and uncovered fruits), *P. cornuti* (3), *Salsola soda* (4; 4a—flower), *Petrosimonia triandra* (5; 5a—flower), and *Sedum caespitosum* (Prodan 1922)

1. Normal plants, *non-halophytes* that avoid salt marshes because the high salt concentration affects plants water supply;
2. *Facultative halophytes* that may support a higher or lower salinity and grow in salty places more because of too high competition dominating nonsaline places;

3. *Obligatory halophytes*, they require for their normal development an abundant amount of salt.

Kovda (1939) divided plants from salt marshes and soils with lower salinity into:

1. *Typical halophytes, succulent Salsola-like species*, characteristic to humid solonchaks, with very high water table; they contain 40–55% salts, where NaCl predominates. For this reason, these species maintain the salinity in upper layers of salt-affected areas;
2. *Semidried halophytes*, from dry solonchaks and other intensely salinized soils, with lower level of water table; they contain 20–30% salts, where chlorides and sulfates are in close proportion. They maintain salinity, but also enrich soil with Ca, Mg, K, SiO_2, to the detriment of sodium;
3. *Halo-xerophytes or dry Salsola-like species*, several *Artemisia* species, and other xerophyte species, which grow on less (or not) salinized solonetzs; they have a mineral content of 10–20%, where P, S, Ca, and K predominate. They contribute to desalinization and de-solonizing of solonetzs;
4. *Poaceae, Fabaceae*, and several *Artemisia* species, characteristic to steppic soils and dry less salinized steppes. They have a mineral content under 10%, where SiO_2, P, Ca, and K predominate. They complete the de-solonizing process and enrichment of complex with calcium.

Ţopa (1939, 1954) offered perhaps the first Romanian work(s) where the "standard" classification of halophytes is explicitly given with appropriate definitions.

Thus, halophytes are classified according to their response to salinity in *obligatory, preferential, supporting, and accidental* halophytes.

1. *Obligatory halophytes* as those plants growing in salty habitats requiring a considerable amount of salt for their development, at least for a short period of the year: *Camphorosma annua, Halocnemum strobilaceum, Salicornia herbacea,* and *Suaeda maritima*;
2. *Preferential halophytes prefer* the saline environments where they find the "*optimal living conditions*": *Atriplex tataricum, Lotus tenuifolius,* and *Tamarix ramosissima*;
3. *Supporting halophytes* endure the salts but do not manage to compete with local vegetation: *Rumex maritimus, Polygonum aviculare,* and *Atriplex hastata*;
4. *Accidental halophytes* occur *accidentally* in the salty habitats but are not able to survive there: *Rumex hydrolapathum, Poa annua,* and *Molinia caerulea.*

These definitions of halophytes and classifications given by Ţopa are very interesting and relevant, because they show the logical and correspondent relation of halophytes with soil salinity. They seem to be an etymological characterization of halophyte classes.

Sennikov (1950) seemed to refine Kovda's classification (1939) and opined that different groups of plants vegetating on saline environments have specific structural and physiological features. Thus, he divided these species into:

1. *Typical halophytes*. Plants exclusively from salt areas that present adaptations for salinized soil: succulent, reduced leaves; they can withstand higher concentration of salts. Their anatomical structure has been nominated as halomorphic. Example of typical halophytes: *Salicornia herbacea, Suaeda maritima, Halocnemum strobilaceum,* and *Salsola soda*. They are adapted to very high osmotic pressure and accumulate in their tissues large amount of salts, especially chlorides and sulfates; for these species, the salinity is necessary and even can stimulate their growth;
2. *Semidried halophytes*: *Petrosimonia crassifolia*, and some species of *Suaeda*, which have a halomorphic structure, with discrete adaptations of xerophytic nature (small trichomes);
3. Halo-xerophytes: *Obione verrucifera*, *O. canum*, from salt areas that moderately dry during summer and have a deeper root system than typical halophytes, succulent (and sometimes) reduced leaves, covered by a layer of water storage hairs that protect stomata;
4. *Xerophytes less halophytic*, characteristic to deep desalinized solonetzs, dried during summer: *Artemisia* spp. and *Camphorosma monspeliaca*. They have a xeromorphic nature, not succulent, but with strong shoots and leaves intensely divided into narrow segments or lacinia.

Genkel (1950, in Gorişina, 1979 and Kuliţiasov 1982) divided halophytes into three categories:

1. *Euhalophytes*, plants that accumulate large amounts of salt in their organs and grow on soils with high salt content: *Salicornia europaea*, *Suaeda maritima*, *Salsola soda*, *Halocnemum strobilaceum*, and *Petrosimonia* spp. They have a higher permeability of cell cytoplasm for salts, lower respiration rates, and reduced enzymatic activity. Due to huge accumulation of salts, osmotic pressure could reach 100–200 atmospheres and present a pronounced succulence. Due to this higher osmotic pressure, they can take up water from salinized soils;
2. *Cryno-halophytes*, plants that excrete salts, develop on soils with low to higher salinity: *Statice gmelini*, *Tamarix gallica*, *Frankenia,* and *Armeria*. Cell protoplasm is permeable for salt and the secretion of salts to exterior is realized by special glands, located on leaves;
3. *Glyco-halophytes*, plants that are not permeable for salts, such as *Artemisia maritima*. Cell cytoplasm has a lower permeability for salts. They belong in fact to freshwater environments but have also a limited capacity to adapt to salinity.

Bucur and collaborators (1957) gave a very consistent bio-ecological classification of halophytes, derived from an intense and elaborate work on halophyte ecology and their relationships with salinity measured in the rhizosphere (Table 1.1). This classification is perhaps among the most consistent and harmonious of all existing worldwide (Grigore 2013). Many systems of classifications are based on arbitrary criteria (see extended comments in Grigore 2008a, b), also taking

Table 1.1 Classification of halophytes according to Bucur and collaborators (1957)

Halophytes (plants vegetating on saline environments)	1. *Euhalophytes*: halophytes strictly adapted to salinity (strictly *obligate* to salinity) are *exclusively preferential* and grow *only* on salinized environments, with the entire or a part of the radicular system, both as seedlings and as mature plants 2. *Neohalophytes*: plants able to adapt to salinity; plants to be adapted to halophytic environment; they are *supporting* and *preferential*, living both on nonsalinized and salinized media, with the entire or a part of the radicular system
Non-halophytes (plants that do not grow on saline environments)	Plants non-adapted to salinized media, non-tolerant to high concentrations of salinity. In relation to concentrations more than 30–40% milligrams of soluble salts, they could be tolerant and preferential

Table 1.2 Equivalence between major Romanian systems of halophytes' classification (Grigore 2012)

Prodan (1939)	Ţopa (1954)	Bucur et al. (1957)		
"First category"	Obligatory	Obligatory	Euhalophytes	Halophytes
"Second category"	Preferential	Facultative halophytes (plants able to adapt to salinity)	Neohalophytes	
	Supporting			
"Third category"	Accidental	Supporting (tolerant to salinity)		Non-halophytes

into consideration the numerical values chosen for describing the thresholds of salinity where halophytes are to be included.

With respect to other major Romanian classifications previously made by Prodan (1939) and Ţopa (1939, 1954), a system to harmonize all these classifications has been figured (Grigore 2012) (Table 1.2).

As Grigore has shown (2013), Bucur's classification (1957) is astonishing, because it is the logic and natural result of scientific activity that is impressive through its vision, conception, and harmonious way in which the obtained data are correlated and interpreted. Bucur investigated the salinity thresholds for over 400 species vegetating in salinized meadows and pastures from Jijia-Bahlui depression; in this way, the obtained results and the derived classification have a high fidelity and consistency.

For establishing the salt affinity of species belonging to plant associations with halophytes, the authors used two investigation methods. The first is based on the variation of plant biomass in accordance with salinity changes in the rhizosphere; this method has been used for establishing salinity tolerance for plants cultivated on salinized areas, without irrigation. The second method is based on the variation of species frequency in the plant communities from salinized meadows.

After applying these methods and obtaining the results (salinity thresholds) for each species, several logical and interesting conclusions have been drawn; these actually offer valuable data about the ecology of species. Thus:

Table 1.3 Hierarchy of euhalophytes, taking into account the soil's salinization degree in the rhizosphere (Bucur et al. 1960)

Euhalophyte	Soil salinity in the rhizosphere (% mg soluble salts)
Very weak	75–95
Weakly/less	95–150
Moderately	150–450
Strongly	450–1400
Very strongly	1400–3400
Excessively	3400–5500

Table 1.4 Hierarchy of neohalophytes, taking into account the soil's salinization degree in the rhizosphere (Bucur et al. 1961)

Tolerant neohalophyte	Soil salinity in the rhizosphere (% mg soluble salts)
Very weak	55–75
Weakly/less	75–95
Moderately	95–150
Strongly	150–450
Very strongly	450–1500
Excessively	1500–3500

1. In several species, the biomass increases according to the increasing soil salinity in the rhizosphere; here two subgroups are described.
2. In other species, the biomass decreases according to the increasing soil salinity nearby the roots; two subgroups are also described.

Briefly, salt plants are divided into:

a. *Obligatory halophytes* (*strictly halophytes*, or halophytes); plants that grow only in saline environments: *Salicornia herbacea*, *Salsola soda*, *Atriplex hastata*, *Plantago schwarzenbergiana*, and *Petrosimonia triandra*;
b. *Facultative halophytes* (adaptable halophytes, plants adaptable to salinity); species that develop both in saline habitats and in normal soils. In saline soils, they have a fragile development and can desiccate faster, during the dry season or severe droughts: *Lepidium ruderale*, *Poa bulbosa*, and *Matricaria chamomilla*;
c. *Halo-phobous*: plants whose biomass decreased according to increasing soil salinity.

Going deeply and having many data at its disposal, Bucur and collaborators proposed some hierarchies within euhalophytes (Table 1.3) and neohalophytes (Table 1.4), in respect of the degree of soil salinization. These specifications are also relevant for the ecological description given by Bucur et al. (1960, 1961).

Bucur et al. (1957), taking into consideration soil humidity as a secondary factor, further divided halophytes into *xero-halophytes, meso-halophytes*, and *hygro-halophytes*. Xero-halophytes are strictly adapted to salinity and drought in the maximal period of plant development. Hygro-halophytes are adapted to salinity and constant humidity in the soil. Meso-halophytes are adapted to soil salinity and humidity.

In the basic classification, other ecological factors are included, resulting in a more detailed classification system. Thus, according to air and soil temperature, euhalophytes and neohalophytes, respectively, can be mega-thermophilous, meso-thermohilous, and micro-thermophilous.

According to the light factor, most of euhalophytes are heliophilous and less sciophilous.

A classification that takes into account the preferences of species for certain chemical elements in the soil is that offered by Şerbănescu (1965):

1. Associations of chloride salt areas: *Bassia hirsuta, Salicornia herbacea, Suaeda maritima, Salsola soda, Halimione pedunculata, H. verrucifera, Aeluropus littoralis, Puccinellia distans, Agropyron elongatum, Taraxacum bessarabicum, Aster tripolium, Cyperus pannonicus, Spergularia marginata, Crypsis aculeata, Petrosimonia triandra, Cerastium anomalum, Juncus gerardii, Beckmannia eruciformis, Trifolium angulatum, Iris halophila, Pholiurus pannonicus, Leuzea salina, Atriplex littoralis, Erysimum repandum,* and *Tamarix ramosissima.*
2. Associations of sodium salt areas: *Carex divisa, Camphorosma annua, C. monspeliaca, Plantago maritima, Lepidium crassifolium,* and *Hordeum maritimum.*
3. Associations of sulfate salt areas: *Artemisia maritima* and *Limonium gmelini.*

1.4 "Obligatory" Halophytes: Perhaps an Inadequate Term

In the common language used in relation to halophytes' definition and classification, lots of papers deal with "obligatory" and "not obligatory" character of halophytes: which species should be considered as "obligatory" halophytes? Moreover, what is the border between "obligatory" and non-"obligatory" character of a halophyte? (see Barbour 1970 iconic question).

Chapman (1975) provides a footnote explaining that obligatory halophytes are those species that reach their optimal growth under conditions of salinities exceeding 0.5% NaCl. This seems to be the "classic" definition that became well established in usual language when generally talking about halophytes.

Nevertheless, talking about the types of halophytes, often their meaning is implicitly and sometimes explicitly inferred (Waisel 1972), in the sense that certain species of halophytes (obligatory) require for their growth high levels of salt. Barbour (1970) suggests that an obligatory halophyte is a species with optimal growth at moderate or high salinity and incapable of growth at low salinity (low salinity meaning in this case more than 2% salt). Although Chapman (1960) and Waisel (1972) suggest that some species (particularly succulent *Chenopodiaceae*) are obligatory halophytes, Barbour argues that there is no evidence supporting that any of coastal or mangrove species is consistent with its definition about obligatory halophyte. Indeed, there are coastal or mangrove species that can grow satisfactorily under normal conditions. The author also notes that species requiring sodium as micronutrient actually need so small amounts of sodium that they can hardly be called obligatory halophytes. Barbour also emphasized that salt tolerance of plants is quite variable and species strictly limited to saline soils are quite rare. Moreover, even salt tolerance often seems to reflect the experimental conditions as well as the self-imposed salinity. Gale et al. (1970) have shown this for *Atriplex halimus*. In low humidity conditions, the growth and yield were maximal at lower salinity of about −5 atm. However, in ambient conditions of high humidity, optimal growth occurs in nonsaline control solution.

Walter (1974) defines halophytes as plants that grow on salinized soils; however, he refers to "*true halophytes, as those plants that accumulate in their organs large amounts of salts*," without being damaged but rather being stimulated at this concentration not "too" high. Therefore, he distinguishes between euhalophytes (true halophytes) and other categories of salt-tolerant plants. He stated that species tolerant to salinity withstand a moderate concentration of salts but develop better on nonsalinized soils. Contrarily, on true halophytes (euhalophytes), growth is stimulated by a certain accumulation of salts. On ordinary soils that contain only small amounts of NaCl, halophytes can take up these small amounts, so their content in salts still remains elevated. Stimulating effect is due to the chloride ion, which intensifies cell turgor; the water uptake is thus increased at the cell level, leading to hypertrophy and, consequently, to increasing succulence. Contrarily, sulfate ions may accumulate in cells, but plants are not succulent. Therefore, a distinction should be made between *chloridic* and *sulfate* halophytes; Walter also delineates *alkaline* halophytes, where accumulated sodium will reach the soil as Na_2SO_4, after the plant decomposition.

From an ecological point of view, however, it would be necessary to demonstrate that halophyte species can successfully complete their life cycle in nonsaline conditions. In fact, ecological tolerance to salt should be defined as the ability of plants to compete and reproduce in particular environments. Unfortunately, this has been demonstrated for a few species, and it should also follow the behavior of halophyte seeds germinating in nonsaline conditions. Schimper (1903) and Warming (1909) have overcome the perception that halophytes would involve the obligatory presence of salt. Schimper (1903) recognized that most halophytes can grow quite well in nonsaline conditions. In this context, it seems unlikely that an obligatory requirement for high levels of salts occurs in angiosperms.

Surprisingly, many typical halophytes are found to vegetate in natural conditions under salinity thresholds that are usually arbitrarily suggested (Table 1.5). The underlined species and their values show important differences within multiple soil samples from the rhizosphere of the same species—a "fact that suggests again the difficulty of managing soil salinity and plant–soil relationships" (Grigore 2008a, b, 2012; Grigore et al. 2010, 2014).

In a dissertation about plants' environments, Hedenberg (1754, published in Hedenberg 1788), supervised by Linné, underlined the importance of salt as an environmental element in the plant life. He gave the example of *Nitraria schoberi* (also a rare Romanian halophyte species, restricted only to Buzau county, Fig. 1.7)

Table 1.5 Values of pH and EC for several halophytes vegetating in Valea Ilenei nature reserve (Grigore and Toma 2014)

Halophyte species	pH	EC (dS/m^{-1})
Halimione verrucifera	8.58	1.54
Suaeda maritima	9.3	4.03
Salicornia europaea	**8.55–8.87**	**8–11.82**
Atriplex littoralis	9.2	2.49
Limonium gmelinii	7.92	2.29
Lepidium crassifolium	**9.04–9.78**	**4.54–10.56**
Artemisia santonicum	8.01	0.57
Aster linosyris	8	0.35
Bolboschoenus maritimus	8.95	2.39
Juncus gerardii	9.05	4.92

Fig. 1.7 *Nitraria schoberi* (Photo courtesy of Ana Cojocariu)

which grew for 20 years under conditions of Uppsala (Sweden) without flourishing, remaining sterile; however, it flourished the next year, when salt was added to the culture medium.

The same idea was expressed by Linné himself in a letter sent in 1759 to François Boissier de La Croix de Sauvages: "*this year, after an assiduous attempt, (I) finally obtained fruiting of Nitraria schoberi (...) whose flowers were not seen in Europe, despite it is found in all gardens; I obtained (flowers) only after salt has been added.*" (D'Hombres-Firmas 1860).

However, discussions about requirements of salts by halophytes are still complicated and controversial. Thus, Weissenbock (1969) believes that the terms "obligatory halophyte" and the "facultative halophyte" should be reconsidered, proposing the use of others, more specific in a physiological sense. The author suggests that the term "facultative" should be applied to plants whose growth is favorably affected by NaCl, but where Na can be replaced by K (*Aster triploum*, *Artemisia maritima, Plantago matritima,* and *Suaeda maritima*). In the same respect, obligatory halophytes would be considered those that are positively affected by NaCl and Cl (*Salicornia herbacea* and *Atriplex vesicaria*).

Ungar et al. (1969) (mentioned by Sharma and Gupta 1986) suggest that many halophytes are perfectly able to grow normally in environments with low salinity, or even nonsaline, and are called facultative halophytes.

Sometimes, though, when the term halophyte is usually being used, the connotation of "obligatory" is implicitly attributed to it (Grigore 2012). Thus, Bucur et al. (1957), when discussing about halophytes, introduced the following remark: "[...] *to distinguish them from those who live only in saline soils and that have been called strict(ly) or obligatory halophytes, or simply, halophyte.*" As discussed above, there was and still exists a tendency to reduce the term *halophyte* to its "*obligatory*" nature.

However, Grigore (2008a, b) suggested that the expression "*preferential halophytes*" is a little bit inconvenient, at least from a semantic point of view; according even to the broadest definition, all halophytes (as their etymology emphasizes) do *prefer* a relatively higher salt concentration in soil solution, as compared to glycophytes. Therefore, "*preferential halophytes*" expression could remain confusing; it suggests an inferior graduation of halophilic scale (as compared to euhalophytes), but, in fact, all halophytes would prefer salts as compared to glycophytes.

Perhaps the "supporting" and "accidental" nominations are more appropriate and seem to correspond to a close reality at plant–soil level.

The "*obligatory*" term, related to halophytes, represents in conclusion an imprecise and equivocally definition.

Sometimes, the distinction between halophytes and other species having a certain degree of salt tolerance is naturally made: "(...) *by halophytes and**other**salt-tolerant plants*" (Goodin et al. 1990, p. 5).

Nevertheless, Grigore (2012) reviewed the semantic field built up about halophyte-related language, both in Romanian and in English as well (Table 1.6).

Starting from a standard definition of halophytes—plants able to grow and complete their life cycle in habitats with soil salinity higher than 200 mM NaCl (Flowers et al. 1986; Flowers and Colmer 2008)—Grigore et al. (2012) experimentally inquire about the *obligatory* character of halophytes. It has been shown that

Table 1.6 Semantic field with different words related to halophytes (Grigore 2012)

Romanian	English
Halofite; plante de sărătură; plante halofile; plante iubitoare de săruri (sare); plante de locuri sărate; plante salifere	Halophytes; salt-tolerant plants; salt plants; high salinity-tolerant plants; salt-loving plants; halophylous plants; halophytic plants; maritime plants

this operational definition, which excludes 99% of all angiosperm species, may be useful, but, obviously, this concentration threshold is rather arbitrary. In fact, there is a continuous spectrum of salt tolerance among plant species, from typical glycophytes (salt-sensitive plants) to extreme halophytes. Another matter of discussion is the possible positive effect of salt on the growth and development of halophytes. Several definitions (Aronson and Le Floc'h 1996; Chapman 1936; Dansereau 1957) suggest that salts—especially NaCl—are compulsorily required during the life cycle of halophytes, due to their stimulating effect upon several biological processes in this type of plants. To emphasize this requirement, a subcategory of halophytes has been described: *euhalophytes*, sometimes called "absolute halophytes," "exclusive halophytes," or "obligatory halophytes"; however, by extrapolation or by misinterpretation of the original definition, the term "euhalophytes" is about to be used for all categories of salt-tolerant plants.

Grigore et al. (2012) conducted a study on two recognized and described as halophytes species (*Plantago crassifolia* and *Inula crithmoides*) and the third one, *Medicago marina*, as a psammophyte, specific for sand dunes. Measurements of growth parameters—number of leaves, plant length, and fresh and dry weights—showed that all three species grew much better on the salt-free and nutrient-rich substrates, peat, and garden soil, than on saline soil and dune sand (Figs. 1.8 and 1.9). These results indicate that salts are not compulsorily required for development of halophytic species and suggest that limitations of water and nutrients, rather than soil salinity per se, are the most important restrictive factors for plant growth in saline habitats. The distribution of halophytes in nature is probably dependent on their limited ability to compete with glycophytes in nonsaline areas while remaining highly competitive under environmental conditions stressful for non-tolerant species.

However, data regarding the effect of salts on halophytes are not homogeneous. In many halophytic species, growth is inhibited by increasing salt concentrations; on the contrary, in several genera—such as *Salicornia*, *Suaeda*, or *Atriplex*—a stimulation of growth in the presence of salt has been observed, although no species has been shown to grow optimally at seawater or higher salt concentrations (Ungar 1991; and references therein). For instance, dry mass production is stimulated by salinity in species that can be regarded as euhalophytes, such as *Aster tripolium* (Baumeister and Schmidt 1962), *Salicornia brachystachya, S. patula* (Grouzis et al. 1977), *S. europaea* (Ungar 1978), *Spartina anglica* (Partridge and Wilson 1987), *Suaeda monoica* (Storey and Wyn Jones 1979), or *S. salsa* (Hekmat-Shoar 1978). In other taxa, which could be considered less halophytic, a decrease in dry mass upon addition

Fig. 1.8 Relative size of representative individual plants of *Inula crithmoides*, after 12 weeks of growth in different substrates: peat (**a**), garden soil (**b**), saline soil (**c**), and dune sand (**d**) (plants of the different species are not shown at the same scale) (Grigore et al. 2012)

Fig. 1.9 Relative size of representative individual plants of *Plantago crassifolia*, after 12 weeks of growth in different substrates: peat (**a**), garden soil (**b**), saline soil (**c**), and dune sand (**d**) (plants of the different species are not shown at the same scale) (Grigore et al. 2012)

of salt has been observed, for example, in *Atriplex gmelini* (Matoh et al. 1986), *A. hastata* (Black 1956), *A. nummularia* (Uchiyama 1987), *A. inflata* (Ashby and Beadle 1957), *A. triangularis* (Drake and Ungar 1989), *A. vesicaria* (Black 1960), or *Spartina alterniflora* (Hopkins et al. 1978).

Nevertheless, even in highly tolerant halophytes such as *Salicornia* species, increased biomass production has been shown to occur in the range from 170 to 340 mM NaCl (data summarized by Ungar 1991), whereas hypersaline conditions in the field, ranging from 500 to 1000 mM total salts, were found to inhibit the biomass production of *S. europaea* (McGraw and Ungar 1981). Similarly, growth of *Inula crithmoides* plants, submitted to increasing salt concentrations for a period of 87 days, was only affected by salinity exceeding 20 dS/m, and the accumulated biomass of plants irrigated with 40 dS/m saline water was nearly half of that of the control plants grown in the absence of salt (Zurayk and Baalbaki 1996).

Most of the examples mentioned above refer to plants treated with salt under controlled artificial conditions. The question remains as to the relative importance of salt stress and other environmental conditions for the distribution of halophytes in nature. Our results suggest that the growth and development of plants present in saline habitats depend not so much on soil salinity per se but rather on other factors such as the availability of water and nutrients.

References

Adam P (1990) Saltmarsh ecology. Cambridge University Press, Cambridge

Aronson J, Le Floc'h E (1996) Restoration ecology of salt-affected, arid and semi-arid lands. In: Choukr-Allah R, Malcolm CV, Hamdy A (eds) Halophytes and biosaline agriculture. Marcel Dekker, New York, pp 55–72

Ashby WC, Beadle NCW (1957) Studies in halophytes: III. Salinity factors in the growth of Australian saltbushes. Ecology 38:344–352

Barbour MG (1970) Is any angiosperm an obligate halophyte? Am Mid Nat 84(1):105–120

Baumeister W, Schmidt L (1962) Über die Rolle des Natrium im pflanzlichen Stoffwechsel. Flora 152:24–56

Black RF (1956) Effect of NaCl in water culture on the ion uptake and growth of *Atriplex hastata* L. Aust J Biol Sci 9:67–80

Black RF (1960) Effects of NaCl on the ion uptake and growth of *Atriplex vesicaria* Howard. Aust J Biol Sci 13:49–66

Bucur N, Dobrescu C, Turcu G, Lixandru G, Teşu C, Dumbravă I, Afusoaie D (1957) Contribuţii la studiul halofiliei plantelor din păşuni şi fâneţe de sărătură din Depresiunea Jijia-Bahlui (partea a I-a). Stud şi Cerc (Biol şi Şt Agric), Acad RPR, filiala Iaşi 8(2):277–317

Bucur N, Dobrescu C, Turcu G, Lixandru G, Teşu C (1960) Contribuţii la studiul halofiliei plantelor din păşuni şi fâneţe de sărătură din Depresiunea Jijia-Bahlui (partea a II-a). Stud şi Cerc (Biol şi Şt Agr), Acad RPR, filiala Iaşi 2(11):333–347

Bucur N, Dobrescu C, Turcu G, Lixandru G, Teşu C (1961) Contribuţii la studiul halofiliei plantelor din păşuni şi fâneţe de sărătură din Depresiunea Jijia-Bahlui (partea a III-a). Stud şi Cerc (Biol şi Şt Agr), Acad. RPR, filiala Iaşi 12(1):169–190

Chapman VJ (1936) The halophyte problem in the light of recent investigations. Quart Rev Biol 11:209–220

Chapman VJ (1942) The new perspective in the halophytes. Q Rev Biol 17:291–311

Chapman VJ (1960) Salt marshes and salt deserts of the world. Plant Science Monographs. Interscience, New York

Chapman VJ (1975) The salinity problem in general, its importance, and distribution with special reference to natural halophytes. In: Poljakoff-Mayber A, Gale J (eds) Plants in saline environments. Springer, Berlin, pp 7–24

Chermezon H (1910) Recherches anatomiques sur les plantes littorales. Ann Sci Nat sér 9 Bot 12:117–313

Contejean C (1881) Géographie botanique. Influence du terrain sur la végétation. Baillière et Fils, Paris

Correll DS, Johnston MC (1970) Manual of the vascular plants of Texas. Texas Research Foundation, Renner

Dansereau P (1957) Biogeography, an ecological perspective. Ronald Press, New York

Drake DR, Ungar IA (1989) The effects of salinity, nitrogen level, and population density on the survival, growth, and reproduction of *Atriplex triangularis* (*Chenopodiaceae*). Am J Bot 76:1125–1135

Duncan WH (1974) Vascular halophytes of the Atlantic and Gulf coasts of North America north of Mexico. In: Reimold RJ, Queen WH (eds) Ecology of halophytes. Academic, New York, pp 23–49

Fernald ML (1950) Gray's manual of botany, 8th edn. American Book Company, New York

Flowers TJ, Colmer TD (2008) Salinity tolerance in halophytes. New Phytol 179:945–963

Flowers TJ, Troke PF, Yeo AR (1977) The mechanism of salt tolerance in halophytes. Ann Rev Plant Physiol 28:89–121

Flowers TJ, Hajibagheri MA, Clipson NJW (1986) Halophytes. Q Rev Biol 61:313–337

Frey RW, Basan PB (1985) Coastal salt marshes. In: Davis RA (ed) Coastal sedimentary environments, 2nd edn. Springer, New York, pp 225–301

Gale J, Naaman R, Poljakoff-Mayber A (1970) Growth of *Atriplex halimus* L. in sodium chloride salinated culture solutions as affected by the relative humidity of the air. Aust J Biol Sci 23 (4):947–952

Goodin JR, Epstein E, Mckell CM, O'Leary JW (eds) (1990) Saline agriculture. Salt tolerant plants for developing countries. National Academy Press, Washington, DC

Gorişina TK (1979) Ecologhia rastenij. Moskv. "Vîsşaia Şkola" (Plant Ecology)

Grigore M-N (2008a) Introducere în Halofitologie. Elemente de Anatomie Integrativă. Ed. PIM, Iaşi

Grigore M-N (2008b) Halofitotaxonomia. Lista plantelor de sărătură din România. Ed. PIM, Iaşi

Grigore M-N (2010) O abordare conceptual-semantică a halofitelor, într-un climat general dominat de salinizare şi insecuritate alimentară. In: Ivănescu L, Zamfirache MM (eds) honorem Prof. Constantin Toma, la a 75-a aniversare. Edit. Graphys, Iaşi, pp 305–323

Grigore M-N (2012) Romanian salt tolerant plants. Taxonomy and ecology. Iasi, Edit. Tehnopress

Grigore M-N (2013) Nicolae Bucur's contribution to create an original system of halophytes classification. An example of holistic ecological vision. Lucr Şt (Hortic), USAMV "Ion Ionescu de la Brad", Iaşi 56(1):19–24

Grigore M-N, Toma C (2010) Halofitele. Aspecte de anatomie ecologică. Edit. Univ. "Al. I. Cuza", Iaşi

Grigore M-N, Toma C (2014) Integrative ecological notes on halophytes from "Valea Ilenei" (Iaşi) nature reserve. Mem Sci Sect Rom Acad 37:19–36

Grigore M-N, Toma C, Boscaiu M (2010) Dealing with halopytes: an old problem, the same continuous exciting challenge. An Şt Univ "Al. I. Cuza" Iaşi, s. II.a (Biol Veget) 56(1):21–32

Grigore M-N, Villanueva M, Boscaiu M, Vicente O (2012) Do halophytes really require salt for their growth and development? An experimental approach. Not Sci Biol 4(2):23–29

Grigore M-N, Ivănescu L, Toma C (2014) Halophytes. An integrative anatomical study. Springer, Cham

Grouzis M, Heim G, Berger A (1977) Croissance et accumulation de sels chez deux salicornes annuelles du littoral médierranéen. Oecol Plant 12:307–322

Guşuleac M (1933) Urme de vegetaţie halofilă în Bucovina. Bul Fac Şt Cernăuţi 7(1–2):329–339
Hedenberg A (1788) Stationes plantarum. Thesis no 54, Upsalia, 1754. In Amoenitates Academicae, vol. IV, editio secunda, Erlangae sumtu IO. Iacobi Palm, pp 64–87
Hekmat-Shoar H (1978) Contribution a l'etude cytotaxonomique de quelques halophytes d'Azerbaidjan (Iran). Rev Cytol Biol Veg Bot 1:97–104
Hopkins CS, Gosselink JS, Parrando RT (1978) Aboveground production of seven marsh species in coastal Louisiana. Ecology 59:760–769
Iversen J (1936) Biologische Pflanzentypen als Hilfsmittel in der Vegetationsforschung. Dissertation. Medd. fra Skalling Laboratoriet, Copenhagen
Jennings DH (1976) The effect of sodium chloride on higher plants. Biol Rev 51:453–486
Khan MA, Duke NC (2001) Halophytes—a resource for the future. Wet Ecol Manag 6:455–456
Kovda VA (1939) Solurile U.R.S.S., vol 1. Partea Europeană a U.R.S.S., Moscova-Leningrad
Kuliţiasov IM (1982) Ecologhia rastenij. Izdat. Moskv. Universit (Plant Ecology)
Lambers H, Chapin SF III, Pons TL (2008) Plant physiological ecology, 2nd edn. Springer, New York
Lawrence GHM (1951) Taxonomy of vascular plants. MacMilian, New York
Le D'Hombres-Firmas B (1860) Lettres inédites de Linné à Boissier de la Croix de Sauvages, Alais, Typographie et Lithographie de A. Veirun
Măianu A, Aksenova I, Albescu I (1965) Toleranţa la salinitate a principalelor plante agricole pe soluri freatic-umede cu salinizare clorurică. An Sect de Pedol ICCA 33:357–373
Marschner H (1995) Mineral nutrition of higher plants, 2nd edn. Academic/Harcourt Brace, London/San Diego
Matoh T, Watanabe J, Takahashi E (1986) Effects of sodium and potassium salts on the growth of a halophyte *Atriplex gmelini*. Soil Sci Plant Nutr 32:451–459
McGraw DG, Ungar IA (1981) Growth and survival of the halophyte *Salicornia europaea* L. under saline field conditions. Ohio J Sci 81:109–113
Oosting HJ (1956) The study of plant communities. W.H. Freeman, San Francisco
Partridge TR, Wilson JB (1987) Salt tolerance of salt marsh plants of Otago, New Zealand. NZ J Bot 25:559–566
Poljakoff-Mayber A, Gale J (1975) Plants in saline environments. Springer, Berlin
Prodan I (1922) Oecologia plantelor halofile din România, comparate cu cele din Ungaria şi Şesul Tisei din regatul SHS. Bul Inf Grăd Bot şi Muz Bot Univ Cluj, 2, 3: 37–52, 69–84, 101–112
Prodan I (1923) Oecologia plantelor halofile din România, comparate cu cele din Ungaria şi Şesul Tisei din regatul SHS. Tipografia 'Ardealul', Cluj
Prodan I (1939) Flora pentru detrminarea şi descrierea plantelor ce cresc în România, II, (ediţia a II-a). Tipografia Cartea Românească, Cluj
Richards LA (1954) Diagnosis and improvement of saline and alkali soils. US Dep Agric Handb 60
Ranwell DS (1972) Ecology of salt marshes and sand dunes. Chapman and Hall, London.
Schimper AFW (1903) Plant geography upon a physiological basis. Clarendon Press, Oxford
Schulze E-D, Beck E, Müller-Hohenstein K (2005) Plant ecology. Springer, Berlin
Şennikov AP (1950) Ecologhia rastenij. Izdat. Sovetskaia Nauka, Moskva (Plant Ecology)
Şerbănescu I (1965) Asociaţiile halofite din Câmpia Română. Com Geol ale Instit Geol St Tehn şi Econ, seria C, Pedologie, Bucureşti 15:1–148
Sharma SK, Gupta IC (1986) Saline environment and plant growth. Agro Botanical, New Delhi
Steiner M (1935) Zur Oekologie der Salzmarschen der nordöstlichen Vereinigten Staaten von Nordamerika. Jahrb f wiss Bot 81:94–202
Stocker O (1928) Das Halophytenproblem. Ergeb Biol 3:265–353
Storey R, Wyn Jones RG (1979) Salt stress and comparative physiology in the *Gramineae*. III. Effect of salinity upon ion relations and glycinebetaine and proline levels in *Spartina* × *townsendii*. Aust J Plant Physiol 5:831–838
Ţopa E (1939) Flora halofitelor din Nordul României în legătură cu cea din restul ţării. Teză prezentată la Facultatea de Ştiinţe din Cernăuţi

Ţopa E (1954) Vegetaţia terenurilor sărate din R.P.R. Natura 6(1):57–76
Uchiyama Y (1987) Salt tolerance of *Atriplex nummularia*. Technical Bulletin of the Tropical Agricultural Research Center no 22. Tropical Agricultural Research Centre. Yatabe, Tsukuba, Ibaraki
Ungar IA (1978) The effects of salinity and hormonal treatments on the growth and ion uptake of *Salicornia europaea*. Bull Soc Bot Fr 3-4:95–104
Ungar IA (1991) Ecophysiology of vascular halophytes. CRC Press, Boca Raton
Van Eijk M (1939) Analyse der Wirkung des NaCl auf die Entwicklung Sukkulenze und Transpiration bei *Salicornia herbacea*, sowie Untersuchungen über den Einfluss der Salzaufnahme auf die Wurzelatmung bei *Aster tripolium*. Rec Trav Bot Neerl 36:559–657
Vilbouchevitch J (1892) L'étude géo-botanique des terrains salants. Bull Sci Soc Bot de France 39 (10):XXVIII–XXXXVI. (Session extraordinaire en Algérie)
Waisel Y (1972) Biology of halophytes. Academic, New York
Walter H (1974) Vegetaţia Pământului în perspectivă ecologică. Edit. Ştiinţifică, Bucureşti
Warming E (1909) Ecology of plants. An introduction to the study of plant communities (English edition prepared by P. Groom, I.B. Balfour). Clarendon Press, Oxford
Weissenbock G (1969) Einfluss des Bodensalzgehaltes auf Morphologie und Ionenspeicherung von Halophyten. Flora (Jena) 158:369–389
Willdenow CL (1792) Grundriss der Kraüterkunde zu Vorlesungen entworfen. Bei haude und Spener, Berlin
Willdenow CL (1799) Grundriss der Kraüterkunde zu Vorlesungen entworfen. Mit v. Ghelenschen Schriften, Wien
Willdenow CL (1805) The principles of botany and of vegetable physiology. Edinburgh University Press, Edinburgh
Zurayk R, Baalbaki R (1996) *Inula crithmoides*: a candidate plant for saline agriculture. Arid Soil Res Rehab 10:213–223

Chapter 2
Saline Environments

The Earth's total surface area covers about 13.2 billion ha, but no more than 7 billion ha are arable and 1.5 billion are cultivated (Massoud 1981). Of the cultivated lands, about 340 million ha (23%) are saline (salt-affected) and another 560 million ha (37%) are sodic (sodium-affected) (Tanji 2002).

Salinity is a common phenomenon and one of the fundamental features of arid and semiarid zones. Basically, there is a close correlation between soil aridity and salinity. Conditions favoring evaporation actually facilitate salt concentration of groundwater in the upper soil layers. It is known that salinity, a major constraining factor for major crops, greatly limits production (Gorham 1995; Shannon 1998; Munns 2002). About 20% of agricultural land and 50% of crops are affected by salt stress (Flowers and Yeo 1995). According to some data, about 23% of world agricultural area (about 1.5×10^9 ha) is saline and about 37% is sodic (Khan and Duke 2001).

The importance of knowledge about saline environments, i.e., habitats in which halophytes vegetate, is relevant for the context of this book from several points of view. First, from a theoretical but very useful point of view, since the language used abroad and in Romanian literature operates with many words and phrases that are often difficult to validate. In addition, there is no absolute system of equivalence regarding different soil classifications. Often, in worldwide literature, be it older or newer, sometimes vague or less precise expressions are being used, which are not easily to be identified at a first glance or by a nonspecialist. Moreover, even a historical evolution of soil science terms can be noticed when consulting various works dealing with salinity and related issues.

In respect of saline soils, there are many systems of classification; at that time, almost every country (the USA, former SSSR, Hungary, France, and Romania) has developed its own system and lots of efforts were made thereafter in order to establish correspondent equivalences (Florea 1963; Szabolcs 1974; Sandu 1984; World Reference Base for Soil Resources 2006).

M.-N. Grigore, C. Toma, *Anatomical Adaptations of Halophytes*,
DOI 10.1007/978-3-319-66480-4_2

Nevertheless, the intention in this chapter is not to review the knowledge about saline soils, in a deep soil science way, but rather to offer several general data about saline environments in their relation with halophytic flora.

For instance, Romanian botanical literature has used and still uses the term "sărătură" (sărături), quite difficult to translate in English; perhaps, "salt marsh," in its broader sense, would fit approximately to this Romanian term.[1]

These salt-affected soils are more or less salinized soils on their entire or partial soil profile. These soils are dominated by the unfavorable or detrimental effects of excess soluble salts or sodium ions from their absorption complex on vegetation (Chiriţă 1955). Nevertheless, in his impressive handbook on soil science, Chiriţă (1955) includes "sărăturile" in the major class of halomorphic and hydrogenetic soils, in the direction of salinization and desalinization with alkaline degradation, next to "related soils." In a footnote, he explains that "sărăturile" term is being used partially inadequately; it refers to soils formed by salinization, desalinization, and alkaline degradation processes. This footnote (p. 590, op. cit.) is subtle, but very important since the author recognizes that this term is common in scientific language and it facilitates the use of a more short, generic expression.

Sandu (1984) defines *sărătura* as a soil whose fertility is intensely affected by the high content of soluble salts in the soil profile, by the presence of exchangeable sodium in the colloidal complex, and by mineralized water table located in the upper layers. The term is linked to agricultural properties of the soil, reflected mainly by its behavior in agricultural work, but also by the behavior of crop plants and their yields. Salinization is the process of increasing the content of soluble salts in the soil, beyond the normal limit of other soils (i.e., more than 0.08–0.1%). Salinization may occur naturally or induced by human activity. Alkalization means the increase of exchangeable sodium content more than normal content from other soils (i.e., more than 5%) and possibly accumulation of carbonate or sodium bicarbonate. The intensity of salinization is evaluated by salinity and alkalization levels at different depth horizons and their location (Lupaşcu et al. 1998).

Saline and alkali soils are those that present (in excess regarding salinity tolerance threshold in crop plants) soluble salts (saline soils), exchangeable sodium in adsorption complex (alkali soils), or both soluble salts and exchangeable sodium (saline-alkali soils). Alkali soil refers only to the presence of exchangeable sodium in excess, compared with alkaline soil, whose pH is greater than 7.2. In connection with these terms, the following may also be used: salinization (accumulation of soluble salts in the soil profile); alkalization (solonetz soil turning), which represents the replacement of divalent cations (Ca^{2+} and Mg^{2+}) from soil adsorption complex with sodium; and solodic soil turning (replacing the exchangeable sodium with hydrogen). Thus, *sărăturarea* (the process of *sărătură* formation) includes salinization and/or alkalization, or both of them (Sandu 1984).

[1]This term will be used throughout the text as a Romanian native one, without being translated, as is the case of Arabic term *sabkha* (flat salt desert), a term accepted nowadays by multidisciplinary scientists (see the series edited by Öztürk et al. 2011; Khan et al. 2014).

As far as we can see until now, this term is not necessarily belonging to soil sciences (see Chiriță's footnote). It is largely used in geobotany and ecology and by botanists, who are not soil scientists, and invented and implemented as an inherent need to express a reality from the field: the intimate relationship between a type of salinized soil and halophytic vegetation. The use of this term was based more on macroscopic observations made by botanists, who are not, in fact, soil scientists. Actually, this term was suggested by botanists and adopted by early soil science handbooks as a familiar term and because of the lack of other specific and precise terms (as we know them nowadays). Turning to soil science field was made only when other terms, more technical and precise, occurred.

Romanian System of Soil Classification (SRCS 1980, Conea et al. 1980) used and naturalized the following terms, many of them included thereafter in botanical works: *solonchak, solonetz, and solod* (mainly derived from former Soviet Union classification system). In the next paragraphs, general information about these type of saline soils will be given, using both Romanian and foreign data.

2.1 Solonchaks (Also Called Saline Soils, Saline Alkali Soils)

These soils are characterized by the accumulation of soluble salts in the upper horizon (0–20 cm) over 1 g/100 g soil with a relatively uniform distribution on the profile by the existence of a salic horizon (Sandu 1984). In the chemical composition of salts, chlorides (NaCl, $CaCl_2$), sulfates ($NaSO_4$, $MgSO_4$), carbonates (Na_2CO_3), bicarbonates ($NaHCO_3$), nitrates ($NaHO_3$, KHO_3), and borates predominate (Sandu 1984). The content of soluble salts on the surface of solonchaks is more than 1–1.5 g/100 g soil depending on the ionic composition and harmfulness of soluble accumulated salts on their profile. Solonchaks surfaces may present salt efflorescence and/or crystals of soluble salts; morphologically, these become noticeable when the amount of salts is higher than 4–5 g/100 g soil. Florea (1963) defines them as soils that contain in their horizon large amounts of soluble salts, generally higher than 1–1.5%. By salinized soils (treated by this author together with solonchaks), he understands soils (of different genetic types) containing in their profile (up to 1–1.5 m) appreciable amounts (over 0.1–0.2%) of soluble salts.

There is a particular arrangement of salts on the vertical structure of solonchaks: in its upper part, chlorides or other soluble salts, then calcium carbonate, and in the lower part, the gypsum is located.

According to the origin and nature of the salinization process, solonchaks are divided into the following types:

- Maritime (coastal and lagoon type)
- Alluvial (on young landforms)
- Continental (which may be wet or meadow like)

- Residual
- Semiresidual (without connection to groundwater)

After the chemical composition of accumulated salts, solonchaks can be divided into chloride, sodic, chloride-sulfatic, sulfate-chloride, nitrate, and even boron type.

From the morphological point of view, solonchaks are divided as follows: wet (soil profile is permanently saturated with mineralized water irrigation); fluffy (they have at the surface a fluffy saline horizon); crust solonchaks; and columnar solonchaks. Romanian System of Soil Classification (Conea et al. 1980) distinguishes the following subtypes of solonchaks (in parentheses are given the similarities): typical (residual, with crust, marine); mollic (humus-like solonchaks); vertic; gleyic (meadow solonchaks); and alkalized (solonized solonchaks) (Sandu 1984). Primary and secondary solonchaks can also be distinguished. Those primaries form on saline parental materials under the influence of natural factors (without human intervention) in places with poor drainage depressions with shallow groundwater, located in arid climates. Secondary solonchaks were formed under the influence of human activity through wasteful irrigation, lakes, and accumulation dams and by insufficient drainage of areas with high potential for salinization. They are formed due to water infiltration into the deep and intensive mobilization of salts on large areas and large thicknesses (Sandu 1984).

Salts found in these soils may have different origins. They may come from sedimentary rock that formed the soil, being accumulated here in a (geological) period on the bottom of sea or lakes; this type of salinization is called residual salinization. In other cases, the salts from the soil come from the atmosphere (dust or rain falling on the ground); this origin—by salt spray—is important only next to seas, oceans, or salt lakes. The most important source of salts is the mineralized groundwater, if located near the soil surface and takes part in soil wetting. In groundwater, salts derive from alteration of primary rocks, from salts existing in sedimentary deposits, and from where water is drained. To these salinization types, another source may be added: water from irrigation (Florea 1963). Salts that reach the ground are subject to a vertical movement, depending on soil hydric regime. Basically, if water status is of the exudative regime (that is characteristic to sectors with shallow groundwater in relatively arid climatic conditions), there is an intense accumulation of salts in the soil. Groundwater rises to the surface by capillarity and during dry periods evaporate, depositing moved salts. Plants also extract water from the soil and favor the accumulation of salts in the upper horizon. In wet periods, descendant currents may occur, which can carry some of the salts in soil depth; during the dry period, however, new amounts of salts are brought to the surface, so finally, a gradual accumulation of salts in the upper horizon and differentiation of salts in various horizons occurs. Conditions favoring the accumulation of salts and thus forming solonchaks are represented by regions without or poor drainage, such as major depressions, lower terraces of rivers, deltas, sea, or lakes littoral. It must be said that with soil salinization, mineralization of groundwater occurs (accumulation of salts in groundwater); these processes take place simultaneously. Fluctuations in the level of groundwater combined with those of ascendant and descendent currents of soil water, during the wet-cool and dry-warm, cause an important dynamic of soluble salts in soil; this implies a variation of water content and, consequently, impact on the osmotic pressure of soil solution and the consistency of the soil. These variations impose development of two types of vegetation: one hygrophilous and weakly halophilous in the spring and another halophilous or halo-xerophilous, during the summer (Chiriţă 1955). Solonchaks have therefore large amounts of soluble salts and thus present typical halophilous vegetation adapted to these environmental conditions; among the most typical plants are *Salicornia europaea, Suaeda maritima, Halocnemum strobilaceum, Salsola soda,* and *Puccinellia distans*. In autumn, the vegetation becomes somewhat reddish, and sometimes small patches may occur without any vegetation cover; these are areas where the salt concentration is too high to allow the existence of plants on their surface.

2.2 Solonetzs (Alkali Soils)

These are soils with a relatively high content of exchangeable Na^+ in the colloidal complex of the illuvial horizon (over 20% of the exchangeable Na from cation exchange capacity) (Florea 1963). Solonetzs are considered those (belonging to different genetic types) that contain exchangeable sodium between 5 and 20% of the cation exchange capacity. Genesis of solonetzs is not yet clearly understood, and it is even controversial: solonetzs would have formed either by desalinization of solonchaks or by salinization and alternative desalinization, accompanied by alkalization (enrichment of colloidal complex with adsorbed sodium) and formation of sodium carbonate (as a result of reaction between the sodium ions and calcium bicarbonates) (Sandu 1984). After desalinization of solonchaks, leaching of soluble salts occurs as well as the decrease in the electrolyte content in the soil solution. Some soil scientists think that to solonetzs formation the influence of magnesium cation can also contribute. Solonetzs formation occurs mainly in wet groundwater regime from steppe soils. Alkali soils contain a high proportion of exchangeable sodium in colloidal complex or salts that are subjected to an alkaline hydrolysis (Na_2CO_3, $NaHCO_3$); they correspond to solonetzs and have a strongly alkaline reaction ($pH > 8.3$) compared to nonsaline, alkali soils.

Classification of solonetzs can be done following several criteria: water regime, the degree of alkalization and salinization, and the thickness of different horizons. According to the hydric regime, solonetzs can be steppic, semi-hydromorphic, and hydromorphic. According to the degree, intensity, and depth where salinization occurs, solonetzs are divided into weak, moderate, strong salinized and solonetzs-solonchacks. According to the degree of soil alkalization, they can be moderately and strongly alkalized. Romanian System of Soil Classification (SRCS, Conea et al. 1980) distinguishes for solonetzs different subtypes and similarities: typical (solonetz), luvic, albic, glossic (solonetzs partially solodized), cambic (alluvial solonetz, residual), salinized mollic (solonetz-solonchak, solonetz-solonchak-like), and gleyic.

Solonetzs are populated by a number of halophytes, such as *Artemisia maritima, Limonium gmelinii, Camphorosma annua,* and *C. monspeliaca.*

2.3 Solods (Alkalized Soils)

They occur rarely in the terminology of soil science; they represent the soils corresponding to the most advanced phase of desalinization of saline soils, where exchangeable sodium from the colloidal complex was replaced by exchangeable hydrogen.

2.4 Saline Soils

It is a subtype of the soil belonging to different subtypes of soils that can be characterized by the presence of a salic horizon.

2.5 Alkalized Soils (Solonized)

It is a subtype of soil belonging to different subtypes of soils, characterized by the presence of an alkalized horizon.

Therefore, *sărătura* is a subtype of soil belonging to different types, being in the same time either salinized and/or alkalized, or both; thus, soil profile is more or less affected by soluble salts, by exchangeable sodium from adsorption complex, as well as by increased osmotic pressure of the soil solution.

Romanian System of Soil Classification (SRCS), revised by the Romanian system Soil Taxonomy (SRTS) (Florea and Munteanu 2003), significantly contributed to a better classification of saline soils and their relation with American or World Reference Base for Soil Resources. SRTS (Florea and Munteanu 2003) includes saline soils in the class of Salsodisols, while SRCS (Conea et al. 1980) maintained them in the former class of halomorphic soils; solonchaks and solonetzs with their complex subtypes are recognized in both classification systems.

According to SRTS (Florea and Munteanu 2003), *solonchaks* are soils with elevated concentration of salts in the upper layer during certain periods of the year. This occurs in regions where evapotranspiration process exceeds greatly the precipitations, at least for a period of time and where the parental material of the soil contains moderate or high amounts of salts. The presence of these salts and the increased osmotic pressure of soil solution or toxicity (induced by several ions) favor installation of typical halophytic vegetation. The involved cations are sodium, calcium, magnesium, and potassium, sodium being the most important. The only common feature of solonchaks is the higher content of salts.

According to SRTS (Florea and Munteanu 2003), *solonetzs* are soils conditioned by the salinity of parental material or water table and poor drainage conditions. They are largely distributed in regions with semiarid climate or dry steppe with very hot and dry summer, on lower surfaces with poor internal and external drainage. Characteristic to solonetzs is the nitric horizon frequently associated with a shallow horizon rich in humus and subjacent salinized horizons. The main chemical characteristics of solonetzs are the higher content of sodium or sodium plus magnesium in the adsorption complex and high pH, usually higher than 9.

American terminology used by US Salinity Laboratory Staff (Richards 1954) classifies the soils affected by salinity into the following operational types:

- *Saline soils* are those having an electrical conductivity of the saturation extract greater than 4 dS/m and an exchangeable sodium percentage (ESP) less than 15. They contain soluble salts in amounts sufficient to interfere with the growth of most crop plants but do not contain enough sodium to significantly alter soil characteristics. Ordinarily, the pH is less than 8. Saline soils are often recognized by the presence of white crusts of salts on the surface. Soil salinity may occur in soils having distinctly developed profile characteristics or in undifferentiated soil material such as alluvium.

  They correspond to "solonchaks" from other classification systems.
- *Saline alkali soils* are those for which the conductivity of the saturation extract is greater than 4 dS/m at 25 °C and the exchangeable sodium percentage is greater than 15. These soils form as a result of the combined processes of salinization and alkalization. As long as excess salts are present, the appearance and properties of these soils are generally similar to those of saline soils. Under conditions of excess salts, the pH readings are seldom higher than 8.5 and the particles remain flocculated.
- *Nonsaline alkali soils* are those for which the exchangeable sodium percentage is greater than 15 and the conductivity of the saturation extract is less than 4 dS/m at 25 °C. This type corresponds to "solonetz" from other classification systems.

American "Keys to Soil Taxonomy" (2010) includes the categories of saline soils in the *Aridisols* order, *Salids* suborder, defined as "other *Aridisols* that have a salic horizon within 100 cm of the soil surface." Further, the key to great groups of Salids includes *Aquisalids* (Salids that are saturated with water in one or more layers within 100 cm of the mineral soil surface for 1 month or more in normal years) and *Haplosalids* ("other salids").

Aquisalids can be further divided into:

- Gypsic Aquisalids: Aquisalids that have a gypsic or petrogypsic horizon within 100 cm of the soil surface;
- Calcic Aquisalids: Other Aquisalids that have a calcic or petrocalcic horizon within 100 cm of the soil surface;
- Typic Aquisalids: Other Aquisalids.

Haplosalids can be further divided into:

- Duric Haplosalids: Haplosalids that have a duripan within 100 cm of the soil surface;
- Petrogypsic Haplosalids: Other Haplosalids that have a petrogypsic horizon within 100 cm of the soil surface;
- Gypsic Haplosalids: Other Haplosalids that have a gypsic horizon within 100 cm of the soil surface;
- Calcic Haplosalids: Other Haplosalids that have a calcic horizon within 100 cm of the soil surface;
- Typic Haplosalids: Other Haplosalids.

The inclusion of saline soils in the major order of *Aridisols* seems very logical when taking into account the physical processes within the soil and especially the resulted effects on plant life. In fact, salt areas are affected by "physiological drought" (Grigore and Toma 2010, 2011) with major constraints on metabolic processes for plants. In fact, problems associated with saline area can be included in three categories (Walter 1974; Lambers et al. 2008; Marschner 1995; Schulze et al. 2005; Fitter and Hay 1987; Grigore et al. 2014):

1. A high salinity is associated with a low soil water potential, giving rise to symptoms similar to those of water stress;
2. Specific ions, especially Na and Cl, may be toxic;
3. High levels of NaCl may give rise to an ion imbalance (predominantly Ca) and lead to deficiency symptoms.

In Romania, according to some data (Sandu 1984), saline and alkali soils and salt-affected areas occupy about 500,000 ha; of course, these data are not absolute, since there are not actualized maps with saline soils and the language is being used ambiguously in relation to saline environments. Interestingly, Florea (1958) divided these salt-affected areas according to a predominant chemical type of salinization.

References

Chiriţă C (1955) Pedologie generală. Ed. Agro-Silvică de Stat, Bucureşti, pp 590–628
Conea A, Florea N, Puiu Ş (1980) Sistemul Român de Clasificare a solurilor. ICPA, Bucuresti
Fitter AH, Hay RKM (1987) Environmental physiology of plants, 2nd edn. Academic/Harcourt Brace Jovanovich, London/San Diego
Florea N (1958) Raionarea preliminară a sărăturilor din R. P R Probl Agric 10(9):48–56
Florea N (1963) Curs de geografia solurilor cu noţiuni de pedologie. Edit Did şi Ped, Bucureşti
Florea N, Munteanu I (2003) Sistemul Român de Taxonomie a Solurilor. Ed. Estfalia, Bucuresti
Flowers TJ, Yeo AR (1995) Breeding for salinity resistance in crop plants: where next? Aust J Plant Physiol 22:875–884
Gorham J (1995) Mechanism of salt tolerance of halophytes. In: Choukr-Allah R, Malcolm CV, Hamdy A (eds) Halophytes and biosaline agriculture. Marcel Dekker, New York, pp 207–223
Grigore M-N, Toma C (2010) Halofitele. Aspecte de anatomie ecologică. Edit. Univ. "Al. I. Cuza", Iaşi

Grigore M-N, Toma C (2011) Halofitele, o categorie ecologică polimorfă. Între seceta fiziologică a solului şi stresul salin. Rev Bot (Chişinău) 2(3):38–46
Grigore M-N, Ivănescu L, Toma C (2014) Halophytes. An integrative anatomical study. Springer, Cham
Khan MA, Duke NC (2001) Halophytes – a resource for the future. Wet Ecol Manag 6:455–456
Khan MA, Böer B, Öztürk M, Al Abdessalaam TZ, Clüsener-Godt M, Gul B (2014) Sabkha ecosystems. Cash crop halophyte and biodiversity conservation, vol IV. Dordrecht, Springer
Lambers H, Chapin SF III, Pons TL (2008) Plant physiological ecology, 2nd edn. Springer, New York
Lupaşcu G, Parichi M, Florea N (1998) Ştiinţa şi ecologia solului. Ed. Univ. "Al. I Cuza" Iaşi
Marschner H (1995) Mineral nutrition of higher plants, 2nd edn. Academic, Harcourt Brace, London/New York
Massoud FI (1981) Salt affected soils at a global scale and concepts for control. FAO Land and Water Develop. Div., Technical Paper, Rome, pp 1–21
Munns R (2002) Comparative physiology of salt and water stress. Plant, Cell Environ 25:239–250
Öztürk M, Böer B, Barth H-J, Breckle S-W, Clüsener-Godt M, Khan MA (eds) (2011) Sabkha Ecosystems. Africa and Southern Europe, vol III. Springer, Dordrecht
Richards LA (1954) Diagnosis and improvement of saline and alkali soils. U.S. Department of Agriculture Handbook no 60
Sandu G (1984) Solurile saline şi alcalice din R.S.R. Ameliorarea lor. Ed. Ceres, Bucureşti
Schulze E-D, Beck E, Müller-Hohenstein K (2005) Plant ecology. Springer, Berlin
Shannon MC (1998) Adaptation of plants to salinity. Adv Agron 60:75–119
Szabolcs I (1974) Salt affected soils in Europe. Martinus Nijhoff-The Hague and Research Institute for Soil Science and Agricultural Chemistry of the Hungarian Academy of Sciences-Budapest
Tanji KK (2002) Salinity in the soil environment. In: Läuchli A, Lüttge U (eds) Salinity: environment-plants-molecules. Kluwer Academic, New York, pp 21–51
Walter H (1974) Vegetaţia Pământului în perspectivă ecologică. Edit. Ştiinţifică, Bucureşti
*** (2010) Keys to soil taxonomy, 11th edn. United States Department of Agriculture. Natural Resources Conservation Service
*** (2006) World reference base for soil resources 2006. A framework for international classification, correlation and communication, Food and Agriculture Organization of the United Nations, Rome

Part II
Major Anatomical Adaptations

Chapter 3
Succulence

It is known, for a long time, that succulence is considered one of the most striking anatomical features involved in salt tolerance in halophytes (Grigore 2008; Grigore and Toma 2007, 2008, 2010a, b, 2014, 2011a, b; Grigore et al., 2011a, b, 2012a, b, 2013, 2014). Succulence—first morphologically noticed—seems to be the first important adaptation observed and discussed within the halophyte group by the early botanists. De Jussieu (1717) described a maritime species collected from Spain seashore—*Kali d'Alicante* (most likely, a *Salsola* species, Fig. 3.1) as having "*cylindrical and succulent*" leaves (. . .) that have at their extremity a "*salty taste*."

Hedenberg in his dissertation on plants environments (1754, published in 1788) included in the group of maritime environments "*salt plants, sub-succulents, constricted*" (referring to general habit or leaves) ("*plantae salsae, subsucculentae, coarctatae*," p. 74). Then, in 1831, Moquin characterizes leaves of *Suaeda* species as being "*sessile, with fleshy, succulent parenchyma*."

At that time, the logical connection between NaCl content in the medium and the degree of succulence was evidenced by Batalin (1886), Lésage (1890), Holtermann (1907), Chermezon (1910), Keller (1925), and Schratz (1934).

Succulence refers, in fact, to an increase in water content in the plant organs and illustrates especially the case of dicotyledonous halophytes and glycophytes exposed to salinity (Poljakoff-Mayber 1975). It was defined as the ratio of total content of water/surface area (Chapman 1942), as an increase in water content/unit leaf area (Jennings 1976; Longstreth and Nobel 1979), as an increase in the percentage of water in the leaf (Handley and Jennings 1977), or as fresh/dry weight (Abd Elbar and El-Maboud 2013).

Together with other adaptive traits, such as reduced leaf area, thick leaves, the higher plasticity of cell walls, and a small number of stomata per unit area, succulence is considered one of the defining features of halophytes (Warming 1897, 1909; Schimper 1903; Stocker 1933; Adriani 1956; Poljakoff-Mayber 1975).

Moreover, the presence of succulence, reduced leaf area, and protective hairs, both in desert plants (xerophytes) and in halophytes, led Schimper (1903) to

M.-N. Grigore, C. Toma, *Anatomical Adaptations of Halophytes*,
DOI 10.1007/978-3-319-66480-4_3

Fig. 3.1 Cylindrical and succulent leaves of maritime species *Kali d'Alicante* (*Salsola ssp.*) (de Jussieu 1717)

formulate the hypothesis of "physiological drought" (discussed in Grigore and Toma 2011a; Grigore et al. 2014). Indeed, starting from the end of the nineteenth century, it was observed that under saline conditions, plant growth was reduced. This was attributed to the lack of water in plant organs, although the plant could grow in wet but saline soils or in salt-enriched solution cultures. The low osmotic potential of the soil solution, resulting from the high concentration of soluble salts, impeded the water uptake by plants. Therefore, the perturbation of water balance in halophytes has been regarded as the main aspect of harmful effects of salt on plants; in addition, a toxic effect of salts has been also attributed. Parallel to this hypothesis, it was assumed that solutions of different salts having the same concentrations would have, in principle, the same effect on plant growth.

This hypothesis has persisted over time, although Osterhout showed since 1906 that diluted seawater was much less harmful to plant growth than equivalent concentrations of each constituent salt from seawater. Lagerwerff and Eagle (1961) came to such conclusions regarding the lower toxicity of mixed salts. Often, it has been observed that the osmotic potential of the leaf sap from plants that grow in saline environments changes in order to maintain a constant gradient of water potential between leaf and soil; this seemed to contradict, at least in part, the hypothesis of physiological drought. Nevertheless, at that time, such a regulatory mechanism has not been considered in the case of roots. Later, when the concept of "free space" has been promoted, Bernstein (1961, 1963) demonstrated the adjustment of osmotic potential also in roots. In these circumstances, it was concluded that water imbalance may not be involved in plant response to salinity, but the harmful effect of salts is rather due to the nature of osmotic adjustment. This finding has been correlated with the observation that plants grown on saline substrates are often succulent but not necessarily less turgid than the control plants. Consequently, the regulation of osmotic potential can be achieved by the uptake of salts (toxic or not), by the release of the K^+ ions in the cells, or by hydrolysis of polysaccharides in small molecules.

Succulence has a dilution effect upon salts accumulated in plant organs and upon toxic ions accumulated in cells, thereby allowing plants to cope with large amounts of salts (Waisel 1972; Grigore 2008). But succulence might also be a tricky strategy for plants, because it can reduce the concentration of calcium and potassium in tissues affected by salt and this can enhance the toxic effect of other ions in the cell. It is known that Na^+ and Cl^- ions are considered to be aggressive osmolytes, due to their ionic small diameter and their high hydrature capacity (Schulze et al. 2005), while salt resistance is partly dependent on the plant's ability to mobilize energy for the removal of sodium from the cell and potassium accumulation (Norkrans and Kylin 1969). In this way, it can be easily understood why succulence is manifold.

However, it is still questionable if succulence is a direct response induced by NaCl or represents a secondary effect, caused by a change in the balance of ions or organic acids. It seems that succulence is a common phenomenon occurring in glycophytes, xerophytes, and halophytes which may suggest that the increase of succulence induced by a strong intensity of light, dryness, and sodium ions is generally based on the same mechanism (Grigore 2008).

Jennings (1968) believed that succulence is caused by a change in ATP metabolism induced by sodium ion transport. Nobody knows certainly if ATP has an effective role in increasing succulence, but it may be involved in the synthesis of new cellular material, or increasing cell extensibility. Such a role would probably explain the role of phosphate in inducing succulence in plant tissues. According to the same author, sodium ability to increase the succulence has a double signification. First, increased succulence has a dilution effect on the ionic content of cells, which may or not reach toxic levels. Second, the sodium may stimulate growth, which tends to reduce the turgor pressure component of the water potential of the cell. The value of this growth response would be to increase the water potential of the leaf cells, without forcing the plant to absorb more ions. Thus, the action of sodium can be regarded as a homeostatic response to the plant toxic ions, so that this negative effect tends to be canceled.

Arnold (1955) suggested that succulence depends on the ratio between absorbed and free ions in plant cells, rather than on absolute amounts of existing sodium, chloride, or sulfate. Succulence seems to be induced only when the accumulation of free ions in a plant organ rises above a critical level. The expression of succulence is, therefore, a consequence of metabolic pathways and can also be induced by some factors, such as radiation, lack of nitrogen, and high light intensity. Sometimes, an increase in the concentration of inorganic ions in mature leaves is also associated with a high succulence (Repp 1939; Biebl and Kinzel 1965).

It is clear, however, that succulence is a major strategy of halophytes in this continuous "negotiating" with higher soil salinity. Halophytes accumulate large amounts of salts in their tissues, thus exhibiting a noticeable succulence. Not incidentally, halophytes secreting salts (cryno-halophytes) developed another major strategy by maintaining a low salt content in their tissues; generally, they do not present succulence (Grigore 2008; Grigore and Toma 2010c).

Succulence was also regarded as a result of the formative action of salts. Salinity and, especially, its type (i.e., composing salts) induce the degree and morphology of succulence.

The data focused on this problem discusses generally the possibility of increasing leaf succulence due to the action of three factors:

1. Salt spray
2. Increasing the salt concentration in the root zone
3. Exposure of plants to the various saline solutions in the culturing media during growth

It is known that a major benefit of a high succulence for halophytes is due to the effect of dilution of the ions in plant cells by the high water content.

Basically, halophytes show a higher affinity for sodium uptake rather than potassium, but halophyte affinity, especially those succulent, is highest for chloride, over all anions. It has been shown that even on sulfur-rich soils of the desert of Namibia (Walter 1937), succulent halophytes accumulate more chloride than sulfur.

Over time, plants were exposed to different types of salts, in order to observe their effects on plant structure. Strogonov (1962) quotes Batalin (1875) as the first scientist who demonstrated induction of succulence in *Salicornia herbacea* grown in NaCl saline substrate. This phenomenon does not occur when plants were grown in nonsalinized substrates, or enriched only with $MgSO_4$, but was rather specific only for NaCl addition. Succulence was attributed to the development of large cells from chlorenchymatic parenchyma and to the presence of a multilayered palisade tissue, otherwise absent in leaves of plants grown on the nonsaline substrate.

Van Eijk (1939) experimentally demonstrated that NaCl plays a major role in building up succulence in *Salicornia europaea*. He worked with NaCl, $CaCl_2$, $MgCl_2$, $NaNO_3$, and $NaSO_4$ salts, or mixtures of salts, and reported their effect on succulence. NaCl was the most important salt involved in inducing succulence. This observation was anticipated by Stocker (1928), who believed that developed succulence was dependent on the action of specific ions rather by modification of osmotic pressure. It seems that chloride plays a more important role than sulfur in this mechanism, a statement supported by Keller (1925), Walter and Steiner (1936), and Williams (1960).

Keller (1925) showed that *Salicornia herbacea* grew best in a medium degree of salinity (Figs. 3.2 and 3.3); he found that the curve of succulence continues to rise

Fig. 3.2 Typical experiment showing that *Salicornia herbacea* optimally grows under salinity conditions (right side plants); its better development is associated with increased succulence (Keller 1925)

Fig. 3.3 Typical experiment showing that *Salicornia herbacea* optimally grows under salinity conditions (left side individuals); its better development is associated with increased succulence (Keller 1925)

with increased concentration of NaCl. A marked increase in succulence was obtained with a small addition of Na or K salts to the nutritive solution. Keller demonstrated that Na_2SO_4 promotes growth, luxuriance, and succulence, but much less than NaCl.

Bickenbach (1932) showed an increased thickness of palisade tissue in *Aster tripolium* and, in addition, an increase of parenchymatic cortex thickness. However, results should not be generalized. Sometimes, even contradictory results may be available for the same species. For example, Shennan and MacRobbie (1987) investigated *Aster tripolium* and did not reveal any increase in succulence at high salinities; they reported that, in fact, the fresh weight/dry weight ratio of leaves even decreased. We think that such different results may be due to natural factors where different species vegetate, to different methods used, and, not least, to a different understanding of a concept such as the succulence. It is true that understanding the same phenomenon in different ways can be confusing, and therefore, a different interpretation of the results might occur. This relative confusion is also underlined by the abovementioned authors. They mentioned Jennings' definition (1976), which refers to succulence as follows: "*by succulence, it is generally meant that leaves of the treated plant are thicker than those of the controls. It is also customary to accept an increase, relative to controls, of the water content per unit dry weight.*" Jennings suggests, therefore, the use of either water content or fresh weight/unit area as a measure of succulence. As defined by Jennings (1976), it would mean that a decrease in dry weight would translate into an increase in succulence, when in fact, the authors did not observe any change in the leaf thickness and water content per unit area. Any increase in the thickness of the leaf could be due to the quick increase of the dry content of the structural material (such as thicker cell walls) than to increase of the moisture content, since at high salinity, fresh weight/dry weight ratio of the leaves decreased, while leaf dry weight per unit area increased and the fresh weight per unit area remained constant.

Results for other *Asteraceae* species, *Jaumea carnosa*, have also shown that under salinity conditions, there is no increase of succulence (St Omer and Schlesinger 1980a), where succulence is defined in this case as a percentage of water content and elongation of the palisade mesophyll cells. There was no correlation between increased leaf succulence and increased NaCl salinity in the root zone. Neither salt spray administration produced a significant increase of succulence. This was increased when plants were moved from NaCl solutions to the initial nonsalinized nutrient solution.

Like Keller (1925), Strogonov (1962) also conducted experiments on *Salicornia*, concluding that in the absence of NaCl, the shoots were thinner, with cortex and pith less developed, as well as conducting tissues; in the presence of salt, the development was normal with all tissues well developed and thicker and succulent shoots. Strogonov has used mixtures of NaCl–Na_2SO_4 in the culture media, preferring this mix because he considered it to be closer to natural conditions. Along with *Salicornia*, he also tested the response of cotton, tomato, sunflower, barley, and beans. It was concluded that the presence of NaCl in the substrate produces succulence in tomato, cotton, and *Salicornia* plants. Excess of Na_2SO_4 causes haloxeromorphism in cotton plants and only a slight succulence in *Salicornia*. Marked xeric effects were recorded in *Salicornia* plants grown in free solutions of salts. In barley, chloride ions induced xeromorphic features.

Similar effects were also recorded in several glycophytes; Hayward and Long (1941) studied the effect of salinity on tomato plants. Under normal conditions, the lamina contains a single-layered palisade tissue and four to five layers of spongy tissue. The authors subjected the plants to three types of salinity: proportional increase of all basic constituents of the nutrient medium, adding NaCl to the basic solution, and salinization with Na_2SO_4. It was observed that with increasing concentrations of equivalent concentrations in the initial nutrient medium, lamina thickness decreased, while under NaCl and Na_2SO_4, leaves became thicker to a slightly different extent. If the salt stress was induced by basic medium, the thickness of the spongy tissue decreased with increasing salinity, while the thickness of the palisade tissue noticeably decreased with increasing salinity (from −0.5 atm. to −1.5 atm.) and then remained constant (at high concentrations of −6 atm.). Under the combined action of NaCl and Na_2SO_4, the thickness of both tissues increased with increasing salinity, but Na_2SO_4 significantly influenced the spongy tissue, while NaCl influenced the palisade tissue.

However, it must be said that, according to these results, the increase in thickness of the leaves caused by high concentrations of NaCl and Na_2SO_4 does not exceed the registered value of the control plants.

Nevertheless, all these results and observations should be manipulated with great caution, as there is certainly a great variability in the response of plants to different types of salinity. In addition, the laboratory experimental conditions can never simulate the environmental factors (Grigore et al. 2011c, 2012c). In nature, salinity is never found to be at a constant level but tends to increase between irrigations, for instance, or during drought periods; contrarily, during rainy periods, or those subjected to irrigation, salinity may decrease considerably.

As evidence of this complex response to salinity, we should mention that Strogonov (1962) reported doubling leaf thickness, including palisade tissue due to chloride salinity. Often, salinity affects different tissues of the plant organs. For example, Boyce (1951) opined that succulence in *Iva imbricata* mainly results from the increase in diameter of the non-chlorenchymatic mesophyll cells.

It is very difficult to say if there are special responses of halophytes or glycophytes to salinity, according to a considered category. Pokrovskaya (1954, 1957) (quoted by Waisel 1972) concluded that in *Statice gmelini* and *Atriplex tatarica*, cell division is inhibited by salinity, but cell growth is stimulated, and consequently, leaf succulence increases. In glycophytes exposed to salinity, both cell division and growth were inhibited; in this way, an explanation for succulence development has been given for these plants. Other explanations have been offered by Stocker (1928), who believed that chloride salts promote swelling of protoplasm in halophytes, thus contributing to their succulence. Tullin (1954) suggested that chloride affects firstly the cell walls, causing some destruction of bonding complex, thereby allowing extensibility of the cell walls and consequently an increase of the diameter of the cell.

Meiri and Poljakoff-Mayber (1970) demonstrated that in bean plants exposed to salinity an increase in leaf surface area and thickness (which depends to some extent on cell division) was recorded. The increase in the surface of the young leaves rapidly decreased by more than 50% after exposure to salinity and continued to decrease gradually over time. Daily increase in thickness was stopped immediately after exposure to the salt, but returned to the original state after 24 h, maintained at a higher value during the experiment than in the leaves of control plants. The increase in total thickness was about 25%. From this thickness, about two-thirds were due to the increase in the size of the cells of the spongy tissue, and one-third was attributed to increasing of palisade cell size. The size of epidermal cells visibly changed. The number of cells per unit area was higher in both epidermises in leaves of plants exposed to salinity. This may suggest that the increase of the size of epidermal cells was affected more by salinity rather than their division.

Another important question is whether monocotyledonous and dicotyledonous plants respond differently to the action of salts. For example, Udovenko et al. (1970) did not show any increase of succulence under salinity in leaves of wheat using a variety rather tolerant to salts—"Federation," thus confirming Strogonov's results (1962) registered for barley. Moreover, in some cases, a decrease in leaf thickness was even observed.

Therefore, succulence can be considered a response of the plant related to the high content of salts, particularly NaCl. It is known that growth of halophytes is usually stimulated by the addition of inorganic salts to the culture medium. For example, *Atriplex spongiosa* and *Suaeda monoica* responded to lowered salinities (Storey and Wyn Lones 1979); in the last mentioned species, content of fresh weight increased with 300% when exposed to a salinity of 500 mol m^{-3} NaCl, although it seems that this level of salinity exceeds the optimum limit for growth, estimated to be approximately 150 NaCl (Waisel 1972; Flowers et al. 1977). This considerable increase in fresh weight was due to the increase in water level tissue, with modification in fresh weight/dry weight ratio. As already stated, this ratio is considered as a good indicator of succulence in plant tissues. For these two species, a correlation between maximum growth and significant succulence was established; therefore, succulence can be considered a good indicator of maximum cell growth due to enlargement of the vacuoles.

In order to characterize succinctly succulents, it should be remembered that thick leaves are due also to increased diameter of mesophyll cells (with fewer chloroplasts than in non-succulent plant leaves) with small intercellular spaces. As discussed previously, mesophyll cells do not respond similarly to the action of salts. From the two layers of chlorenchymatic tissue (external and internal chlorenchyma) of *Suaeda monoica*, only the internal one develops after saline treatment, while the external one (from subepidermal level) remains about at the same thickness (Waisel 1972) (Fig. 3.4). Usually, increasing succulence is accompanied by a reduction of the leaf area per unit volume.

Stomata of succulent leaves are often sunken at epidermis level, and their number is usually lower. In Table 3.1, several data regarding stomata number are

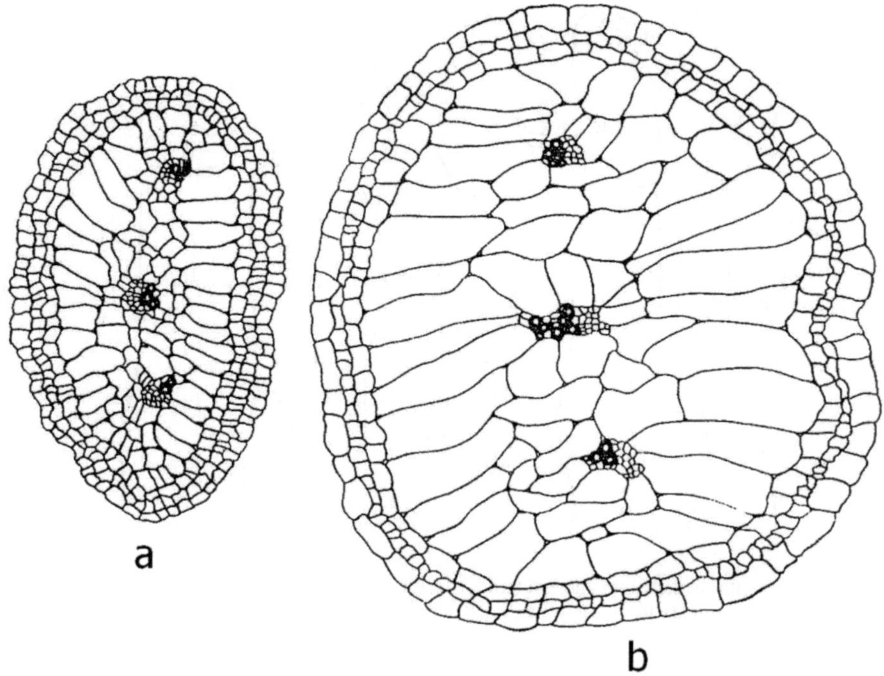

Fig. 3.4 Cross section through the lamina of *Suaeda monoica*: plants grown in nonsaline (**a**) and saline (**b**) conditions (Waisel 1972)

Table 3.1 Number of stomata on leaves of several halophyte species

	Number of stomata/mm^2		
Species	Upper epidermis	Lower epidermis	Mean
Plantago maritima	117	212	
Lepidium crassifolium	150	165	
Camphorosma ovata	120	120	
Triglochin maritima	77	103	
Aster pannonicus	47	67	
Suaeda maritima	38	50	
Alhagi maurorum	107	136	
Prosopis farcta	58	70	
Nitraria retusa	–	–	64
Arthrocnemum glaucum	–	–	57
Suaeda monoica	–	–	48–75
Suaeda fruticosa	–	–	11–23

After Repp (1939) and Shmueli (1948), adapted after Waisel (1972)

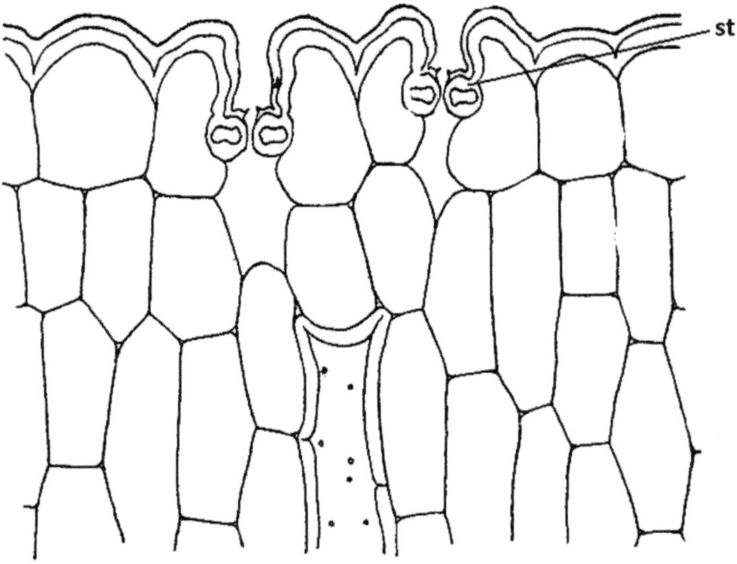

Fig. 3.5 Stomata sunken into the epidermis, in the leaf of *Arthrocnemum macrostachyum* (*st* stomata) (Chermezon 1910)

given for several species of halophytes from Neusiedller region (Austria) (Repp 1939) and from species vegetating on the seashore of the Dead Sea, Israel (Shmueli 1948).

Location of stomata—sunken under the surface of the epidermis—was usually regarded as a xeromorphic feature in halophytes; Chermezon (1910) highlighted this character in many species, such as *Arthrocnemum macrostachyum* (Fig. 3.5). However, this feature should not be generalized, nor regarded as a universal adaptation of halophytes.

Lesage (1890) conducted a very interesting study regarding the modifications of leaves in halophytes (maritime plants); he dealt with this issue both in natural (working with ecotypes—*varieté maritime* and *varieté terrestre*) and in experimental conditions. He compared therefore the leaf modifications in plants collected near to the sea (maritime plants) and their ecotypes from inland nonsaline area; to strengthen his results, he validates them in an experimental frame. It is worth mentioning his study, since it looks like a very modern and complex study of experimental plant ecophysiology.

Lesage (1890) stated that:

1. The same species vegetating on the seashore is expected to have succulent leaves.
2. If it is moved toward interior (on inland, nonsaline areas), plants would develop thin, less succulent leaves.
3. If it usually vegetates in inland areas and if it is moved on seashore, it would acquire thicker leaves than those from inland areas.

He clearly shows that leaf succulence is mainly the result of increasing mesophyll volume, by increasing the volume and length of palisade cells and increasing the number of palisade tissue layers.

Consequently, Lesage (1890) demonstrated that *Eryngium maritimum* (Fig. 3.6) and *Aster tripolium* (Fig. 3.7) have more succulent leaves when collected from

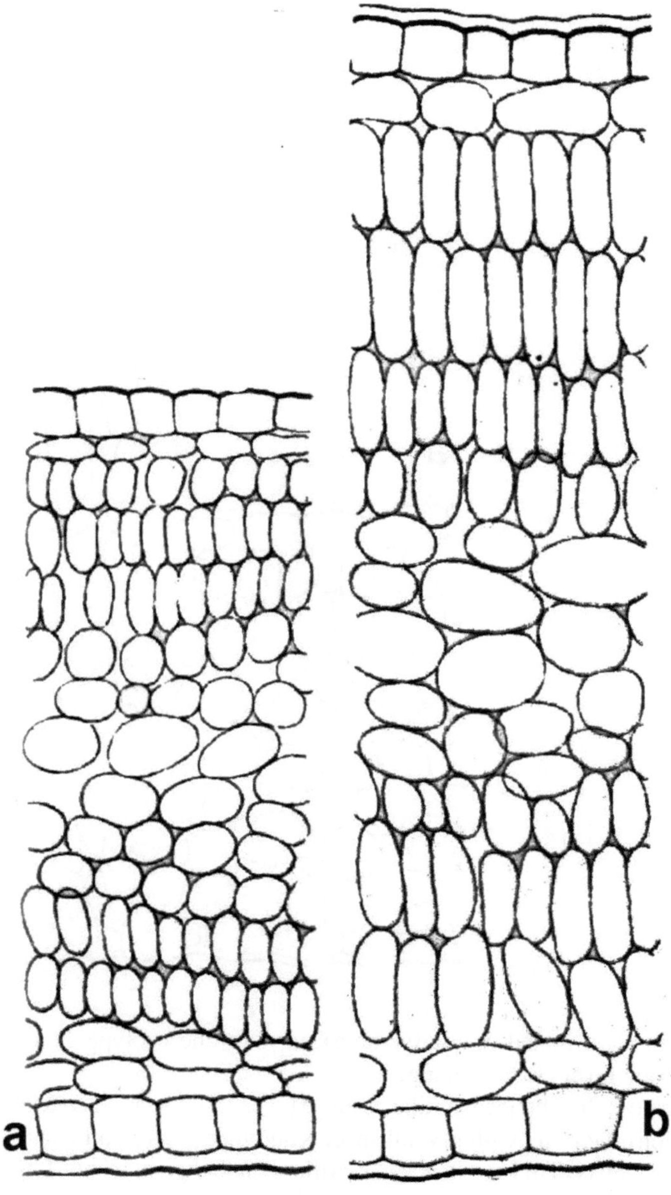

Fig. 3.6 Cross sections through the leaves of *Eryngium maritimum*: (**a**) plants collected from inland, nonsaline areas and (**b**) maritime plants (Lesage 1890)

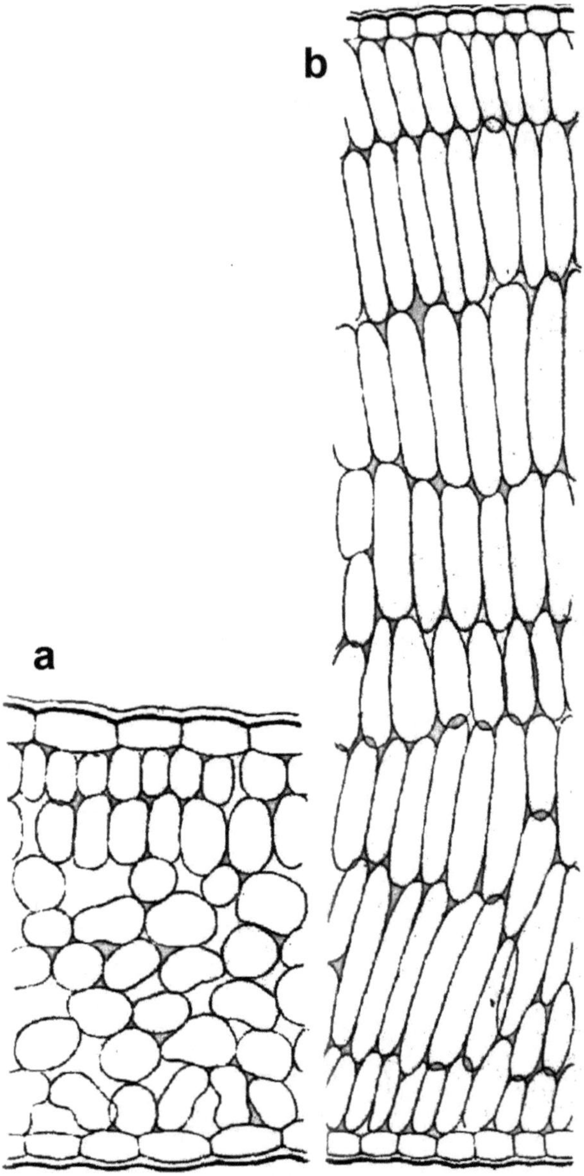

Fig. 3.7 Cross sections through the leaves of *Aster tripolium*: (**a**) plants collected from inland, nonsaline areas and (**b**) maritime plants (Lesage 1890)

littoral than from inland, nonsaline areas: in maritime ecotypes, the leaves have more palisade layers with prolonged cells.

Furthermore, he obtained similar results (Figs. 3.8, 3.9, and 3.10) in experimental conditions, working with several species and many salt treatments.

Finally, in respect of succulence in maritime (halophytes) plants, Lesage came to several interesting conclusions:

Fig. 3.8 Cross sections through the cotyledons of *Lepidium sativum*: (**a**) plants grown on garden soil and (**b**) plants irrigated with seawater—plants approximately after 15 days of salt exposure (Lesage 1890)

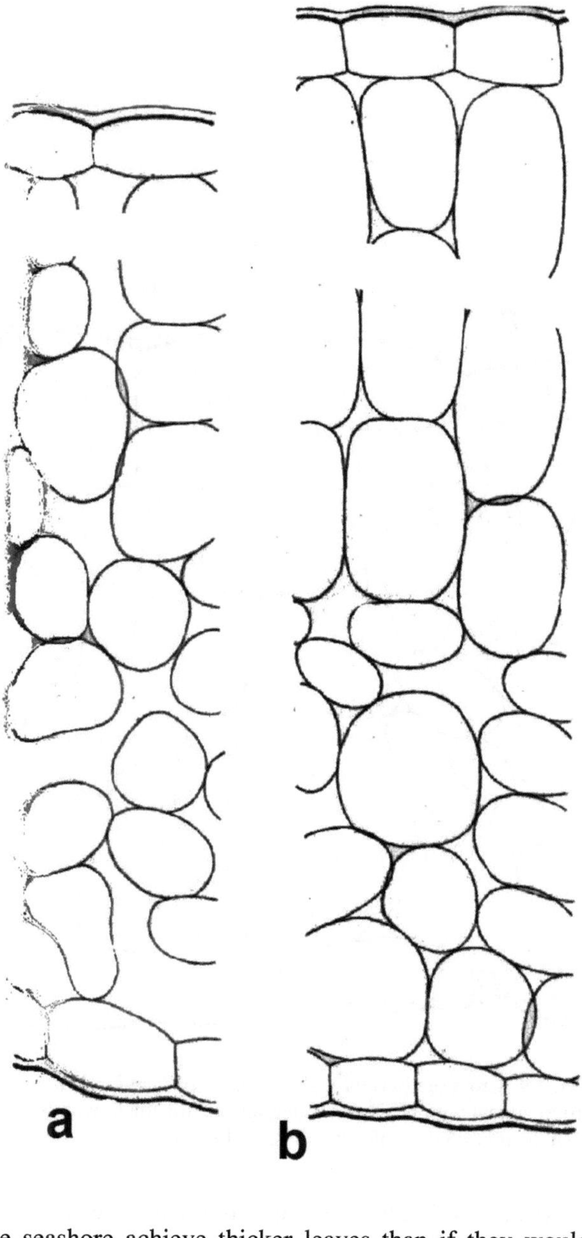

1. Plants living near to the seashore achieve thicker leaves than if they would vegetate in inland areas; however, not all the plants would behave in a similar way.
2. In plants that successfully withstand maritime influence, palisade cells are very developed. When the thickness of leaves considerably increases, then the palisade cells greatly prolong; in addition, the number of mesophyll layers may or may not remain unchanged.

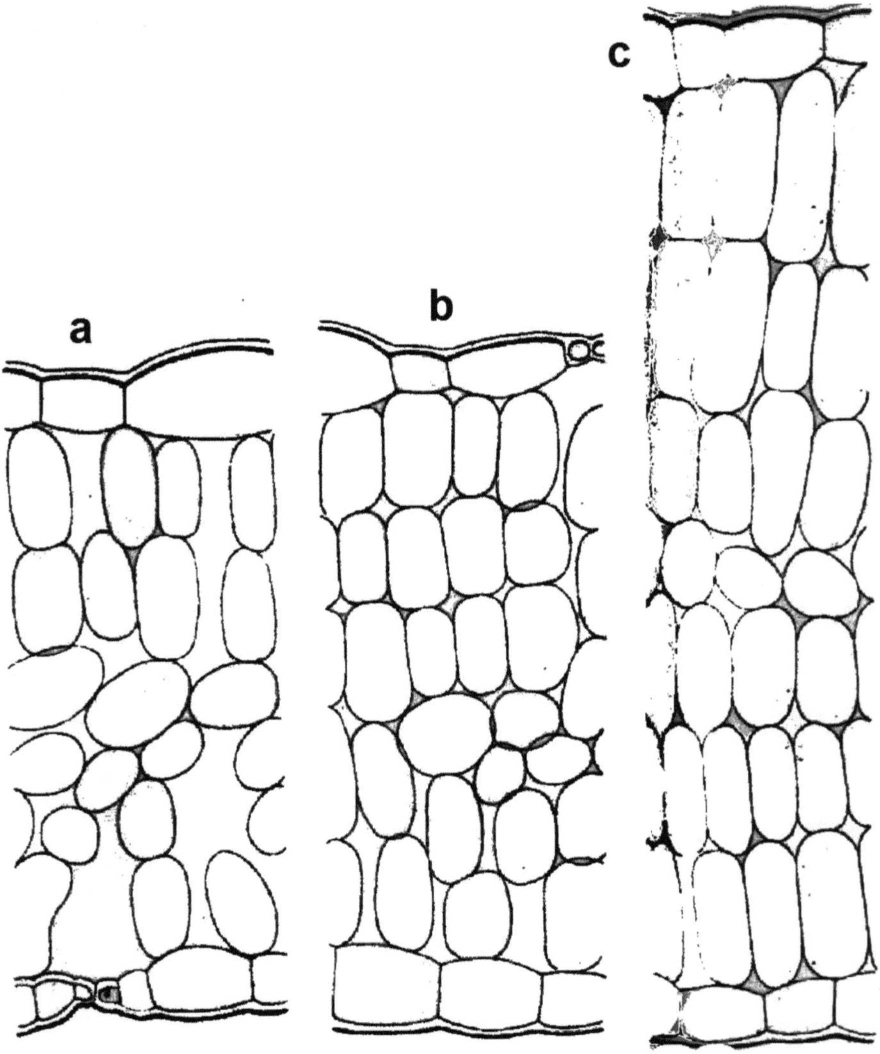

Fig. 3.9 Cross sections through the leaves of *Lepidium sativum*: (**a**) plants grown on garden soil, (**b**) treatment with NaCl and water, and (**c**) treatment with NaCl and garden soil—plants approximately after 45 days of salt exposure (Lesage 1890)

3. Meatus is significantly reduced in the leaves of littoral plants.
4. Chlorophyll is less abundant in the cells of plants growing on seashore.
5. Succulence, development of palisade tissue, and reduction of meatuses and chlorophyll may be induced experimentally in cultures with varying salt concentration.

As already underlined, plants grown in a medium enriched with NaCl have a lower leaf surface but with larger cells, while plants treated with Na_2SO_4 have small

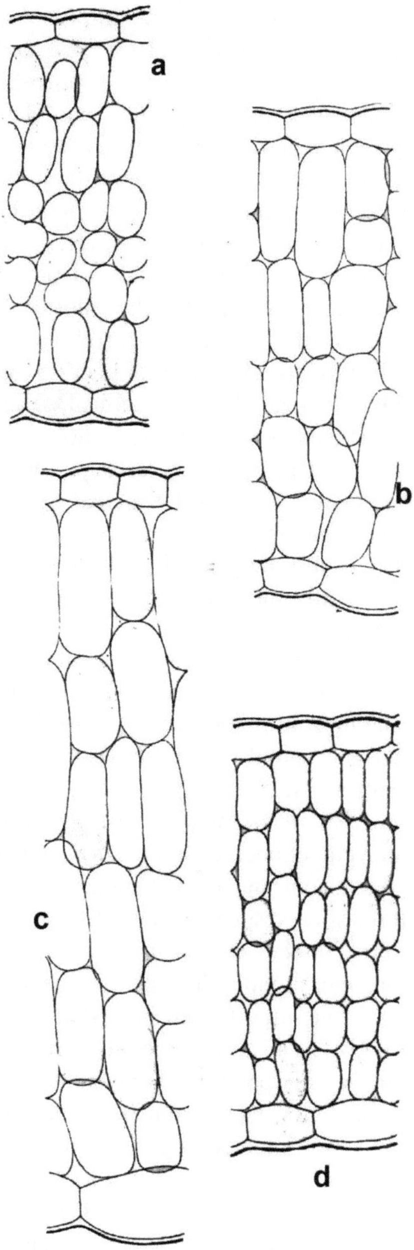

Fig. 3.10 Cross sections through the leaves of *Lepidium sativum*: (**a**) plants grown on garden soil, (**b**) treatment with NaCl and water, (**c**) plants irrigated with seawater, and (**d**) treatment with NaCl and garden soil— plants approximately after 60 days of salt exposure (Lesage 1890)

leaf cells but leaves have a larger area (Waisel 1972). Apparently, the sodium sulfate is more involved in the increasing cell division, and chloride salts influence more the cell enlargement rather than cell division.

Jennings (1968) suggested that succulence is one of the mechanisms that plants use to cope with toxic concentrations of ions. This mechanism has been also investigated in mangrove plants, where high concentrations of salts in their leaves can be avoided due to the dilution effect produced by increasing the water content of the cells. Bowman (1921) revealed a higher succulence in *Rhizophora mangle* grown in seawater than the same species cultivated with freshwater. Reinders-Gouwentak (1953) pointed out that the leaves of *Sonneratia* were frequently succulent and succulence was due to a distinct water storage hypodermic layer. In addition, in the leaves immersed in tidal water, the hypodermic layer was three to five times as thick as leaves from the higher levels of the same tree. The same author states that the hypodermic layer was almost absent in trees grown in freshwater in botanical gardens. Reinders-Gouwentak (1953) believed that succulence in *Sonneratia* is related to the chloride content in the water.

Nevertheless, several correlations may be established between succulence and other structural features or phenomena. For example, Longstreth and Nobel (1979) have correlated the effects of salinity on leaf anatomy with implications on photosynthesis in beans, cotton, and *Atriplex patula*. Succulence induced by salinity may decrease resistance to absorbed CO_2, and thus, the photosynthetic rate would increase by increasing the internal surface of the leaf, per unit area, where gas exchange can occur per unit area. Sometimes, high concentrations of NaCl substrate generally reduced the photosynthetic rate, and sometimes, it can be practically unaffected by elevated salinity in some species vegetating in saline habitats.

For such correlations, several clarifications may be useful; at optimal illumination, photosynthesis is generally limited by the diffusion rate of CO_2 into the leaf. The most important components that control this diffusion are stomatal resistance and mesophyll resistance. Using the surface area/leaf mesophyll ratio, the mesophyll resistance can be further subdivided into effects of foliar structure and the inherent resistance to the diffusion of CO_2 of mesophyll cells. In the abovementioned experiments, Longstreth and Nobel (1979) came to the following conclusions: the leaf succulence increased with increasing NaCl concentration in all three studied species. Mesophyll thickness also increased with elevated salinity due to increased size of palisade cells and increasing the number of spongy cells. The diameter of palisade cells from beans and cotton remained approximately constant under the saline treatment. The diameter of spongy cells tends to increase along with salinity in all three species. Increasing the length of palisade cells and increased number of spongy cells layers are related to a high ratio between the surface of mesophyll cells and unit area of leaves for beans and cotton, while in *A. patula*, the same ratio slightly ranged with salinity because palisade cells grow in both diameter and length.

This is just one example that highlights the important consequences of succulence and the other phenomena that may occur in the plant organs.

Modifications related to succulence involve a wide range of anatomical and ultrastructural features, as it can be seen in Table 3.2.

Succulence is also widespread within mangrove species, which have to cope with high values and variations of salinity levels.

Schimper (1891, 1898) has shown that the leaves of most mangrove species contain water storage tissues; they are in the form of a hypodermis in *Rhizophora* and *Avicennia* species and occur as a central layer of cells, in *Sonneratia*.

Table 3.2 Effects of salinity on structural and ultrastructural features of halophytes

Anatomical or ultrastructural feature	Effect of salinity	Species	References
Leaf thickness	Increase	*Atriplex nummularia*	Greenway (1968)
		A. hastata	Black (1958) and Mendoza (1971)
		A. hortensis	Handley and Jennings (1977)
		A. patula	Longstreth and Nobel (1979)
		A. spongiosa	Sorey and Wyn Jones (1979)
		Suaeda monoica	Lacerda et al. (2006)
		Vigna unguiculata	
Leaf number per plant	Decrease	*Jaumea carnosa*	St. Omer and Schlesinger (1980b)
		Suaeda maritima	Yeo şi Flowers (1980)
Leaf area per plant	Increase	*Atriplex hastata*	Black (1958)
		Suaeda maritima	Clipson (1984) (quoted by Flowers et al. (1986)
		Atriplex halimus	Gale and Poljakoff-Mayber (1970)
		Jaumea carnosa	St. Omer and Schlessinger (1980b)
Leaf epicuticular wax	Increase	*Agrostis stolonifera*	Ahmad and Wainwright (1976)
		Suaeda maritima	Hajibagheri et al. (1983)
		Arachis hypogaea	Rao et al. (1981)
Leaf cuticle	Increase	*Suaeda maritima*	Hajibagheri et.al. (1983)
Number of stomata per unit area	Decrease	*Salicornia hortensis*	Strogonov (1964)
		Suaeda maritima	Siadat-Pour (1978) (quoted by Flowers et al. 1986)
		Jaumea carnosa	St. Omer and Schlessinger (1980b)
		Kandelia candel	Qiu et al. (2007)
Leaf cell size	Increase	*Suaeda maritima*	Yeo and Flowers (1980)
		Jaumea carnosa	St. Omer şi Schlessinger (1980a)
		Atriplex hastata	Mendoza (1971)

Stace (1966) questioned the possibility of using epidermal characters in several phylogenetic problems and made a detailed study of leaf anatomy of several mangrove species from *Combretaceae*, *Rhizophoraceae*, and *Avicenniaceae*; he also provided data about epidermal characteristics in *Bruguiera* and *Avicennia*. He concluded that these anatomical characters are of xeromorphic nature. All species had many common epidermal features, including a thick cuticle, straight cell walls, and the presence of water storage tissue and hydathodes. Almost all investigated species had sunken stomata or stomata surrounded by dense hairs. All taxa, except

Avicennia and *Conocarpus*, lacked lateral veins and had lesser epidermal veins, a feature generally associated with the development of water storage tissue.

In Table 3.3, the most important anatomical characters of the species investigated by Schimper (1891), which can be correlated with succulence, are synthetized.

Table 3.3 Comparison of several anatomical features in several mangrove species

	Anatomical features		
Species	Stomata position in respect with epidermis	Mesophyll	Water storage tissue position within leaf structure
Lumnitzera racemosa (Fig. 3.11d–g)	Apparently on the same level with epidermis (Fig. 3.12)	Two layers of palisade tissue beneath each epidermis; spongy tissue absent (Fig. 3.13)	Centrally placed: Approximately 7–11 layers of more or less isodiametric cells.
Rhizophora mucronata (Fig. 3.14)	Sunken (Figs. 3.15 and 3.16)	Four layers of palisade and approximately eight to ten layers of spongy below upper hypodermis (Fig. 3.15)	Probably under upper hypodermis (more developed) and under lower hypodermis (less developed)
Ceriops candolleana (Fig. 3.11a–c)	Sunken (Fig. 3.17)	Differentiated in palisade and spongy tissues	Hypodermic position
Bruguiera gymnorrhiza (Fig. 3.18)	Sunken (Fig. 3.19)	Differentiated in palisade and spongy tissues	Hypodermic position (unilayered)
Sonneratia acida	Sunken (in natural conditions; on leaves collected from species from Buitenzorg Botanical Gardens—stomata seem to be at the same level as epidermal cells—Figs. 3.20 and 3.21)	Palisade tissues located beneath lower and upper epidermis—mucilage cells distinguished between palisade cells (Fig. 3.22)	Central position—ramified sclereids can be noticed in its mass
Carapa moluccensis	Not sunken (Fig. 3.23)	Differentiated in palisade and spongy tissues (Fig. 3.24)	Central position—under hypodermis
Scyphiphora hydrophyllacea		Differentiated in palisade and spongy tissues	Hypodermic position
Aegiceras majus (Fig. 3.25)	Sunken (Fig. 3.26)	Differentiated in palisade and spongy tissues (Fig. 3.27)	Hypodermic position
Avicennia tomentosa and *A. officinalis*		Differentiated in palisade and spongy tissues	Hypodermic position
Acanthus ilicifolius		Differentiated in palisade and spongy tissues	Hypodermic position

Based on Schimper (1891)

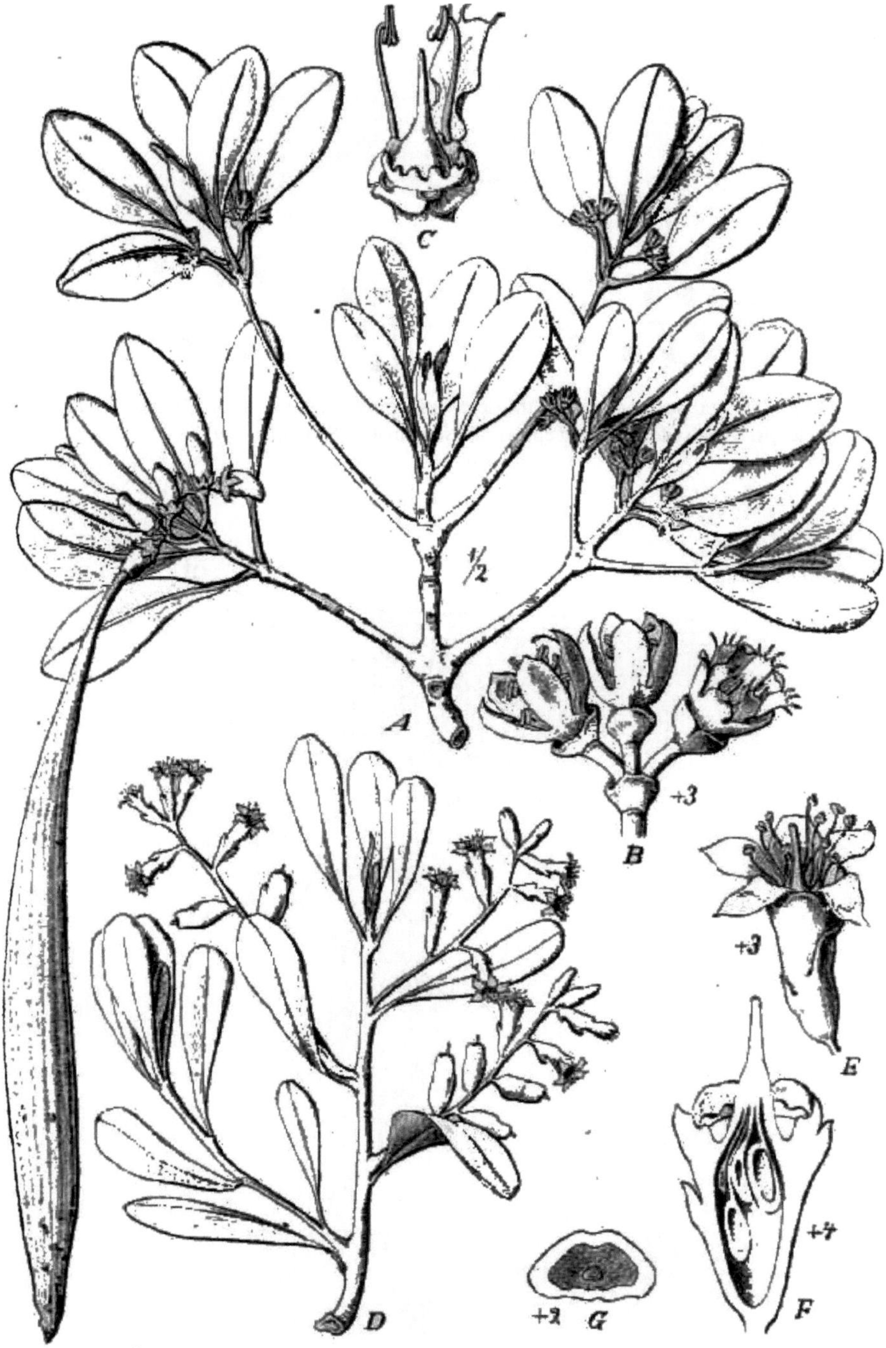

Fig. 3.11 *Ceriops candolleana* (**a–c**) and *Lumnitzera racemosa* (**d–g**) (**a**—flowering and fruiting branch; **b**—branch bearing flowers; **c**—gynaecium and a part from stamen; **d**—flowering branch; **e**—flower; **f**—gynaecium in longitudinal section; **g**—fruit and seed in cross section) (Engler and Drude 1921)

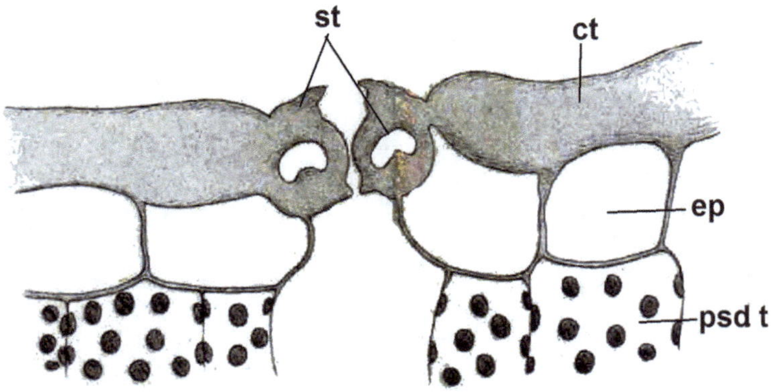

Fig. 3.12 Stomata position within the leaf epidermis of the leaf of *Lumnitzera racemosa* (*ct* cuticle, *psd t* palisade tissue, *ep* epidermis, *st* stomata) (Schimper 1891)

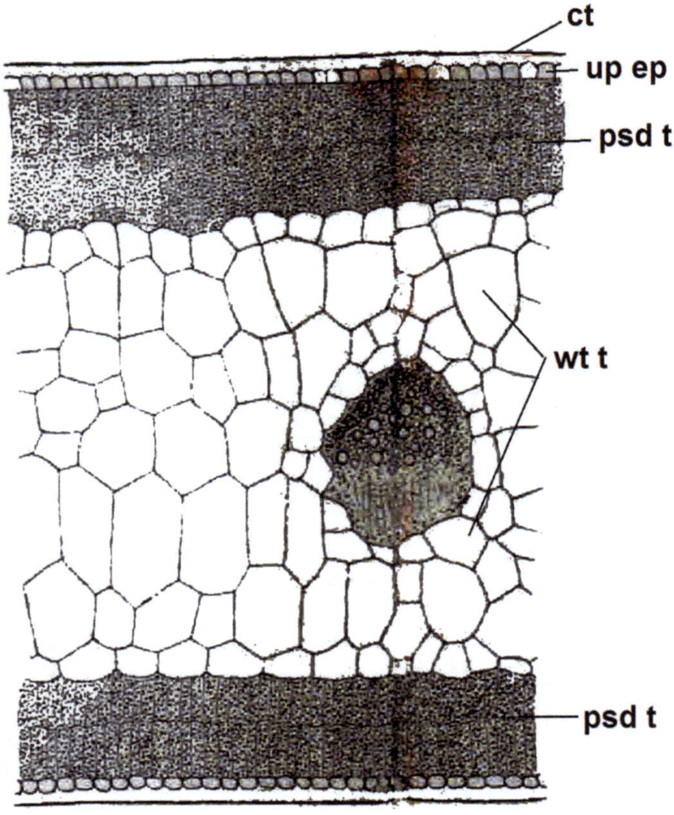

Fig. 3.13 Cross section through the leaf of *Lumnitzera racemosa* (*ct* cuticle, *psd t* palisade tissue, *wt t* water storage tissue, *up ep* upper epidermis) (Schimper 1891)

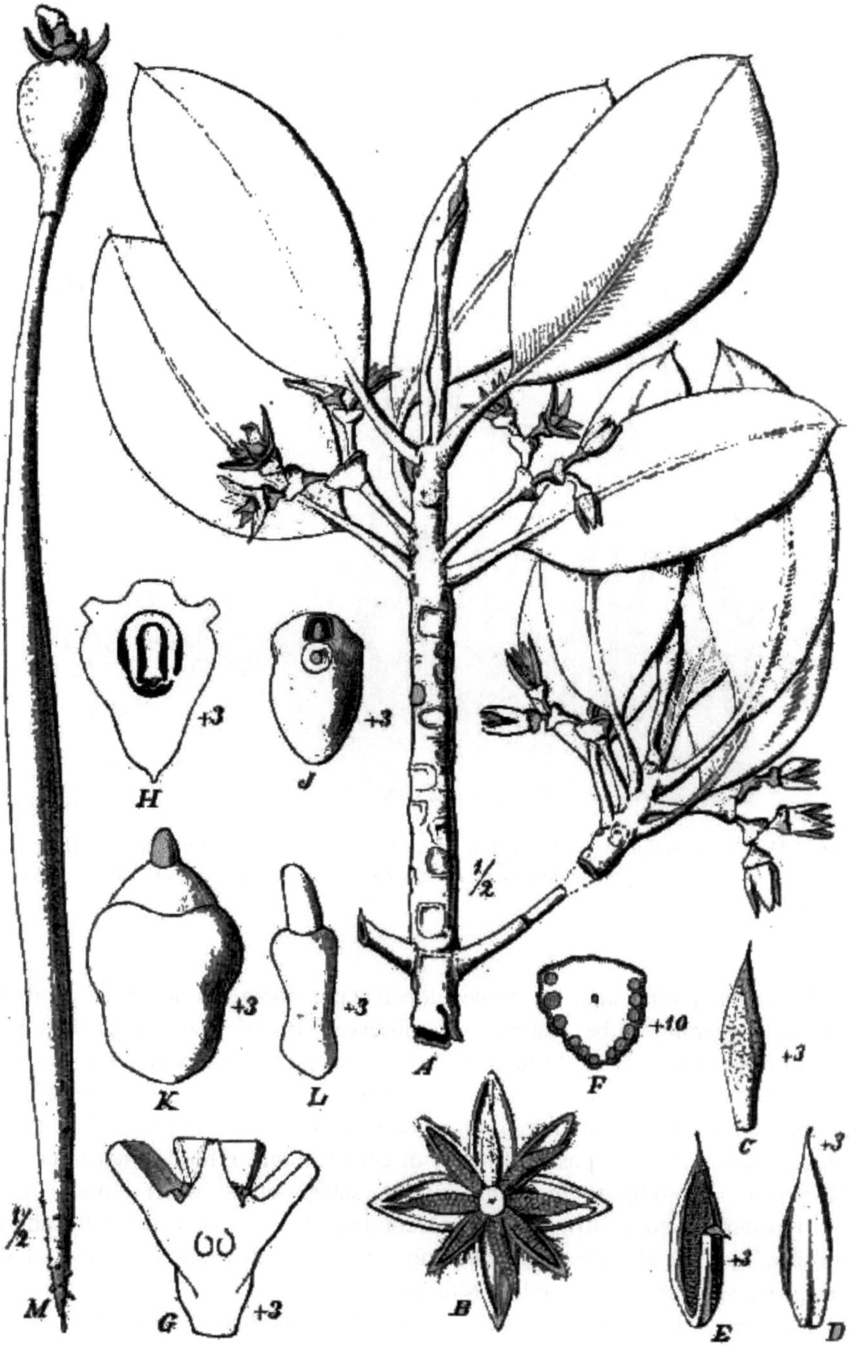

Fig. 3.14 *Rhizophora mucronata*: (**a**) flowering branch, (**b**) flower, (**c**) stamen, front view, (**d**) stamen, view from behind, (**e**) stamen with an opened anther, (**f**) stamen, cross section, (**g**) longitudinal section through forming fruit and receptacle with calyx, (**h**) longitudinal section through the young (half) fruit, (**i**) young fruit, with the tip of hypocotyl noticed in the enlarged micropyle, (**k**) young fruit, evidencing the endosperm and hypocotyl, (**l**) dissected embryo, (**m**) fruit with embryo (Engler and Drude 1921)

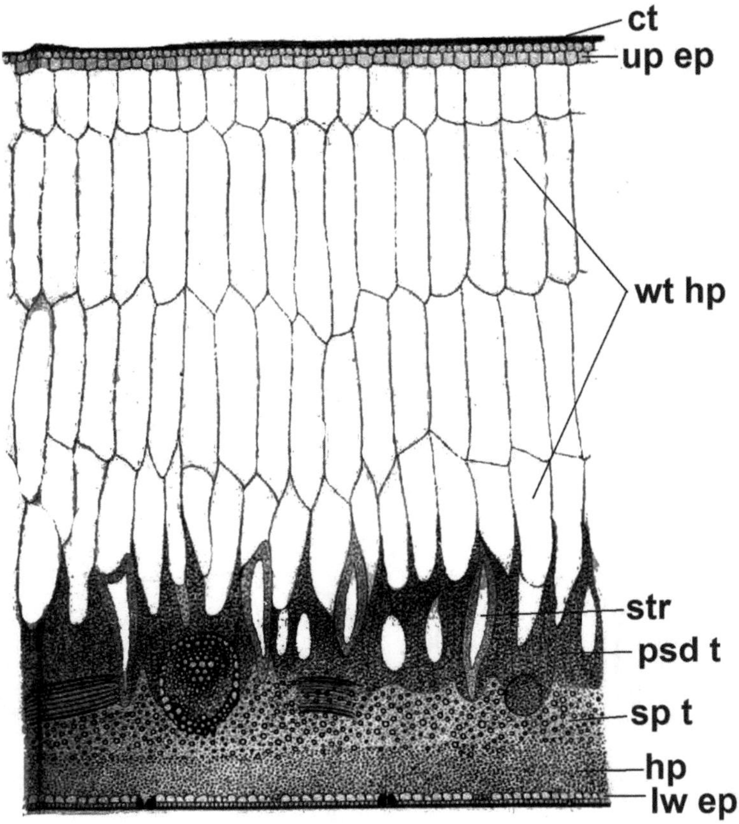

Fig. 3.15 Cross section through the leaf of *Rhizophora mucronata* (*ct* cuticle, *psd t* palisade tissue, *wt hp* water storage hypodermis, *up ep* upper epidermis, *lw ep* lower epidermis, *str* stereids, *sp* spongy tissue) (Schimper 1891)

In *Avicennia officinalis*, a well-developed water storage tissue and a hypodermis with cells seeming to be aqueous were observed (Baylis 1940–1941) (Fig. 3.28). Most of the cross section through the leaf is occupied by the water storage tissue, which fills about half of the volume of the leaf. Water storage function of hypodermis corresponding to the lower epidermis is being assumed, although it is suggested that it is composed of layers of cells similar to those of aqueous tissue. Moreover, according to the same author, subepidermal water storage tissues (in hypodermic position) appear to be common features of leaves of mangrove species. They are also present, for example, in seven of the eight species of Indian mangrove species investigated by Mullan (1931).

The succulence was also well expressed in another species of mangrove, *Rhizophora mangle*, as a result of treatment with NaCl salt (Werner and Stelzer 1990). Moreover, increasing succulence was one of the most evident effects of

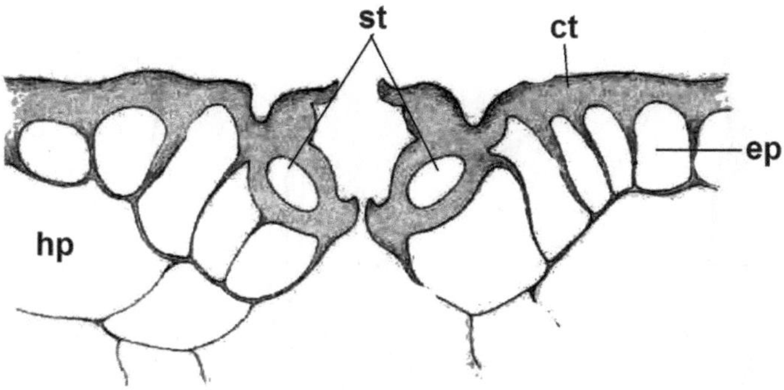

Fig. 3.16 Stomata position within the leaf epidermis of the leaf of *Rhizophora mucronata* (*ct* cuticle, *ep* epidermis, *st* stomata, *hp* hypodermis) (Schimper 1891)

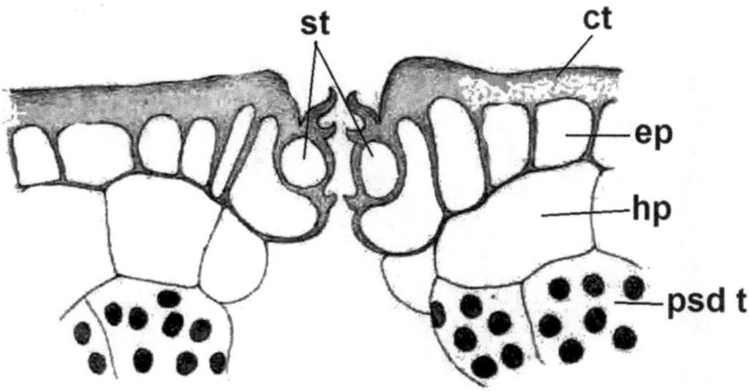

Fig. 3.17 Stomata position within the leaf epidermis of the leaf of *Ceriops candolleana* (*ct* cuticle, *ep* epidermis, *st* stomata, *hp* hypodermis, *psd t* palisade tissue) (Schimper 1891)

NaCl action; in hypostomatic leaves, the number of stomata has been reduced and the average of stomatal pore size increased, while the average of transpiration rate did not vary significantly, due to saline treatment. The leaves of plants subjected to salinity were thicker than those of control plants, due to hypodermic water tissue and mesophyll cells, which seemed slightly larger and more turgid. By comparison, the leaves of control plants and cells of the spongy tissue and especially those of palisade tissue are more compact, with smaller intercellular spaces. In addition, with regard to the location of the various ions in the tissues, the authors have found the highest concentrations of Na^+ and Cl^- in the vacuoles of both hypodermis.

Reduced number of stomata per unit area is considered a characteristic for succulence induced by salinity and appears to be balanced by a greater stomatal opening. An increased rate of CO_2 uptake in plants treated with a saline solution corresponds to a better growth of plants and might be due to a combination of

Fig. 3.18 *Bruguiera gymnorrhiza*: (**a**) lower part of the trunk, with prop roots, (**b**) flowering branch, (**c**) flower, (**d**) floral diagram, (**e**) petal with two stamens, (**f**) lower part of a petal with stamens, (**g**) fruit with embryo (Engler and Drude 1921)

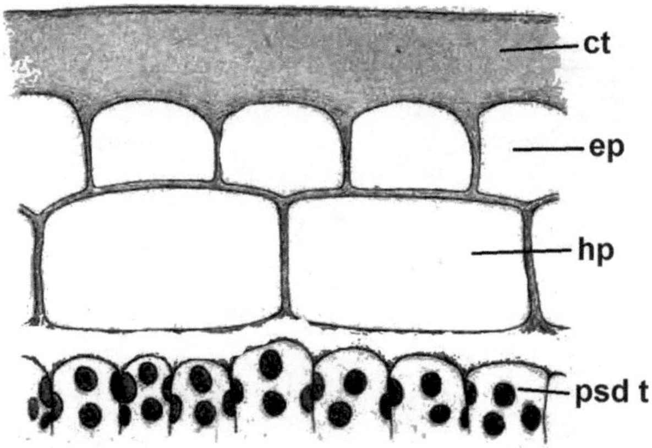

Fig. 3.19 Cross section through the leaf of *Bruguiera gymnorrhiza* (*ct* cuticle, *psd t* palisade tissue, *ep* epidermis, *hp* hypodermis) (Schimper 1891)

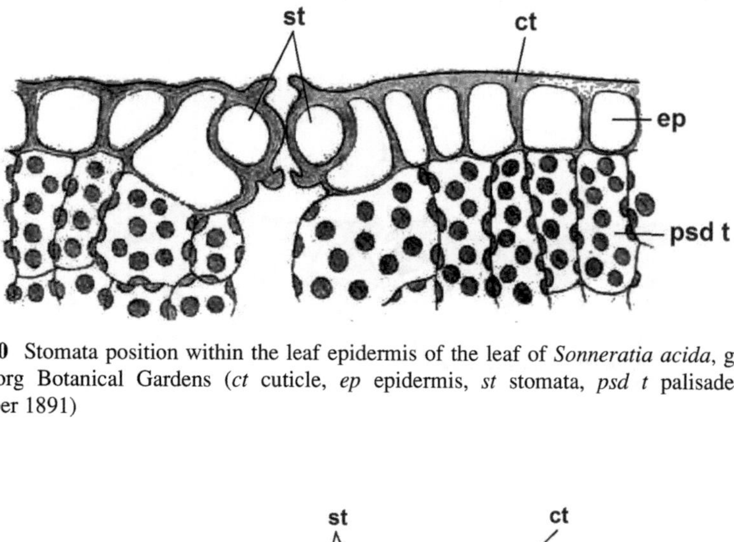

Fig. 3.20 Stomata position within the leaf epidermis of the leaf of *Sonneratia acida*, grown in Buitenzorg Botanical Gardens (*ct* cuticle, *ep* epidermis, *st* stomata, *psd t* palisade tissue) (Schimper 1891)

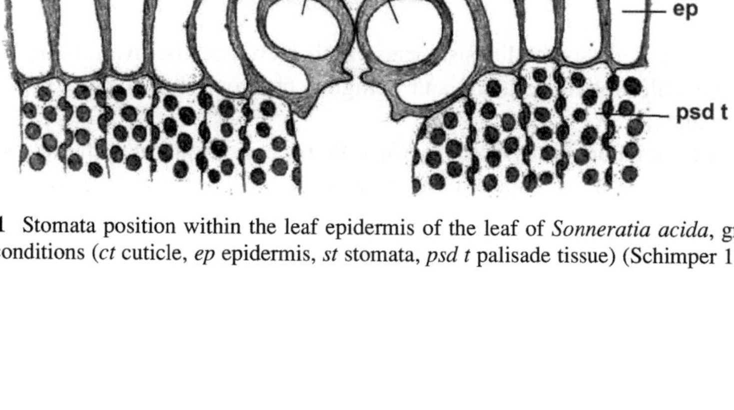

Fig. 3.21 Stomata position within the leaf epidermis of the leaf of *Sonneratia acida*, grown in natural conditions (*ct* cuticle, *ep* epidermis, *st* stomata, *psd t* palisade tissue) (Schimper 1891)

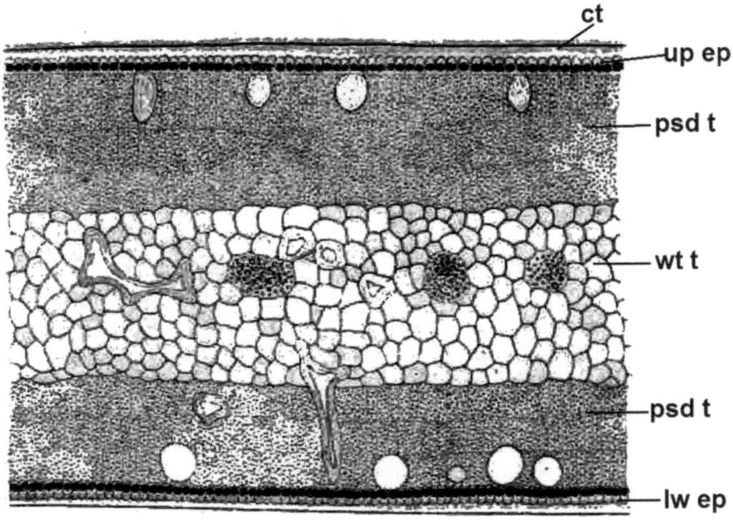

Fig. 3.22 Cross section through the leaf of *Sonneratia acida* (*ct* cuticle, *psd t* palisade tissue, *ep* epidermis, *hp* hypodermis, *up ep* upper epidermis, *lw ep* lower epidermis, *wt t* water storage tissue) (Schimper 1891)

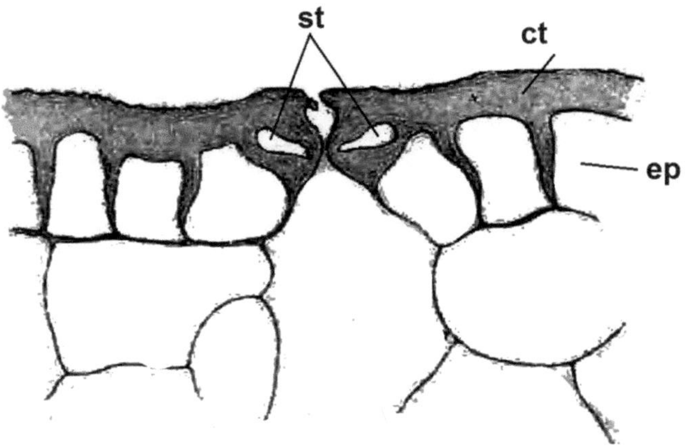

Fig. 3.23 Stomata position within the leaf epidermis of the leaf of *Carapa moluccensis* (*ct* cuticle, *ep* epidermis, *st* stomata) (Schimper 1891)

factors: decreased stomatal resistance, which increases the availability for CO_2, increasing chlorophyll content, and a higher chlorophyll a/b ratio in leaves of treated plants.

Succulence induction by NaCl in *Rhizophora*, in association with high water use efficiency, as well as experiments conducted with *Aegiceras majus*—another

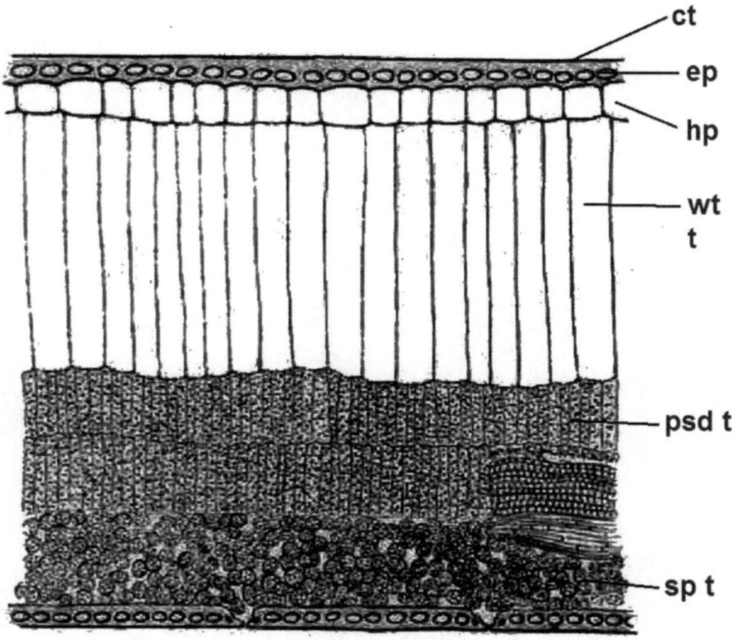

Fig. 3.24 Cross section through the leaf of *Carapa moluccensis* (*ct* cuticle, *psd t* palisade tissue, *ep* epidermis, *hp* hypodermis, *wt t* water storage tissue, *sp t* spongy tissue) (Schimper 1891)

species of mangroves—where it was observed that NaCl induces increased PEP carboxylase activity (Joshi et al. 1974), suggests that *Rhizophora* may be a C_3-CAM plant.

These details support again the profound implications of succulence in the complexity of plant–salinity interrelationships.

A study on two species of mangrove, *Avicennia germinans* and *Conocarpus erectus* from northern Venezuela (Smith et al. 1989), also pointed out several histo-anatomical features related to local environmental factors (rainy and dry seasons) and a certain functionality of these adaptations, where succulence plays a prominent role.

The leaves of *A. germinans* did not show significant differences in terms of succulence in young or mature leaves, or in both rainy and dry seasons. In contrast, in *C. erectus*, succulence was generally higher in mature leaves than in young ones, as reflected in the lower ratio between dry and fresh weight in older leaves. Succulence was slightly higher in the dry season compared to the rainy one, but obviously, the most succulent leaves were of those branches from individuals exposed to saltwater from seashore. Differences of succulence in these isolateral leaves were assigned rather to cell length perpendicular to the leaf than changes in the cell number. The relationship between the total thickness and length of the leaf cells of the four layers of cell mesophyll—delineated by the abaxial epidermis and

Fig. 3.25 *Aegiceras majus*: (**a**) general appearance, (**b**) flower part, (**c**) bud, (**d**) sepal, (**e**, **f**) stamen, (**g**) loculament) (Pax 1897)

vascular bundles—shows that the inner layers of cells are responsible for the differences in the leaf succulence. Moreover, apart from *Conocarpus,* isolateral leaf structure with succulence derived from the significant development of mesophyll in the leaf center is also a characteristic of other mangrove species, such as *Laguncularia, Lumnitzera,* and *Sonneratia* (Walter and Steiner 1936; Biebl and Kinzel 1965; Stace 1966). This anatomical type is different from that referring to dorsiventral (heterofacial) leaves as in *Aegialitis, Avicennia,* and *Rhizophora*, where succulence results from the development of a thick hypodermis (Walter

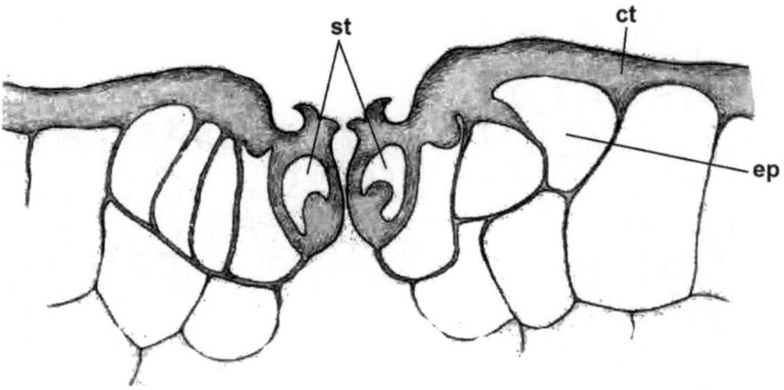

Fig. 3.26 Stomata position within the leaf epidermis of the leaf of *Aegiceras majus* (*ct* cuticle, *ep* epidermis, *st* stomata) (Schimper 1891)

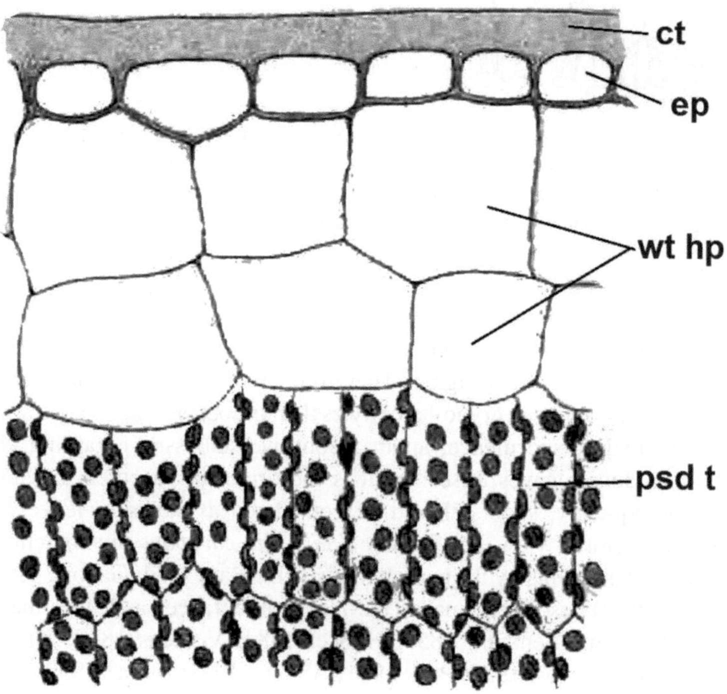

Fig. 3.27 Cross section through the leaf of *Aegiceras majus* (*ct* cuticle, *psd t* palisade tissue, *ep* epidermis, *hp* hypodermis, *wt t* water storage tissue) (Schimper 1891)

and Steiner 1936; Stace 1966). It is known that leaf succulence has been associated both with leaf aging (Walter and Steiner 1936; Biebl and Kinzel 1965) and with soil salinity (Camilleri and Ribi 1983); most likely, it appears to be rather the result of the increased cell size, than cell division.

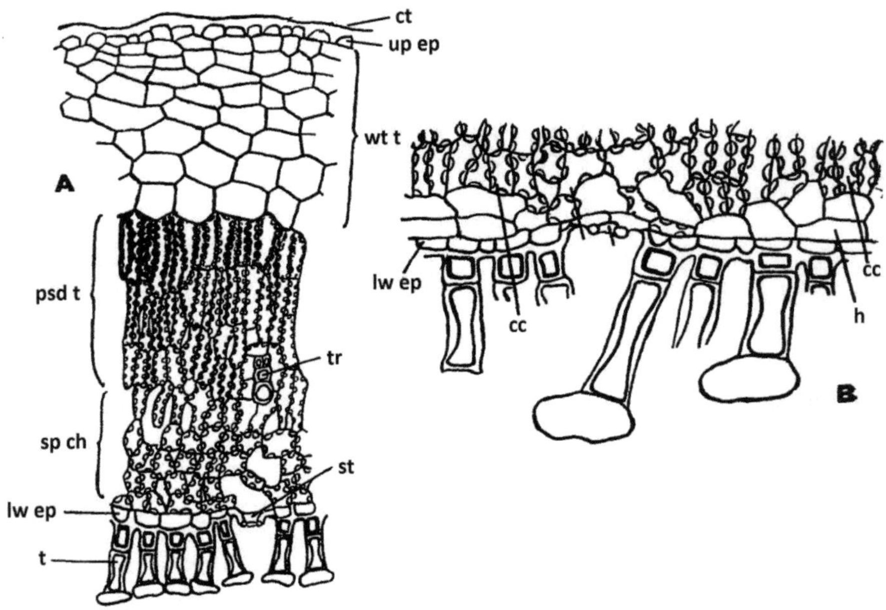

Fig. 3.28 Cross section through the lamina of mature leaf of *Avicennia officinalis* (**a**) and lower epidermis and adjacent tissues (**b**) (*ct* cuticle, *cc* compact chlorenchyma, *h* hypodermis, *lw ep* lower epidermis, *up ep* upper epidermis, *sp ch* spongy chlorenchyma, *psd t* palisade tissue, *wt t* water storage tissue, *t* tomentum, *tr* tracheids) (Baylis 1940–1941)

Salt glands in *Avicennia* may explain why this species is not as succulent as *Conocarpus* because the effectiveness of these structures is known to control and adjust the salt content in plant organs. Therefore, there are two distinct mechanisms, and succulence varies less in cryno-halophytes than in species that do not excrete salts.

Closely related to succulence, another controversial issue regarding chenopod species with reduced leaves arises: what is the origin of succulent tissues (fleshy organs)? Consequently, shoots should be considered as stems or leaves? From a morphological point of view, species like *Salicornia europaea* are sometimes characterized as leafless plants by the early botanists (De Bary s884; Ganong 1903; Cross 1909), while others described small scale-like protuberances as leaves that may be fused with the stem (Bentham 1858; Hooker 1884; Volkens 1887; Jepson 1923; Schischkin 1936; Evenari 1938; Peck 1941; Abrams 1944; Muntz 1959).

Succulent halophytes are classified generally into two categories: those with succulent leaves and those with succulent stems. From a morphological point of view, the distinction between the two organs is problematic.

Some authors (De Fraine 1912; Halket 1928; James and Kyhos 1961) regarded these succulent organs as leaves by their origin, while others (Fahn and Arzee 1959; Fahn 1963) consider them as stems, their cortex being succulent. From strictly a histo-anatomical perspective, our opinion is that the nature of these fleshy organs is foliar, an idea supported by the typical lamina structure, with the epidermis, palisade tissue with two or three layers, and a water storage parenchyma surrounding the central cylinder (Grigore et al. 2014). Moreover, the presence of numerous vascular bundles at the limit between the palisade and the lacunose tissues would be proof of the fact that the entire structure belongs to a pair of opposite leaves, tightly bonded to the external side of the stem.

Anderson (1974), who anatomically investigated *Salicornia virginica*, refers to this succulent tissue as being a leaf, showing that, in fact, the so-called leafless shoot is, in fact, a stem with two appressed leaves, surrounding each internode. In cross section, there is a single-layered papillate epidermis composed of small cells, with external walls highly cutinized. Beneath the epidermis, there is a palisade tissue, composed of two to three cell layers. The remaining leaf tissue is a parenchyma with large intercellular spaces. Inside the leaf tissue, considered by several authors as stem tissue, there is a layer of cells with external, radial, and inner thickened cell walls. Casparian strips have not been evidenced at this level, but it seems to have other structural characteristics of an endodermis. Beneath endodermis, there is a pericycle, which gives rise not only to adventitious roots but to the vascular and cork cambia.

The foliar origin of these succulent tissues is also supported by Duval-Jouve (1868). Based on the morphological and anatomical data, he stated that this succulent tissue comes from an increase of decurrent leaves. His point of view was also supported by a shedding phenomenon of this tissue. Dangeard (1888) and Monteil (1906), working with *Salicornia* and *Arthrocnemum* species, have assumed that this succulent tissue would represent the fused sheaths of opposite leaves.

De Fraine (1912), in his study on the *Salicornia* genus, came to the conclusion that this cortex is of foliar origin, derived from the decurrent growth of leaves. His observations were based on the following considerations:

1. Similarity of the tissue of succulent cortex to that of the leaves
2. Venation system of the cortex derives from the anastomosing lateral branches of the leaf strands
3. Shedding of the assimilatory cortex in the fall as a result of suberization of inner layers of cells
4. Similarity in the development of cotyledons and hypocotyl to that of the subsequent leaves and internodes

Keller (1951) suggests that in *Salicornia* species, the cortex derived from the fusion and adnation of the leaves to the stem. His findings were based on the study of some anomalies, i.e., seedlings with three cotyledons and unifoliate nodes, and asymmetrical arrangement of the internodal fleshy cortex in some plants grown in experimental plots.

Leisle (1949), studying the anatomy and ecology of halophytes and xerophytes with reduced leaves, specifies that, in *Anabasis aphylla*, this tissue is derived from the fusion of opposite leaves and argues as follows:

1. In the seedlings of *Anabasis*, the lowest nodes, adjacent to the cotyledons, have quite prominent leaves, while in the more distant, leaves are gradually reduced.
2. From the free leaf tips to the base of fleshy internodal tissue, similar palisade tissue and water storage tissue appear.

3. Different length of the leaves in various species of *Anabasis* (*A. micradena, A. brevifolia,* and *A. salsa*) is interpreted as a result of leaf fusion.

Cooke (1911), studying the anatomy of the *Salicornia australis* species, refers to this succulent formation as actually representing the leaf base surrounding the stem (Fig. 3.29b). According to Cooke, here are some of the reasons for calling leaf base what appears to be and has been described as "cortex":

1. The vascular system in this cortex-like portion resembles that of a leaf; the position of the vascular bundles is comparable to that in the basal portion of peltate leaves.
2. Except for the median branch, the network has no connection with the stem.
3. The bundles end blindly in mesophyll.
4. There is no difference between the palisade cells in the leaf and leaf base.
5. The water tissue of the leaf base is completely similar to the leaf mesophyll.
6. Below the leaf base, the stem loses its palisade tissue; it is a very short portion of each internode, i.e., the portion covered by the leaves of the next node below.

Cross section of the internode before leaf base disappears shows:

1. Epidermis, a single layer of cells, the outer walls of which have developed a cuticle.
2. Palisade parenchyma and scattered tracheids.
3. Water storage tissue, the internal limit of which is the endodermis.
4. Portions of fibro-vascular bundles scattered about in the water tissue. These are regarded as the vascular bundles of the leaf bases.
5. Central cylinder with a well-marked pericycle. According to the quoted author, this is the only portion of this section which can be called "stem." In this are embedded the collateral fibro-vascular bundles.

This latter point of view, which basically reduces the stem to a central cylinder, is also supported by Anderson (1974) and Chermezon (1910), who investigated *Salicornia fruticosa* (Fig. 3.30). Furthermore, he considers the layer of cells bordering inside the water tissue as consisting of separate tangentially flattened cells, which actually would represent the upper epidermis of the leaf fused with the stem.

Finally, returning to Cooke's observations, only sections made under the leaf shows the following:

1. An epidermis with thin-walled cells, which are much smaller than those of the leaf.
2. Hypodermis, consisting of two-layered cells.
3. Cortex with parenchymatous thin-walled cells, resembling those of water tissue, only much smaller. Chloroplasts are few in number in comparison with those of the chlorenchyma in the leaf and leaf base.
4. Central cylinder.

There are no cuticle, no stomata, no palisade tissue, no scattered tracheids, and no fibro-vascular bundles except in central cylinder.

The stomata are sunken in the epidermis (Fig. 3.29a), a feature—as already stated—common to other halophyte species.

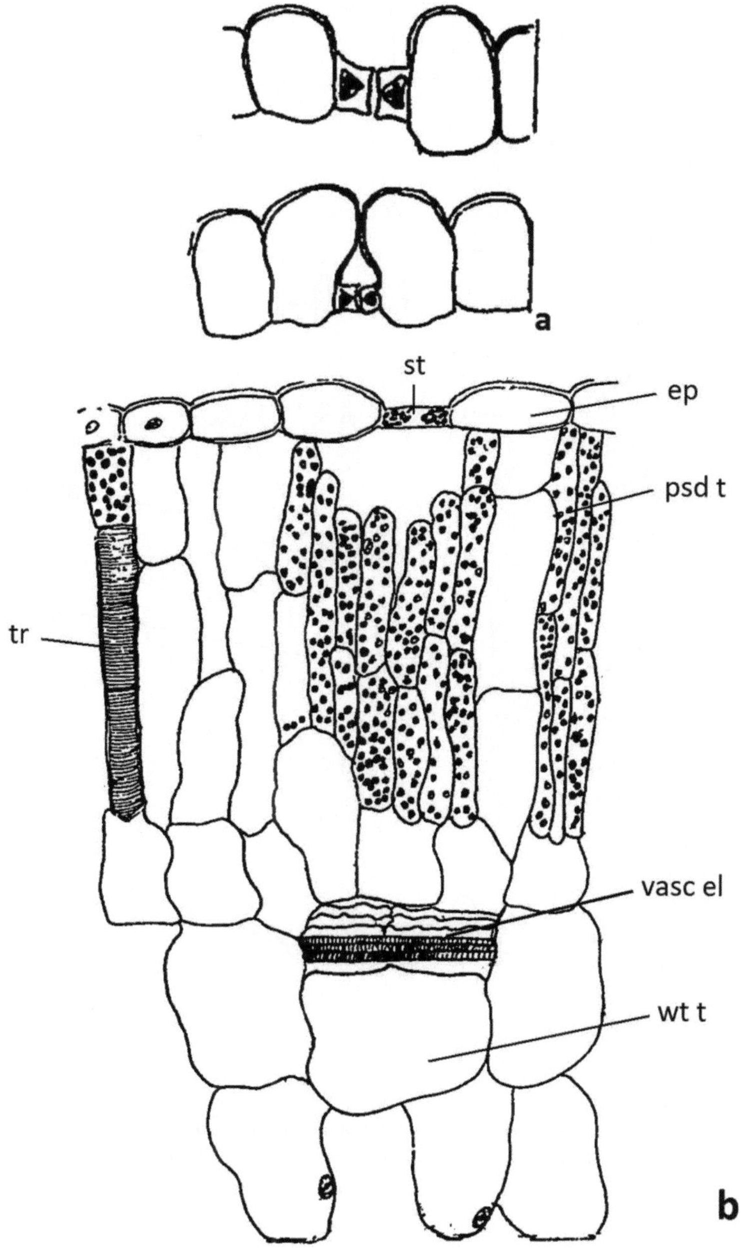

Fig. 3.29 *Salicornia australis*: (**a**) stomata sunken into epidermis and (**b**) cross section through leaf base (*ep* epidermis, *psd t* palisade tissue, *tr* tracheids, *vasc el* vascular elements, *wt t* water storage tissue) (Cooke 1911)

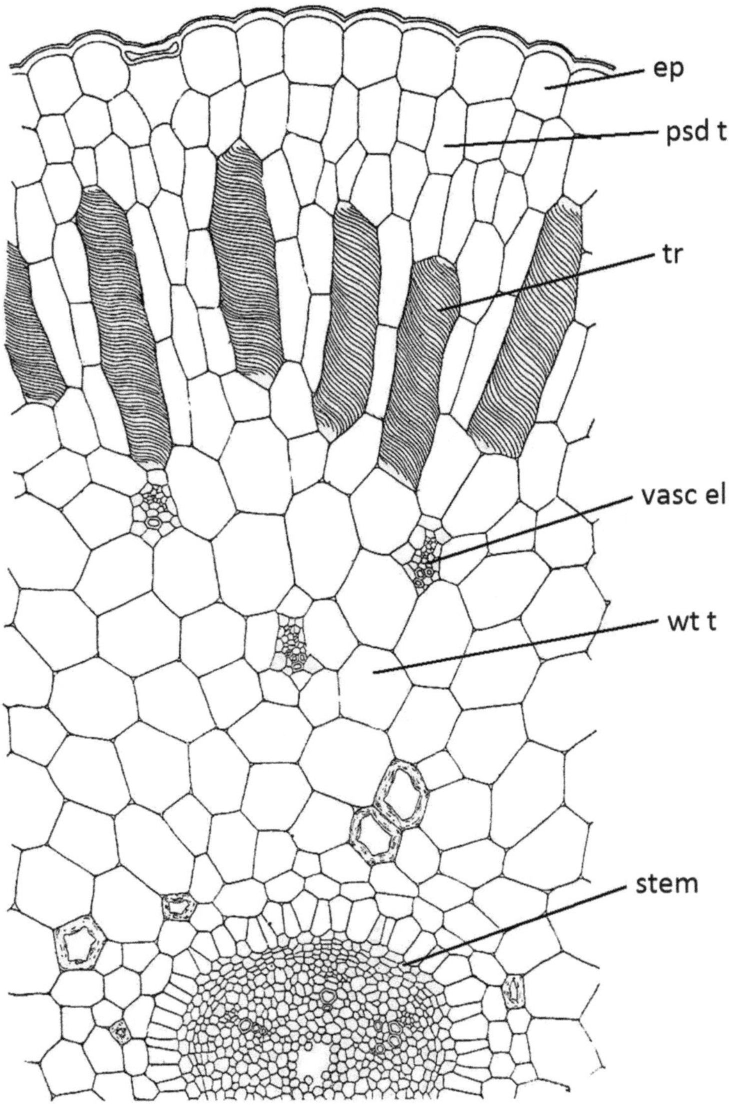

Fig. 3.30 Cross section through a young segment of *Salicornia fruticosa* shoot (*ep* epidermis, *psd t* palisade tissue, *wt t* water storage tissue, *tr* tracheids, *vasc el* vascular elements) (Chermezon 1910)

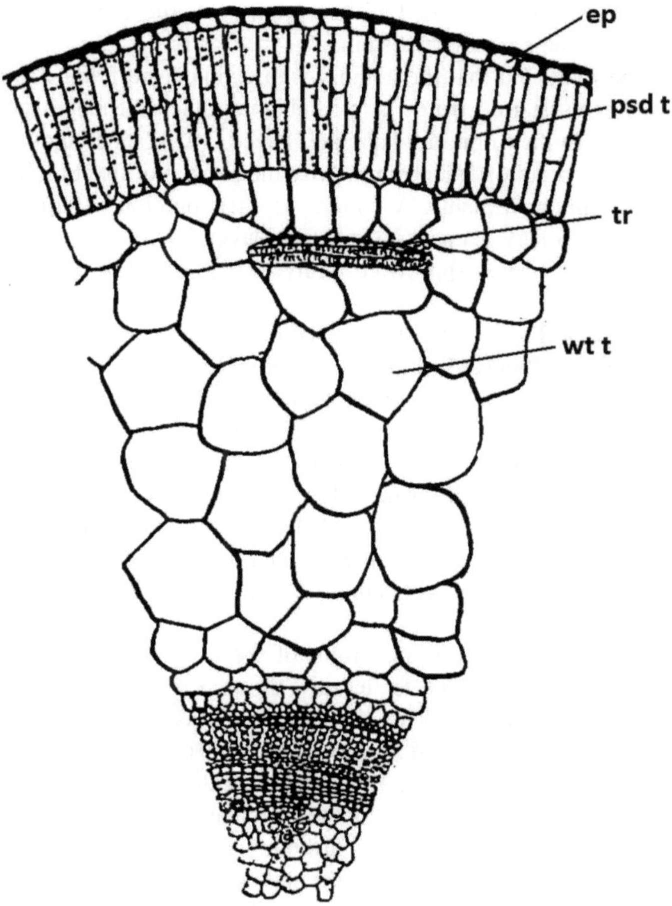

Fig. 3.31 Cross section through the "*stem*" of *Salicornia australis* (*ep* epidermis, *psd t* palisade tissue, *tr* tracheids, *wt t* water storage tissue) (Cross 1909)

A surprising issue in all this controversy is the lack of a uniform language and a constant characterization of structures evidenced at the level of these fleshy tissues.

For instance, the same topographical picture is sometimes described as a cross section through the lamina, or through the leaf base, or segment or through stem. Cross (1909), who studied the same species, *Salicornia australis* (Fig. 3.31), characterized it as a plant where "*leaves are* ***entirely*** *absent*" (our emphasis), although the cross section through the "*succulent stem*" is quite similar to that presented by Cooke (1911).

After reviewing several descriptions of these succulent tissues and the arguments used by each author for supporting them as being of foliar nature, data contradicting this idea will be provided in the next paragraphs. These data refer to the authors claiming that these tissues located at the periphery of the central cylinder would actually represent the cortex of the stem.

It should be reminded that the authors who claimed the foliar nature of these succulent segments started from the resemblance of chlorenchymatic tissues to those of the leaf. In their opinion, Fahn and Arzee (1959) stated that this does not provide consistent proof of foliar origin, since the transfer of function from organ to the organ is common in plants. According to their opinion, assimilatory tissues appear in stems of many plants with reduced leaves, such as *Genista sphacelata*, *Retama* spp., and *Spartium* spp. More rarely, it is possible to modify stems into phylloclades.

According to De Fraine (1912), the shedding of the cortex provides its foliar nature. Volkens (1887) described the separation of the cortex as caused by a deep-lying cylindrical phellogen. It is well known when such a phellogen arises in a stem, the external tissues are separated as a hollow cylinder. Fahn and Arzee (1959) thought that this type of shedding is a characteristic of many stems and cannot be therefore used as an argument for the foliar origin of the cortex. In the articulated *Chenopodiaceae*, the cortex and reduced leaves of each internode are removed together as a unit but separately from each internode in part. This shedding from each internode is due to the constriction of each internode at its base, where the cortex is quite narrow.

Another argument for the foliar origin of these tissues was given by Leisle (1949), referring to the gradual decrease of the free portions of the leaves from seedlings of *Anabasis aphylla*. This was interpreted as a result of the fusion and adnation to the stem, but Fahn and Arzee (1959) believe that this may be considered merely as stages of leaf reduction. They show that this can be also seen in young plants of *Acacia spp.*, where normal composed leaves, phyllodes, and intermediate structures are found on the same plant. Furthermore, if adnation had taken place, an increased number of vascular strands in the stele should be found, as a result of the inclusion of leaf strands. Therefore, they reject these arguments and conclude that, in fact, succulent tissues located outside the central cylinder of the plant must be regarded as a true cortex.

Nevertheless, succulence was, as already underlined, an adaptive structural feature early recognized by botanists within anatomical sets of halophyte strategies.

Warming in his consistent study referring to structural characteristics in halophytes (1897, but also prepared by several previous studies, 1890, 1891), evidenced succulence in various species; actually, many of investigated species have succulent or narrow-reduced leaves (Fig. 3.32), a xeromorphic adaptation (Grigore et al. 2014).

Warming (1897) found water storage tissue in the leaf structure of many halophytes: *Tournefortia gnaphalodes* (Fig. 3.33), *Scaevola plumieri* (Fig. 3.34), *Borrichia arborescens* (Fig. 3.35), *Philoxerus vermiculatus* (Fig. 3.36), *Remirea maritima* (Fig. 3.37), *Euphorbia buxifolia* (Fig. 3.38), and *Haloxylon ammodendron* (Fig. 3.39).

Chermezon (1910) made an impressive study on structural features in littoral plants, investigating approximately 134 species belonging to various botanical

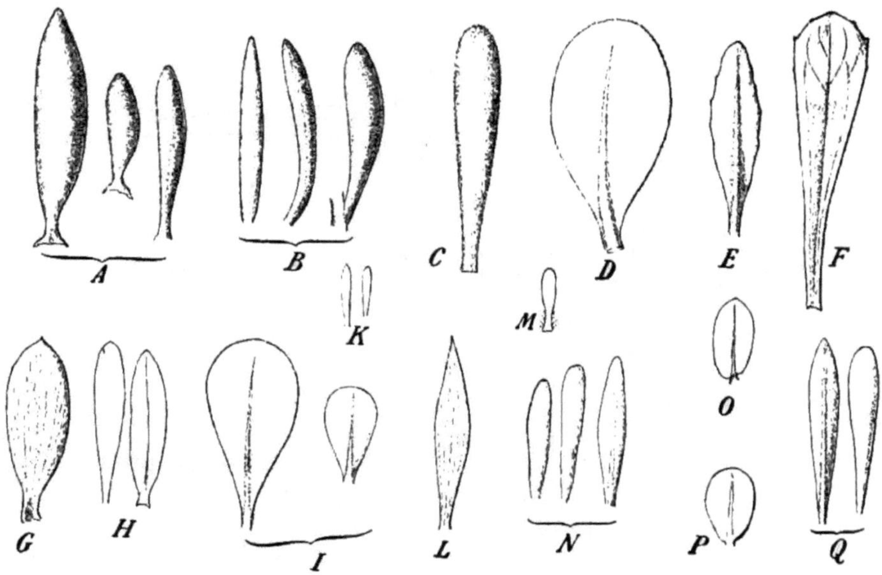

Fig. 3.32 General form of leaves of halophyte species: *Sesuvium portulacastrum* (**a**), *Batis maritima* (**b**), *Tournefortia gnaphalodes* (**c**), *Scaevola plumieri* (**d**), *Cakile aequalis* (**e**), *Acicarpha spathulata* (**f**), *Borrichia arborescens* (**g**), *Philoxerus vermiculatus* (**h**), *Baccharis dioica* (**i**), *Alternanthera muscoides* (**k**), *Ernodea littoralis* (**l**), *Pectis humifusa* (**m**), *Heliotropium curassavicum* (**n**), *Euphorbia buxifolia* (**o**), *Portulaca oleracea* (**p**), and *Suriana maritima* (**q**) (Warming 1897)

families and having different ecological spectra. Many of them are typical halophytes, where succulence (nominated by the French botanist as *carnosité*) has been evidenced. Apart from anatomical descriptions of littoral plants, he also made several interesting conclusions regarding the occurrence of succulence in halophytes. He underlined that succulence is not a general feature in all investigated taxa, and it has a very high variability character, depending on leaf form and especially on ecological requirements. Plants from salt marshes (*stricto sensu*) often display the most striking succulence, whereas plants from other habitats (also subjected to reduced salinity influence) present leaf succulence to a lesser extent. Nevertheless, he correlated the degree of succulence with the intensity of environmental salt, albeit he realized that this correlation does not have an absolute value. He adopted and used a rather flexible definition of succulence, as resulted from the hypertrophy of leaf cells; therefore, all cells increase in their size, and the epidermal and mesophyll cells become thicker. When these cells are not thicker, then succulence is mainly due to the development of water storage tissue.

Within littoral plants, he found a correlation between saline environments (and salinity degree, consequently) and the development of succulence. Plants from salt marshes have the most striking succulence (*Spergularia marginata, Inula crithmoides, Plantago crassifolia, Suaeda fruticosa, S. maritima, Salsola soda,*

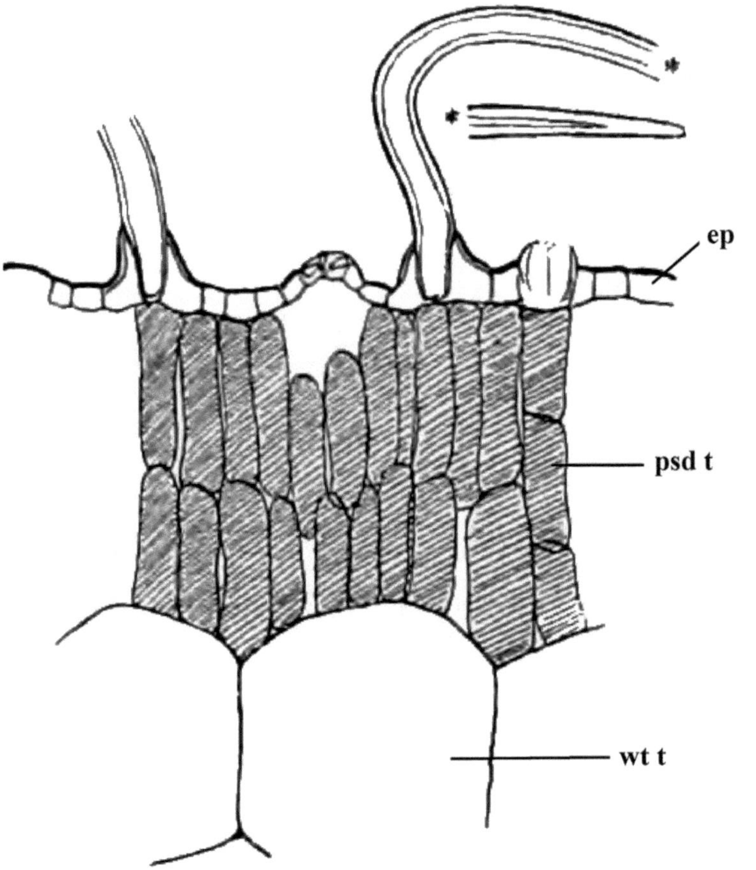

Fig. 3.33 Cross section through the leaf of *Tournefortia gnaphalodes* (*ep* epidermis, *psd t* palisade tissue, *wt t* water storage tissue) (Warming 1897)

Salicornia herbacea, and *Arthrocnemum macrostachyum*—many of the drawings shown in the lines below), while others are less succulent: *Aster tripolium, Artemisia gallica, Plantago maritima, Atriplex portulacoides,* and *A. littoralis.* Plants vegetating in rocky shores have leaves less succulent than those of salt marshes: *Crambe maritima, Silene maritima, Crithmum maritimum,* and *Beta maritima.* Finally, desert plants from saline environments are often succulent: *Zygophyllum album, Tetradiclis eversmanni,* and *Halocnemum strobilaceum.*

Within the succulent species he investigated, the following should be mentioned: *Cakile maritima* (Fig. 3.40, also investigated by Toma et al. 1979), *Honckenya peploides* (Fig. 3.41), *Silene maritima* (Fig. 3.42), *Spergularia lobeliana* (Fig. 3.43), *Zygophyllum album* (Fig. 3.44; Elhalim et al. 2016), *Tetradiclis eversmanni* (Fig. 3.45), *Crithmum maritimum* (Fig. 3.46), *Inula crithmoides*

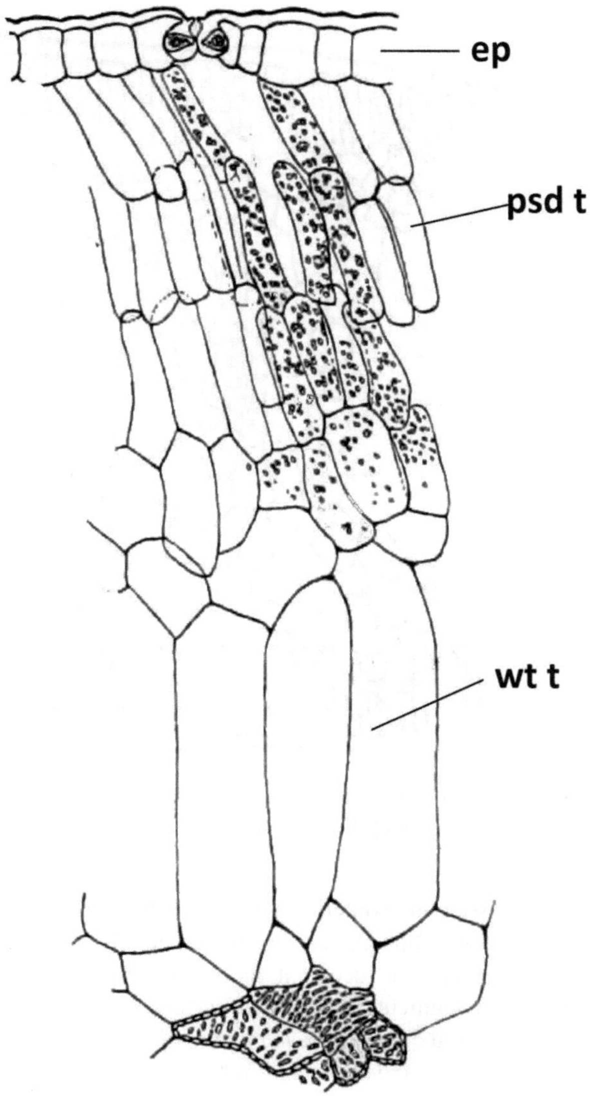

Fig. 3.34 Cross section through the leaf of *Scaevola plumieri* (*ep* epidermis, *psd t* palisade tissue, *wt t* water storage tissue) (Warming 1897)

(Fig. 3.47), *Artemisia crithmifolia* (Fig. 3.48), *A. gallica* (Fig. 3.49), *Heliotropium curassavicum* (Fig. 3.50), and *Plantago crassifolia* (Fig. 3.51).

Succulence has been evidenced in many halophyte species collected from Romanian saline environments (Grigore 2008; Grigore and Toma 2008, 2010a, b; Grigore et al. 2012a, b; Grigore et al. 2014) and from Mediterranean salt marshes (Grigore et al. 2011a, b, 2012b, 2013, 2014).

In a recent monograph, Grigore et al. (2014) reinforced the idea that succulence is a typical adaptive feature found in halophytes. It has been evidenced as a water

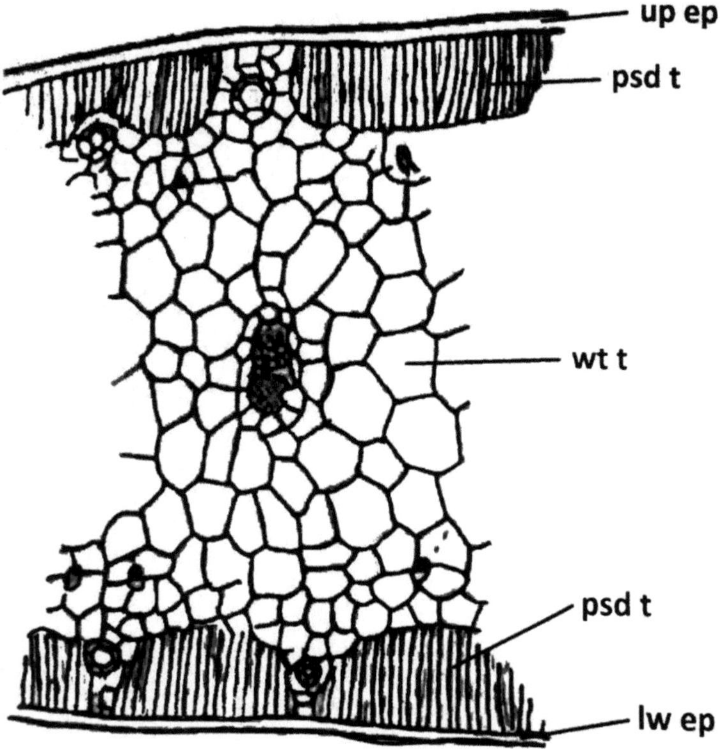

Fig. 3.35 Cross section through the leaf of *Borrichia arborescens* (*up ep* upper epidermis, *lw ep* lower epidermis, *psd t* palisade tissue, *wt t* water storage tissue) (Warming 1897)

storage tissue and sometimes as a well-developed palisade tissue in many investigated species. Succulence occurs especially in *Chenopodiaceae* species. Here, several groups of succulent halophytes have been proposed. Halophytes with articulated segments: *Salicornia europaea* (Figs. 3.52 and 3.53), *S. ramosissima* (Figs. 3.54 and 3.55), *Halocnemum strobilaceum, Sarcocornia fruticosa* (Figs. 3.56 and 3.57), and *Arthrocnemum macrostachyum* (Fig. 3.58). These are apparently leafless species, where water tissues confer the special appearance of nominated taxa. Another group includes species with small, cylindrical or flattened laminas, in which water storage tissue occupies a large part of the leaves: *Suaeda maritima* (Fig. 3.59), *S. spicata* (Figs. 3.60 and 3.61), *Bassia hirsuta* (Fig. 3.62), *B. sedoides* (Figs. 3.63 and 3.64), and *Halimione portulacoides* (Fig. 3.65). Succulence in these[1] species can be correlated with the C_3 photosynthetic pathway. Other groups of succulent chenopods include C_4 species; here, the water storage tissue has a central position within the leaves, being delineated at the exterior by the two typical

[1]On the micrograph explanation, RO designates that the species have been collected from Romania, while ESP, from Spain.

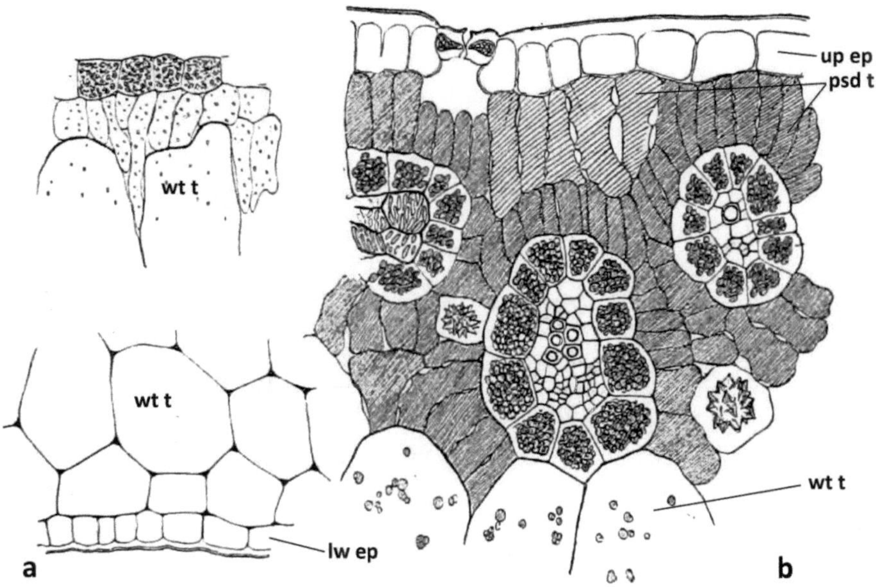

Fig. 3.36 Cross section through the leaf of *Philoxerus vermiculatus*: (**a**) detail of water tissue and (**b**) position of water storage tissue within leaf section (*up ep* upper epidermis, *lw ep* lower epidermis, *psd t* palisade tissue, *wt t* water storage tissue) (Warming 1897)

chlorenchymatic tissues evidenced in C_4 species: *Petrosimonia oppositifolia* (Fig. 3.66), *P. triandra* (Figs. 3.67 and 3.68), *Camphorosma annua* (Figs. 3.69 and 3.70), *Salsola oppositifolia* (Figs. 3.71 and 3.72), *S. kali* (Fig. 3.73), *Suaeda splendens* (Fig. 3.74). Succulence has been also found in two *Plantago* species, *P. crassifolia* (Fig. 3.75) and *P. tenuiflora* (Figs. 3.76 and 3.77), in *Inula crithmoides* (Fig. 3.78), *Spergularia media* (Figs. 3.79 and 3.80), *Crithmum maritimum* (Fig. 3.81), and Iranian Bassia species (C_4 species): *B. turkestanica* (Fig. 3.82), *B. pilosa*(Fig. 3.83), and *B. stellaris* (Fig. 3.84).

There are also several halophytes species with central storage water tissue, both C_3 and $C_{4.}$ Several drawings with cross sections through halophyte organs can help to obtain an accurate picture of tissues involved in achieving succulence appearance (Fig. 3.85—*Salicornia herbacea*, C_3; Fig. 3.86—*Suaeda maritima*, C_3; Fig. 3.87—*Plantago maritima*¸C_3; Fig. 3.88—*Spergularia media*, C_3; Fig. 3.89—*Petrosimonia triandra*, C_4; Fig. 3.90—*Salicornia europaea*, C_3; Fig. 3.91—*Salsola kali*—C_4; Fig. 3.92—*Kochia hirsuta*—C_3).

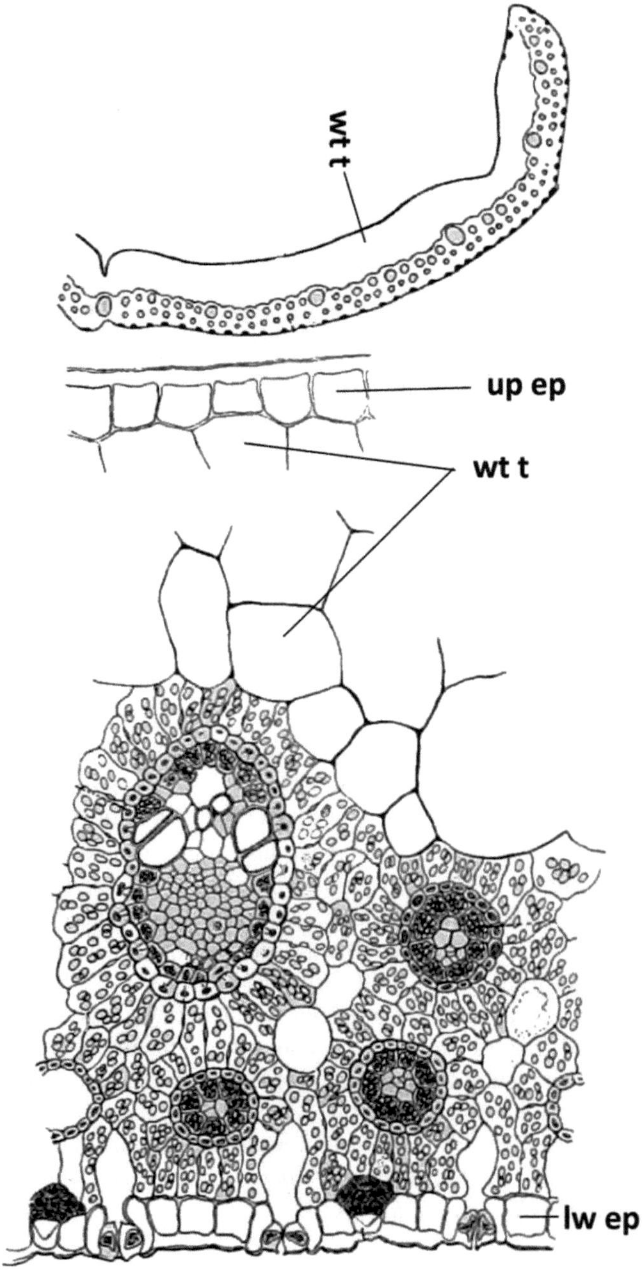

Fig. 3.37 Cross section through the leaf of *Remirea maritima* (*top*—position of water storage tissue within leaf section; bottom—detail) (*up ep* upper epidermis, *lw ep* lower epidermis, *wt t* water storage tissue) (Warming 1897)

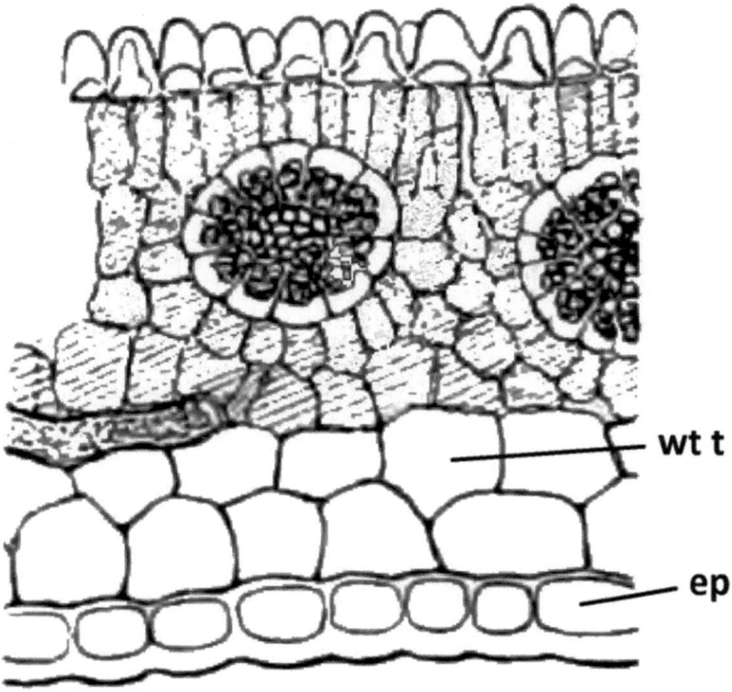

Fig. 3.38 Cross section through the leaf of *Euphorbia buxifolia* (*ep* epidermis, *wt t* water storage tissue) (Warming 1897)

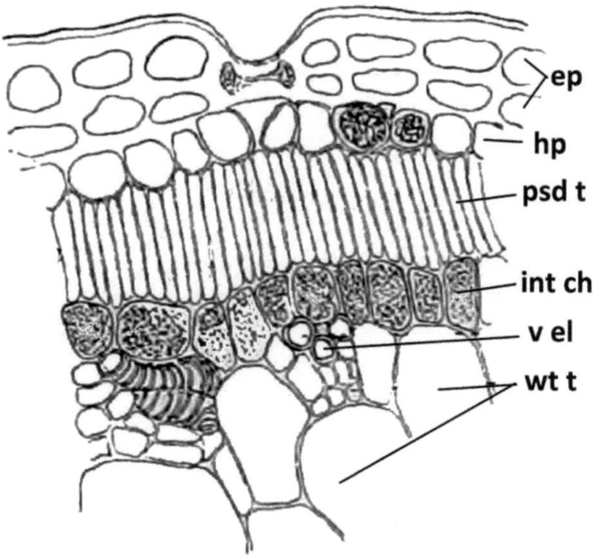

Fig. 3.39 Cross section through the assimilating shoot of *Haloxylon ammodendron* (*ep* epidermis, *int ch* internal chlorenchyma, *v el* vascular elements, *wt t* water storage tissue) (original drawing from Warming 1897)

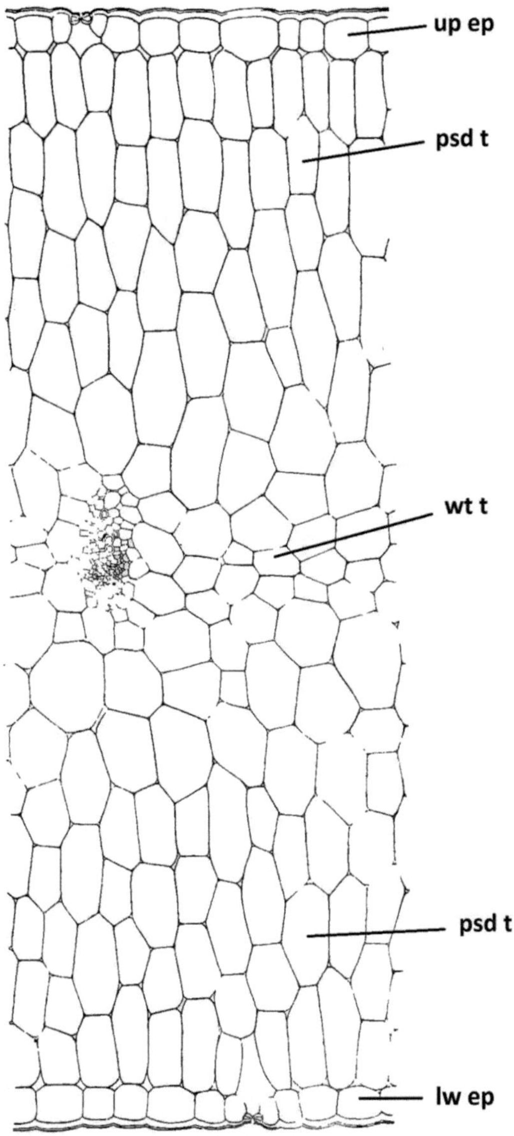

Fig. 3.40 Cross section through the leaf of *Cakile maritima* (*up ep* upper epidermis, *lw ep* lower epidermis, *psd t* palisade tissue, *wt t* water storage tissue) (Chermezon 1910)

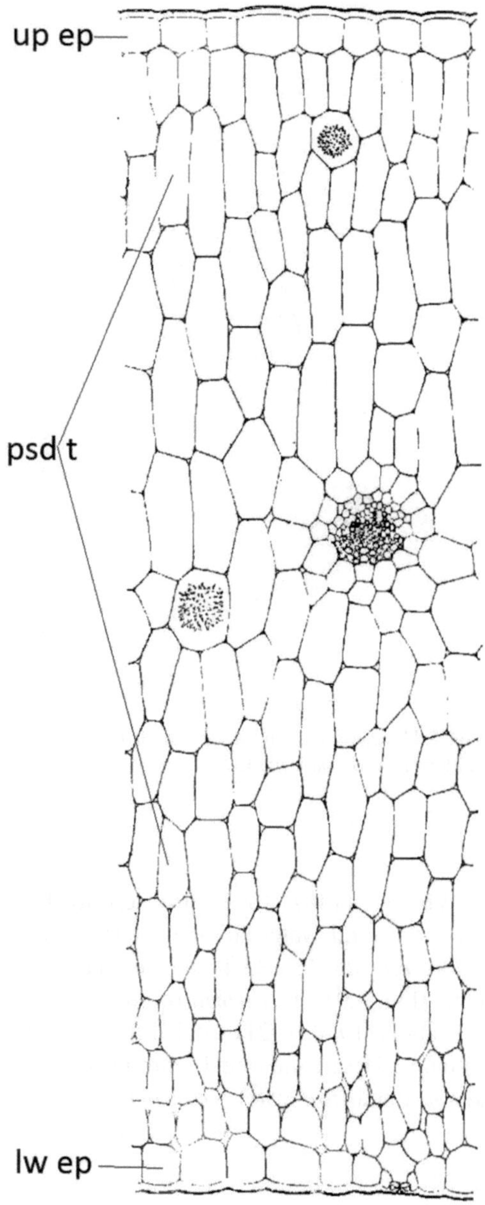

Fig. 3.41 Cross section through the leaf of *Honckenya peploides* (*up ep* upper epidermis, *lw ep* lower epidermis, *psd t* palisade tissue) (Chermezon 1910)

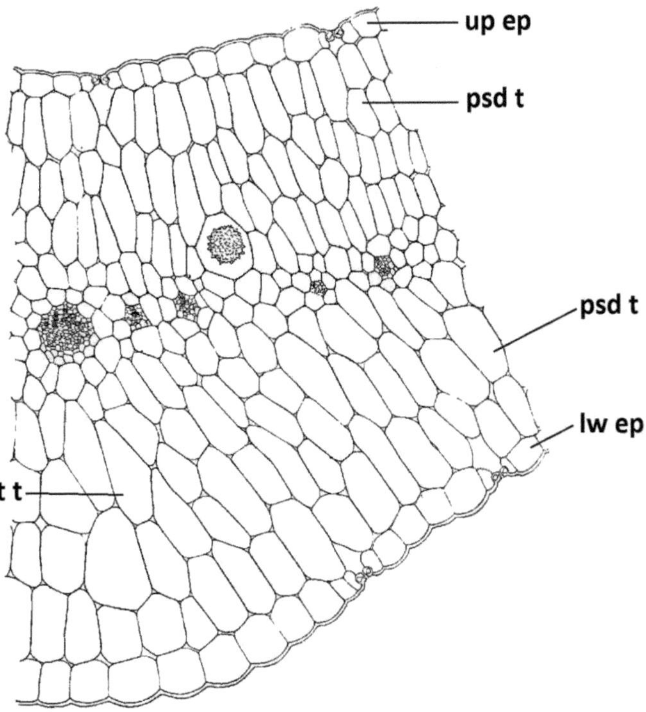

Fig. 3.42 Cross section through the leaf of *Silene maritima* (*up ep* upper epidermis, *lw ep* lower epidermis, *psd t* palisade tissue, *wt t* water storage tissue) (Chermezon 1910)

Volkens (1887, 1893) also described several C_4 halophytes from deserts that have a central water storage tissue, *Salsola longifolia* (Fig. 3.93) and *Haloxylon schweinfurthii* (Fig. 3.94), and very large water-storing cells located within palisade cells (Fig. 3.95) in *Nitraria retusa*.

Sabnis (1920, 1921) studied the physiological anatomy of some plants growing in Indian Desert and found central water storage in two C_4 halophytes: *Zygophyllum simplex* and *Haloxylon recurvum*.

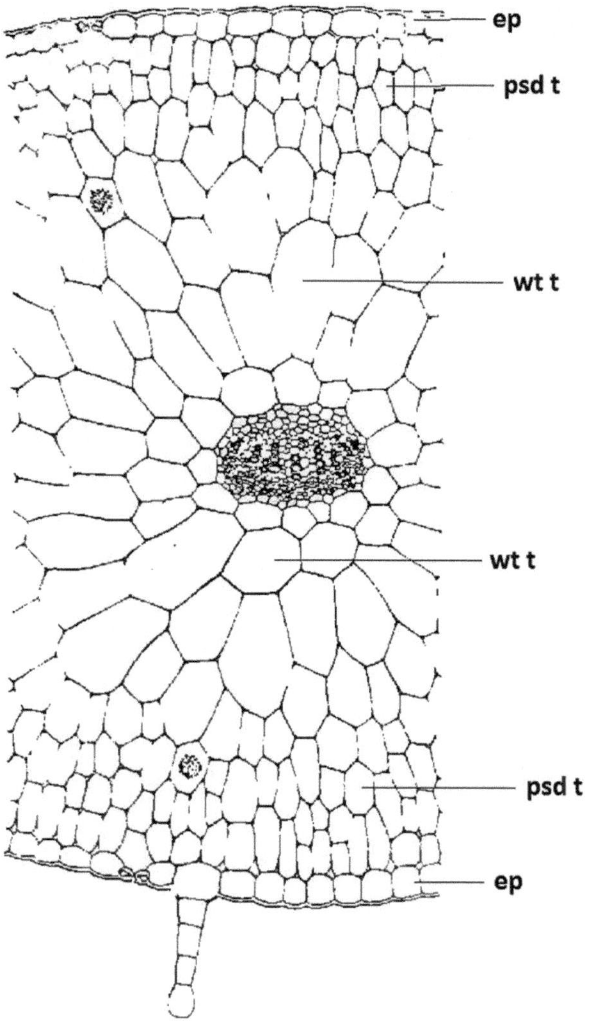

Fig. 3.43 Cross section through the leaf of *Spergularia lobeliana* (*ep* epidermis, *psd t* palisade tissue, *wt t* water storage tissue) (Chermezon 1910)

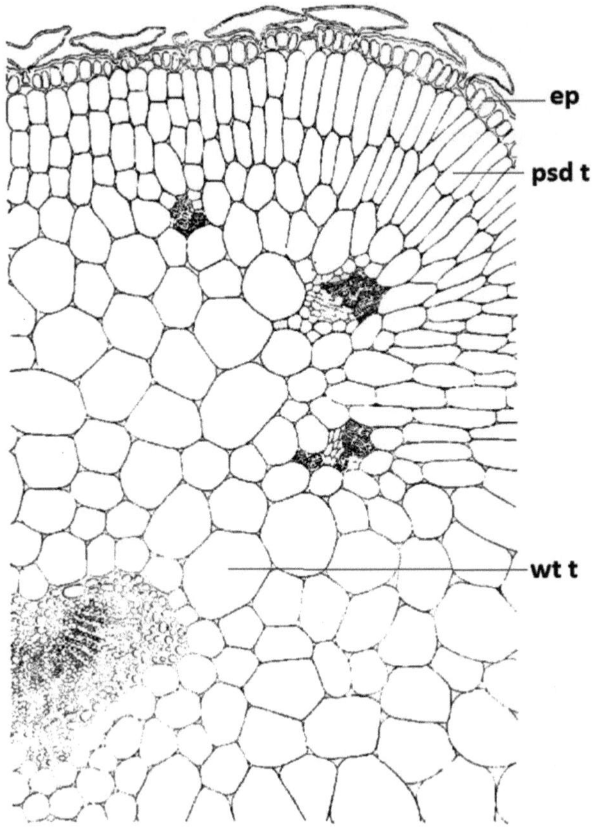

Fig. 3.44 Cross section through the leaf of *Zygophyllum album* (*ep* epidermis, *psd t* palisade tissue, *wt t* water storage tissue) (Chermezon 1910)

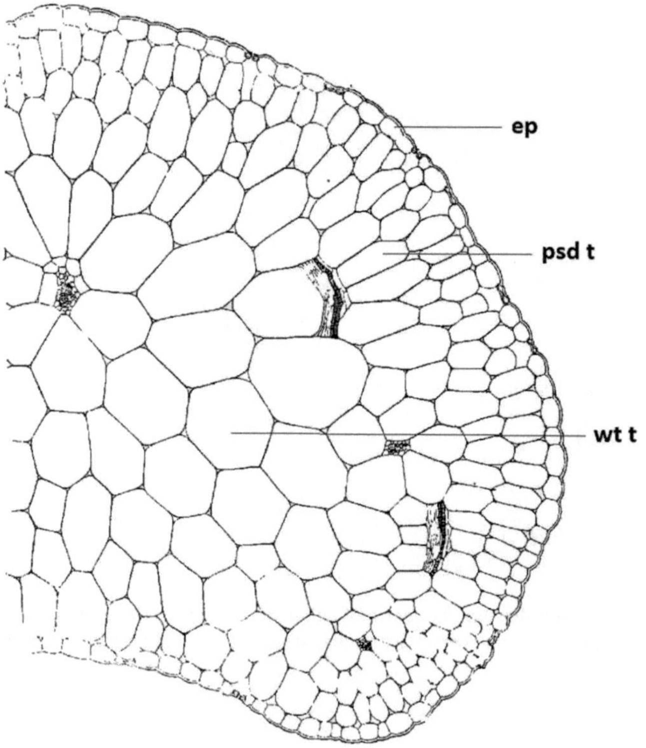

Fig. 3.45 Cross section through the leaf of *Tetradiclis eversmanni* (*ep* epidermis, *psd t* palisade tissue, *wt t* water storage tissue) (Chermezon 1910)

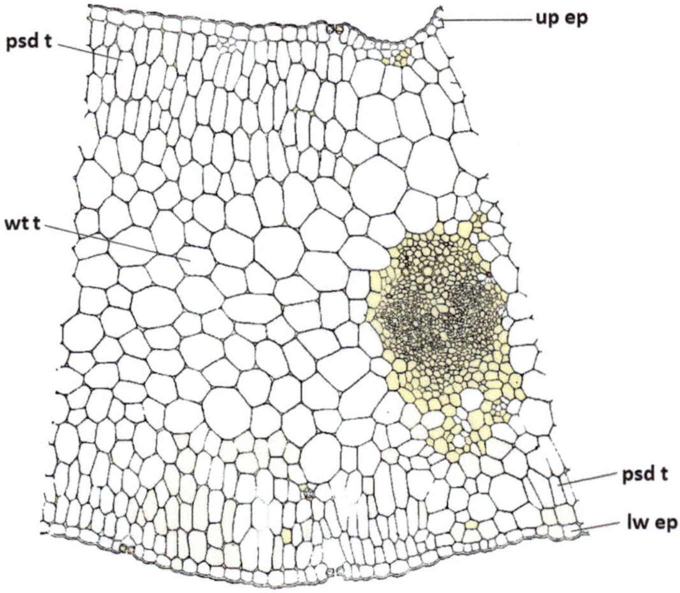

Fig. 3.46 Cross section through the leaf of *Crithmum maritimum* (*up ep* upper epidermis, *lw ep* lower epidermis, *psd t* palisade tissue, *wt t* water storage tissue) (Chermezon 1910)

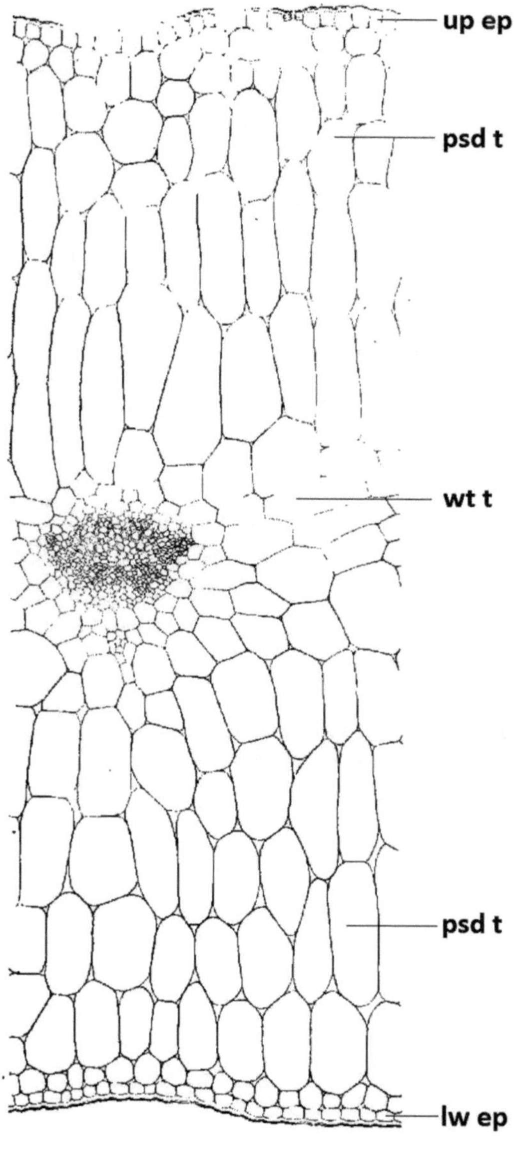

Fig. 3.47 Cross section through the leaf of *Inula crithmoides* (*up ep* upper epidermis, *lw ep* lower epidermis, *psd t* palisade tissue, *wt t* water storage tissue) (Chermezon 1910)

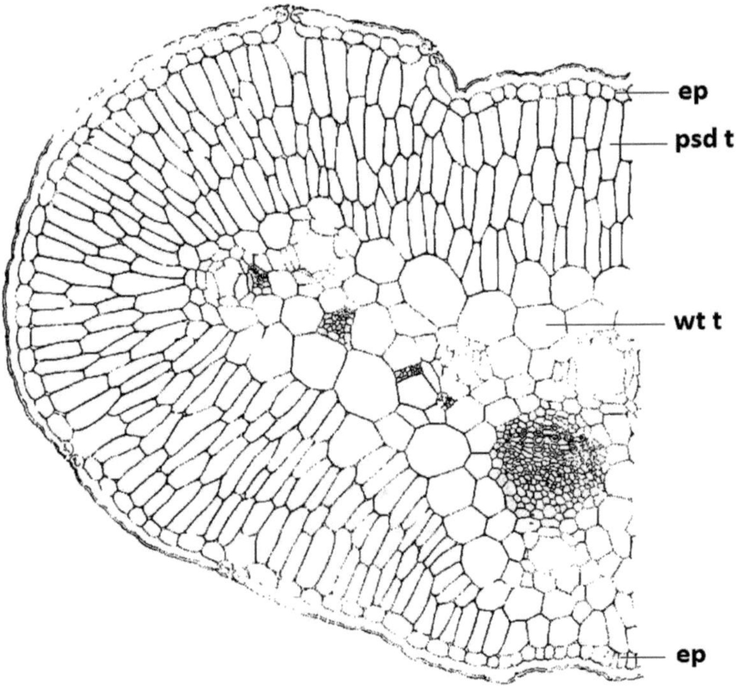

Fig. 3.48 Cross section through the leaf of *Artemisia crithmifolia* (*ep* epidermis, *psd t* palisade tissue, *wt t* water storage tissue) (Chermezon 1910)

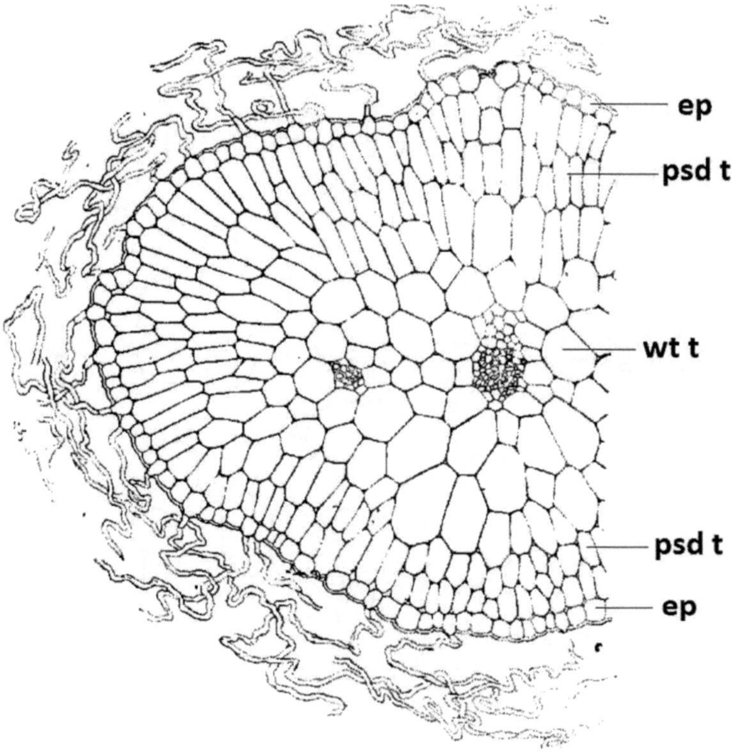

Fig. 3.49 Cross section through the leaf of *Artemisia gallica* (*ep* epidermis, *psd t* palisade tissue, *wt t* water storage tissue) (Chermezon 1910)

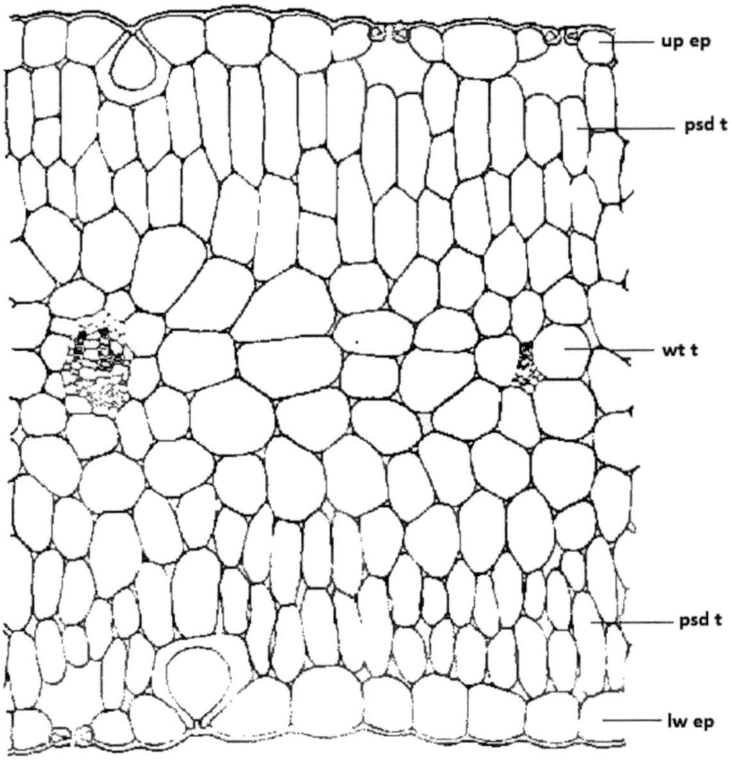

Fig. 3.50 Cross section through the leaf of *Heliotropium curassavicum* (*up ep* upper epidermis, *lw ep* lower epidermis, *psd t* palisade tissue, *wt t* water storage tissue) (Chermezon 1910)

Fig. 3.51 Cross section through the leaf of *Plantago crassifolia* (*ep* epidermis, *psd t* palisade tissue, *wt t* water storage tissue) (Chermezon 1910)

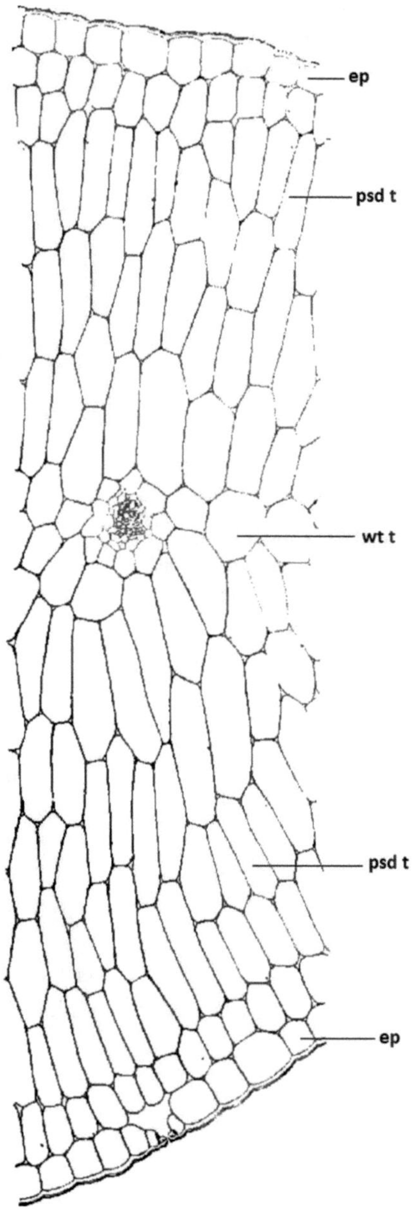

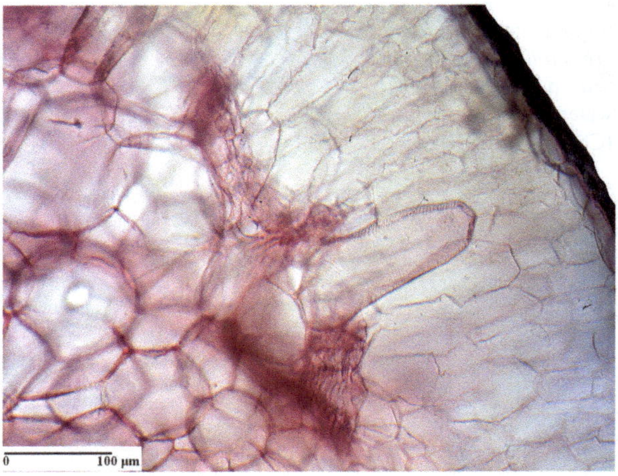

Fig. 3.52 Cross sections through the fleshy segments of *Salicornia europaea* (RO)

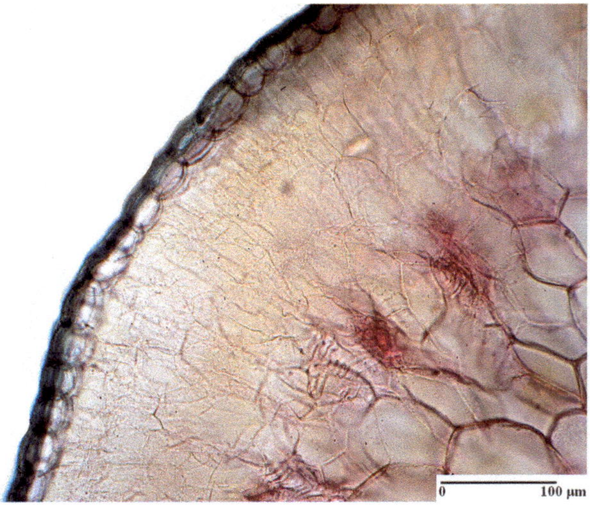

Fig. 3.53 Cross sections through the fleshy segments of *Salicornia europaea* (RO)

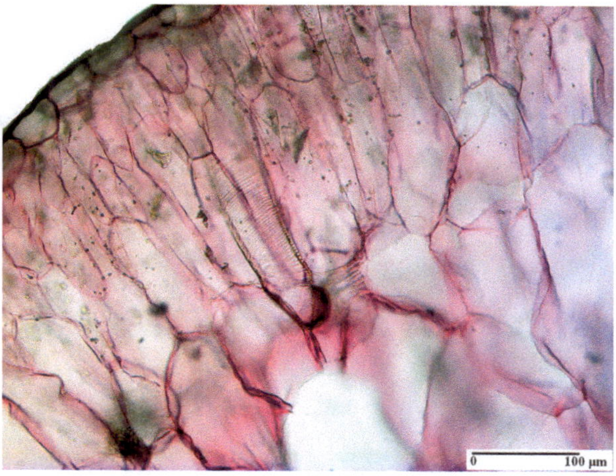

Fig. 3.54 Cross sections through the fleshy segments of *S. ramosissima* (ESP)

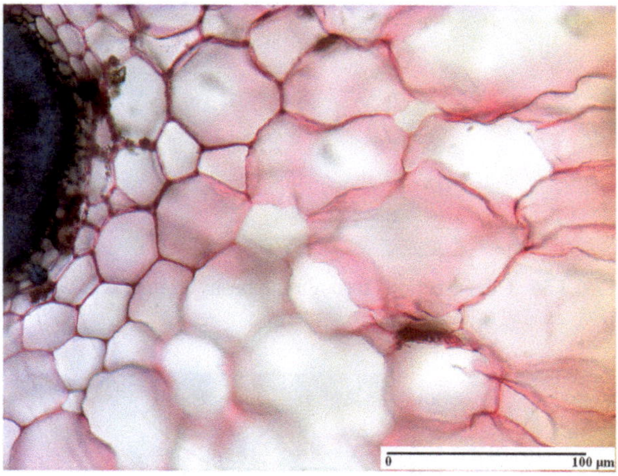

Fig. 3.55 Cross sections through the fleshy segments of *S. ramosissima* (ESP)

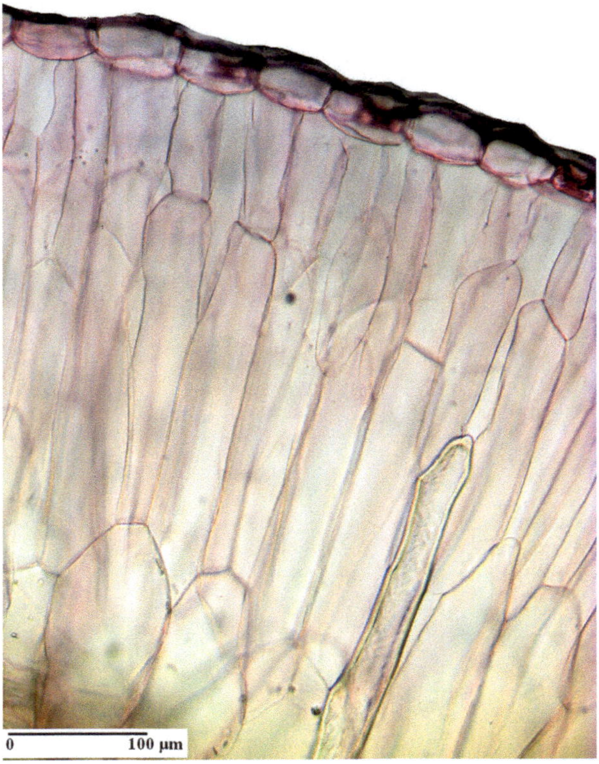

Fig. 3.56 Cross sections through the fleshy segments of *Sarcocornia fruticosa* (ESP)

Fig. 3.57 Cross sections through the fleshy segments of *Sarcocornia fruticosa* (ESP).

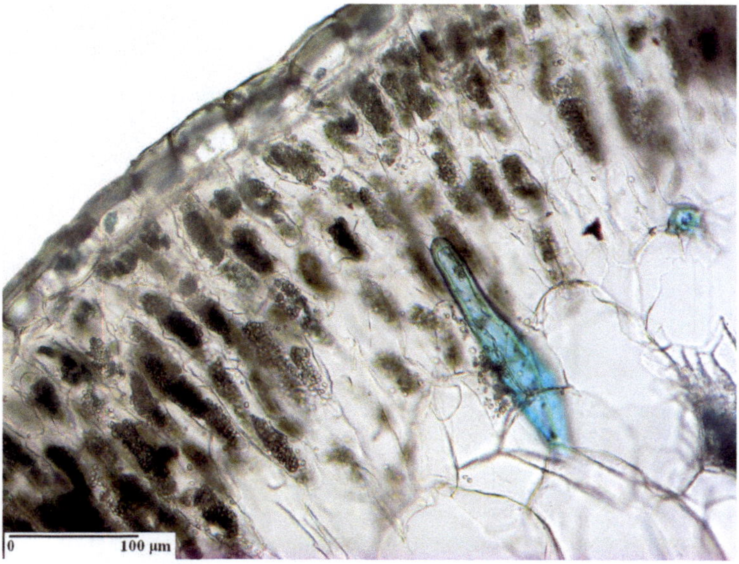

Fig. 3.58 Cross sections through the lamina of *Arthrocnemum macrostachyum* (ESP)

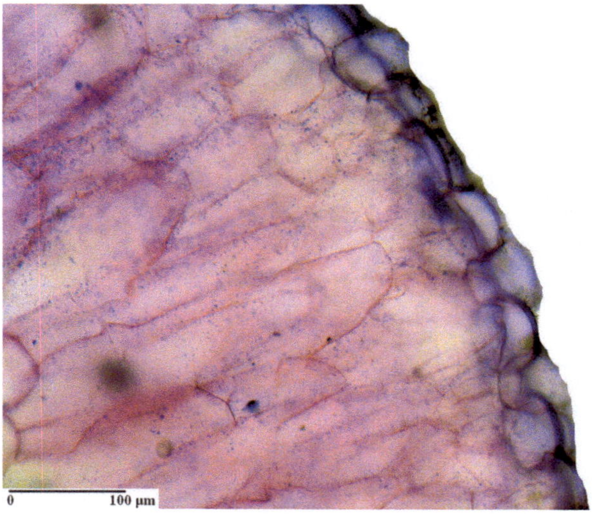

Fig. 3.59 Cross sections through the lamina of *Suaeda maritima* (RO)

Fig. 3.60 Cross sections through the lamina of *S. spicata* (ESP)

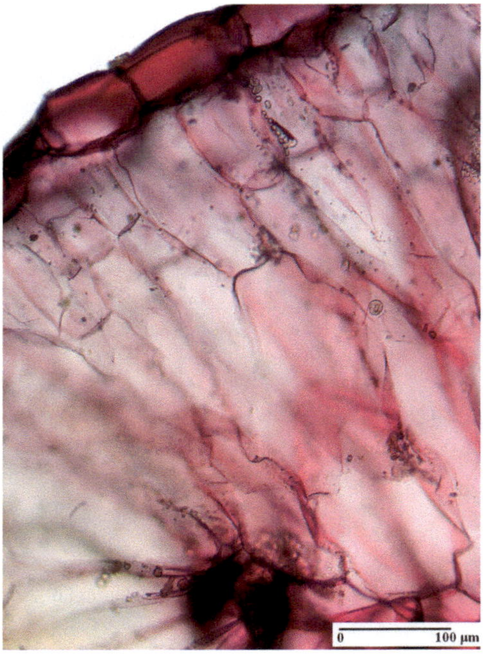

Fig. 3.61 Cross sections through the lamina of *S. spicata* (ESP)

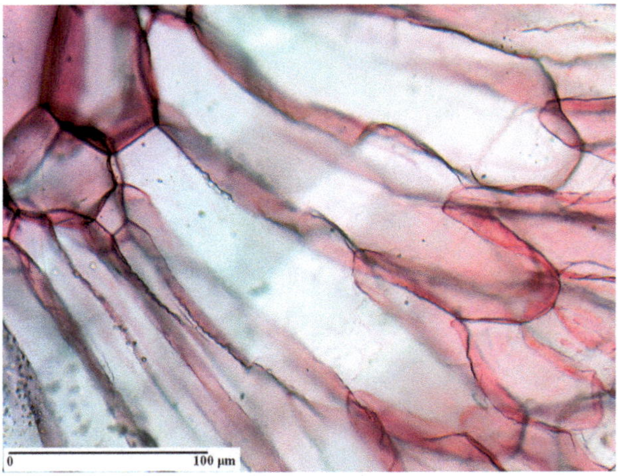

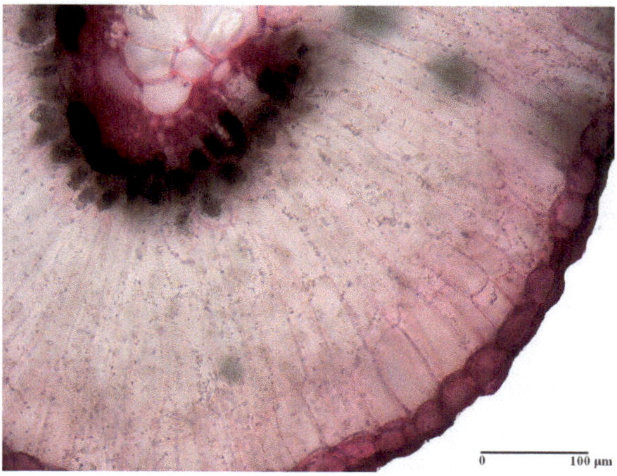

Fig. 3.62 Cross sections through the lamina of *Bassia hirsuta* (RO)

Fig. 3.63 Cross sections through the lamina of *B. sedoides* (RO).

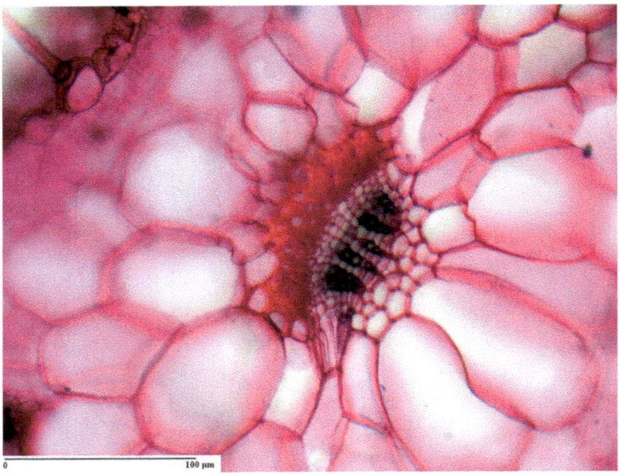

Fig. 3.64 Cross sections through the lamina of *Bassia sedoides* (RO)

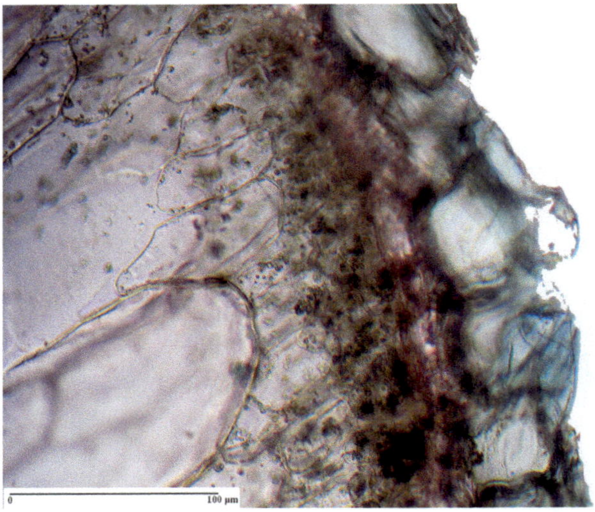

Fig. 3.65 Cross sections through the lamina of *Halimione portulacoides* (ESP)

Fig. 3.66 Cross sections through the lamina of *Petrosimonia oppositifolia* (RO)

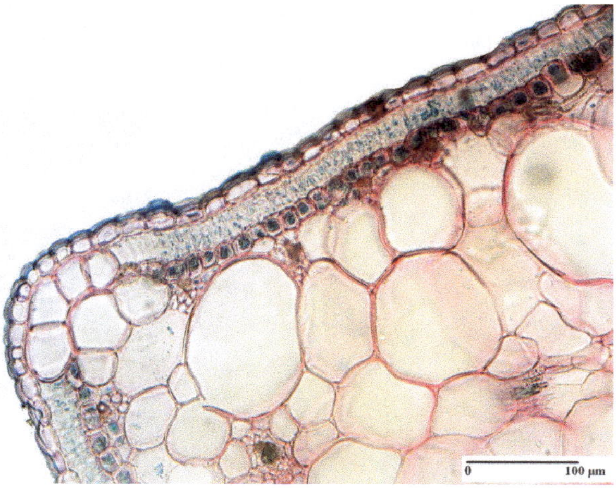

Fig. 3.67 Cross sections through the lamina of *P. triandra* (RO)

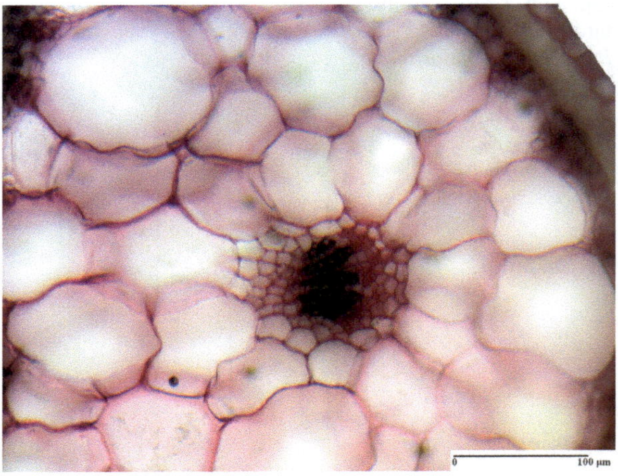

Fig. 3.68 Cross sections through the lamina of *P. triandra* (RO)

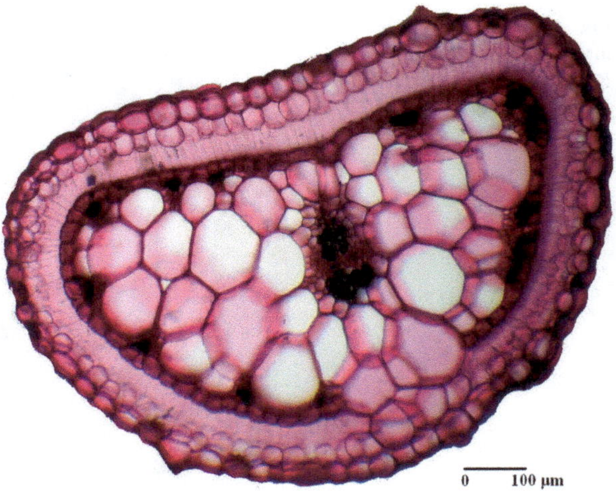

Fig. 3.69 Cross sections through the lamina of *Camphorosma annua* (RO).

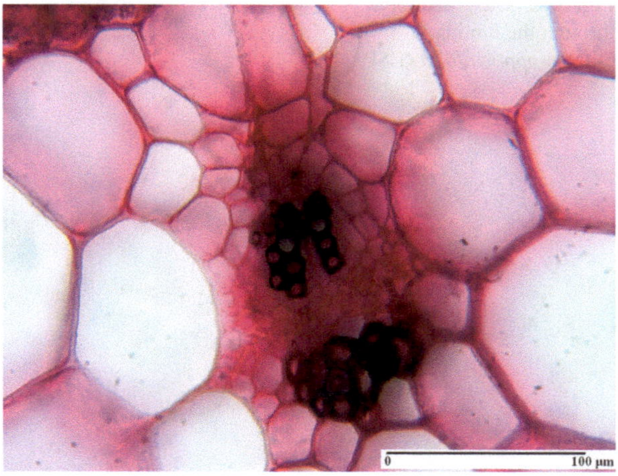

Fig. 3.70 Cross sections through the lamina of *Camphorosma annua* (RO)

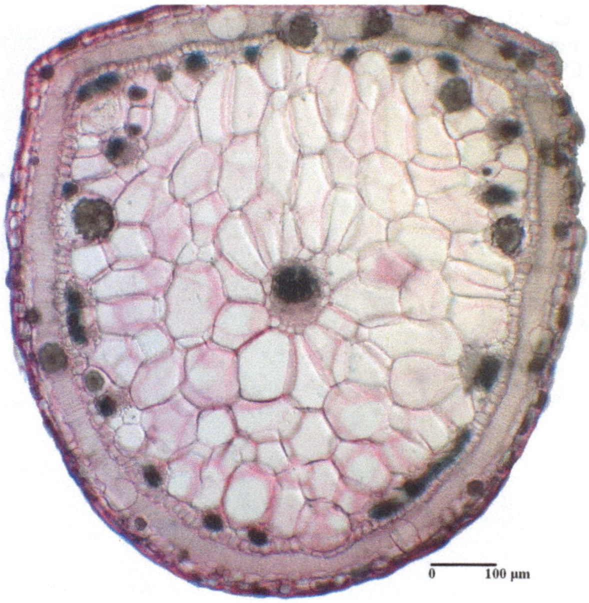

Fig. 3.71 Cross sections through the lamina of *Salsola oppositifolia* (ESP)

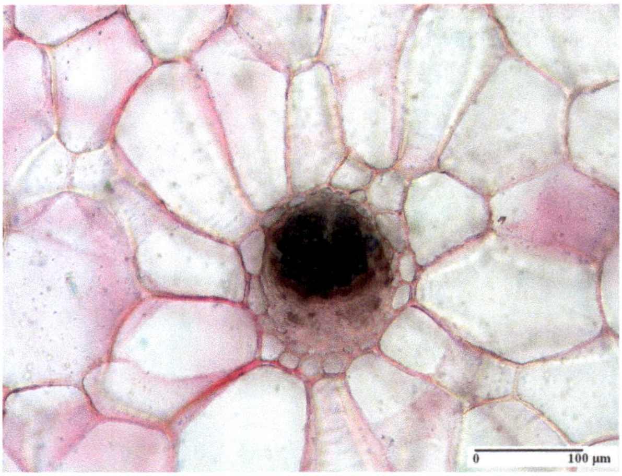

Fig. 3.72 Cross sections through the lamina of *Salsola oppositifolia* (ESP)

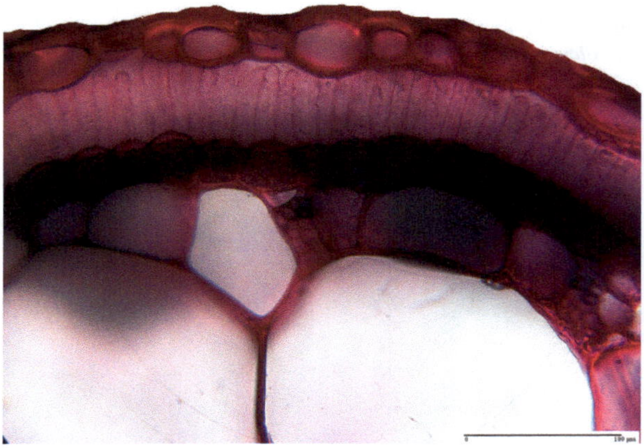

Fig. 3.73 Cross sections through the lamina of *S. kali* (ESP)

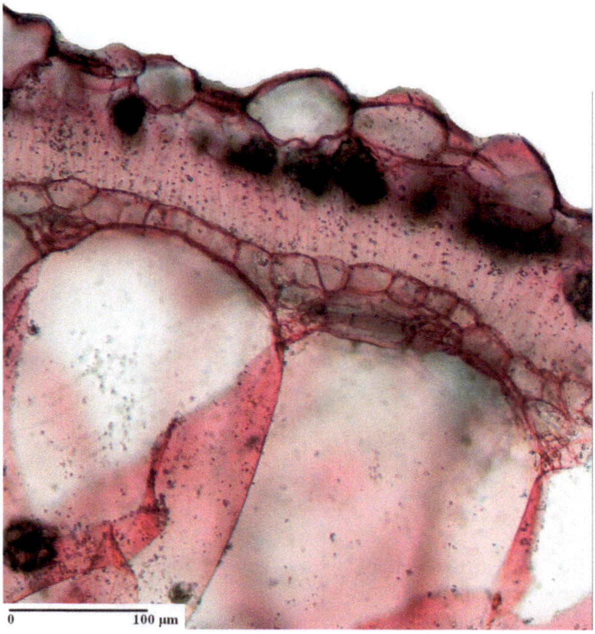

Fig. 3.74 Cross sections through the lamina of *Suaeda splendens* (ESP)

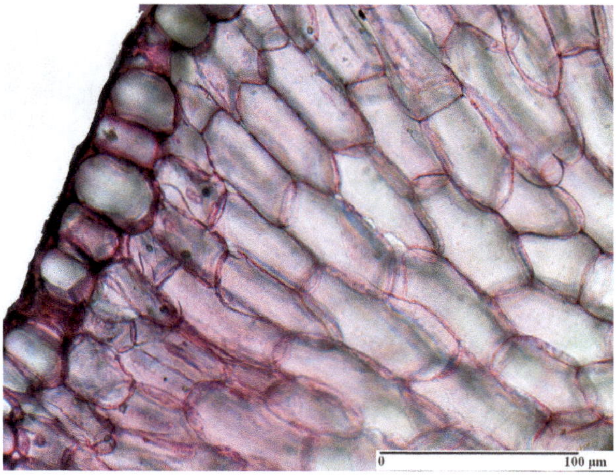

Fig. 3.75 Cross sections through the lamina of *Plantago crassifolia* (ESP)

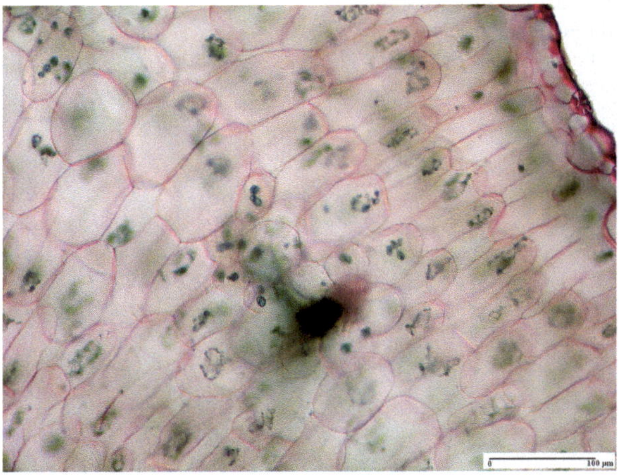

Fig. 3.76 Cross sections through the lamina of *Plantago tenuiflora* (RO)

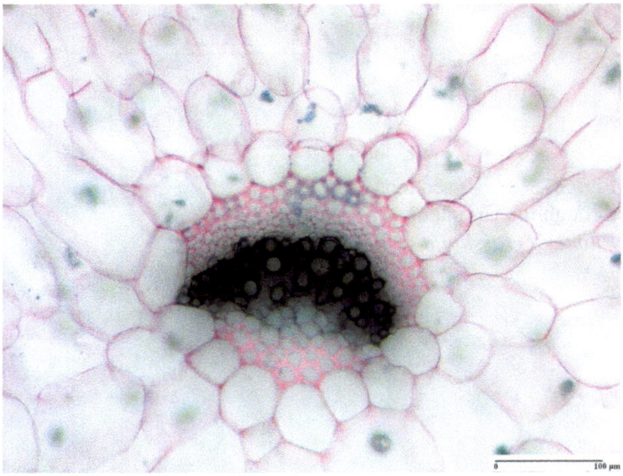

Fig. 3.77 Cross sections through the lamina of *Plantago tenuiflora* (RO)

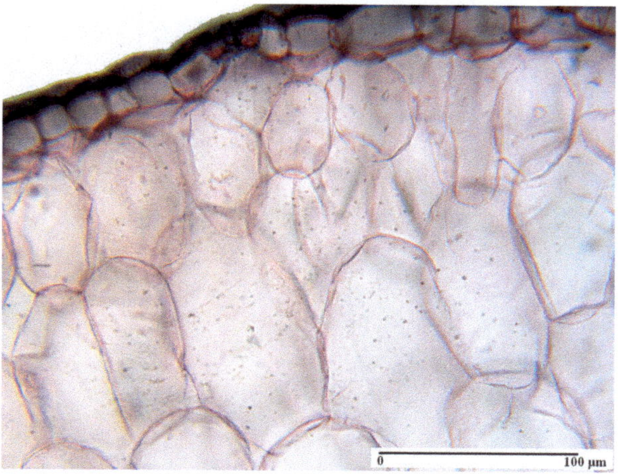

Fig. 3.78 Cross sections through the lamina of *Inula crithmoides* (ESP)

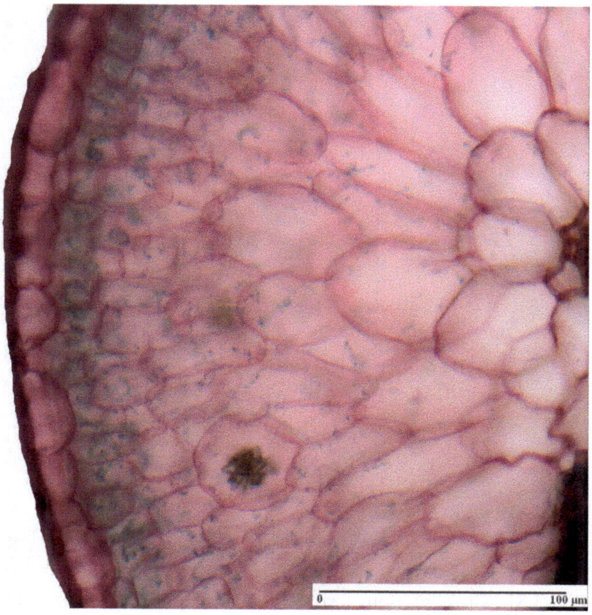

Fig. 3.79 Cross sections through the lamina of *Spergularia media* (RO)

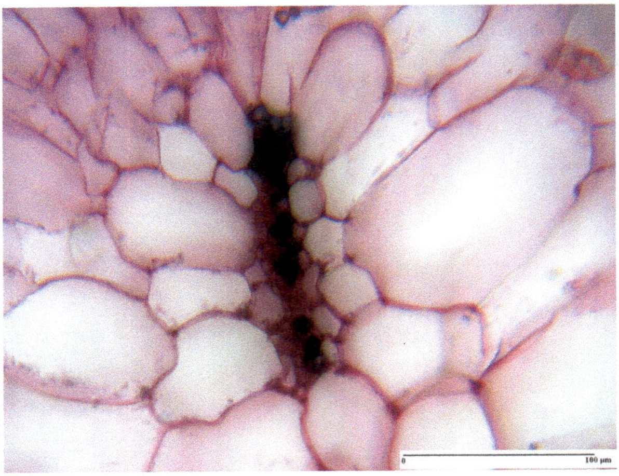

Fig. 3.80 Cross sections through the lamina of *Spergularia media* (RO)

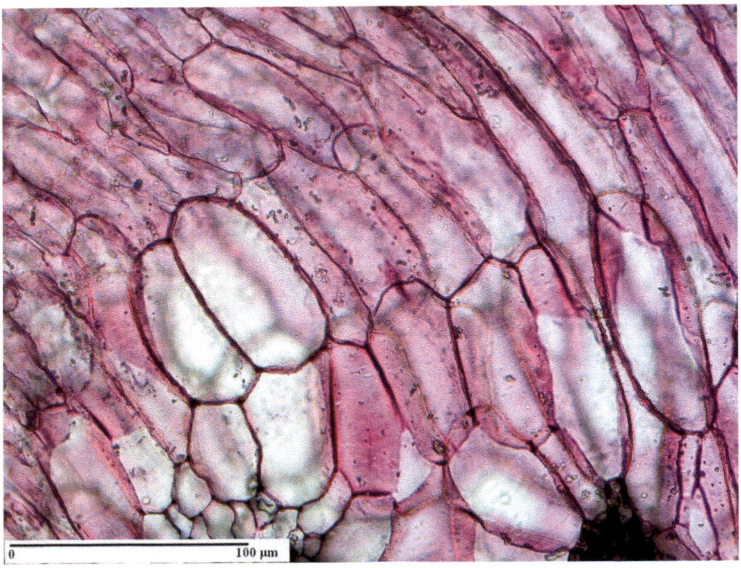

Fig. 3.81 Cross sections through the lamina of *Crithmum maritimum* (ESP)

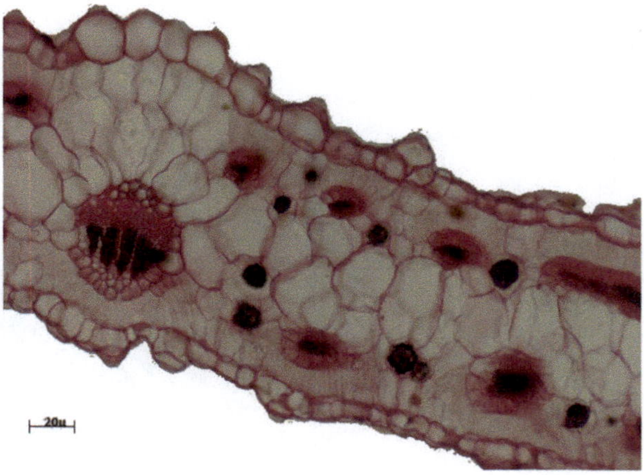

Fig. 3.82 Cross sections through the lamina of *B. turkestanica* (courtesy of Somayeh Safiallah)

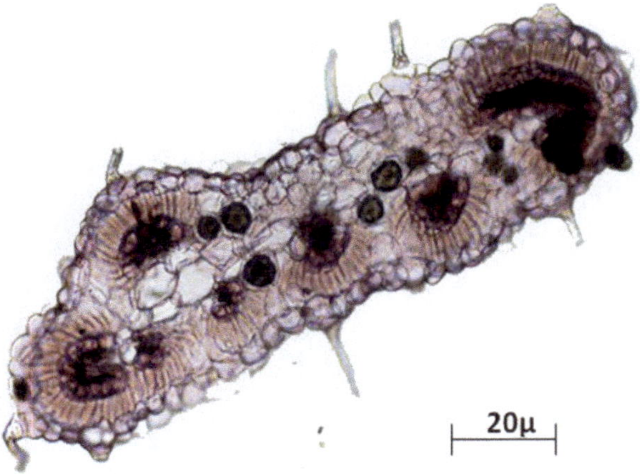

Fig. 3.83 Cross sections through the lamina of *B. pilosa* (courtesy of Somayeh Safiallah)

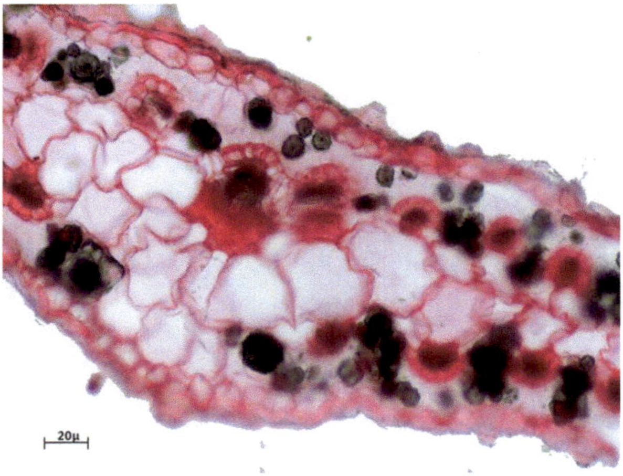

Fig. 3.84 Cross sections through the lamina of *B. stellaris* (courtesy of Somayeh Safiallah)

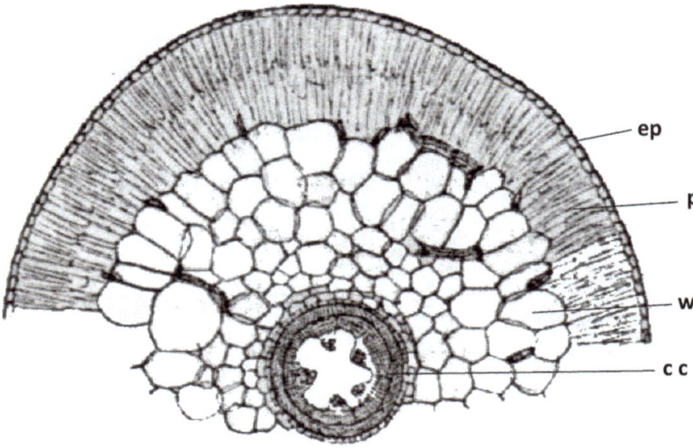

Fig. 3.85 Cross section through a fleshy segment of *Salicornia herbacea* (*ep* epidermis, *c c* central cylinder, *psd t* palisade tissue, *wt t* water storage tissue) (Warming 1890)

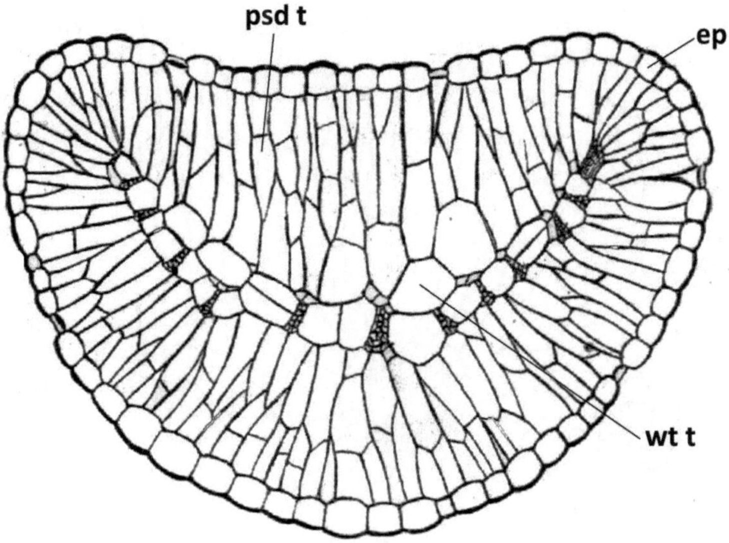

Fig. 3.86 Cross section through a fleshy segment of *Suaeda maritima* (*ep* epidermis, *psd t* palisade tissue, *wt t* water storage tissue) (Warming 1890)

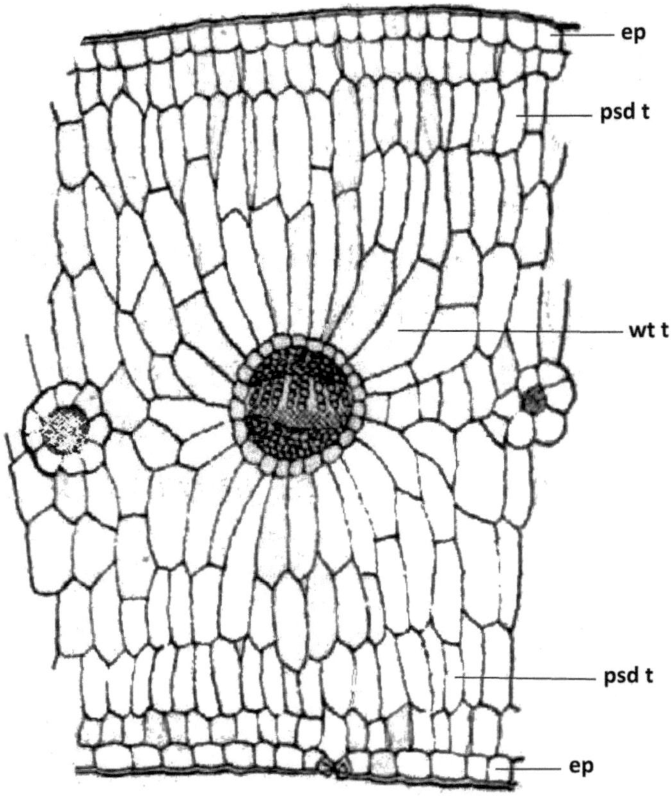

Fig. 3.87 Cross section through a lamina of *Plantago maritima* (*ep* epidermis, *psd t* palisade tissue, *wt t* water storage tissue) (Warming 1890)

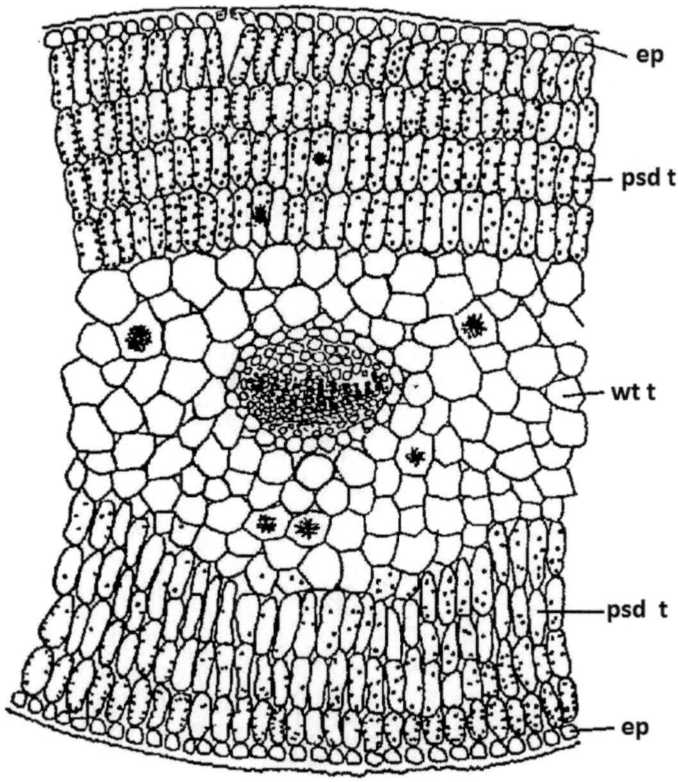

Fig. 3.88 Cross section through a lamina of *Spergularia media* (*ep* epidermis, *psd t* palisade tissue, *wt t* water storage tissue) (Cross 1909)

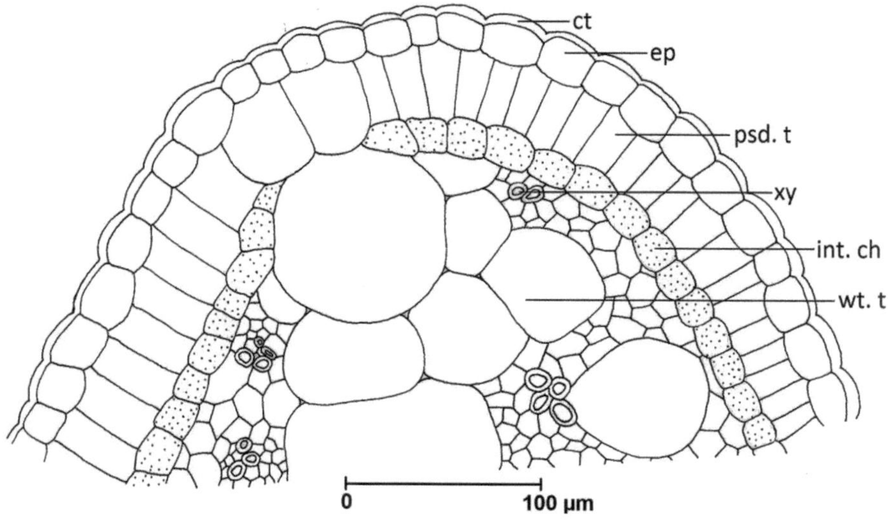

Fig. 3.89 Cross section through the lamina of *Petrosimonia triandra* (*ep* epidermis, *int ch* internal chlorenchyma, *psd t* palisade tissue, *xy* ylem, *wt t* water storage tissue) (original)

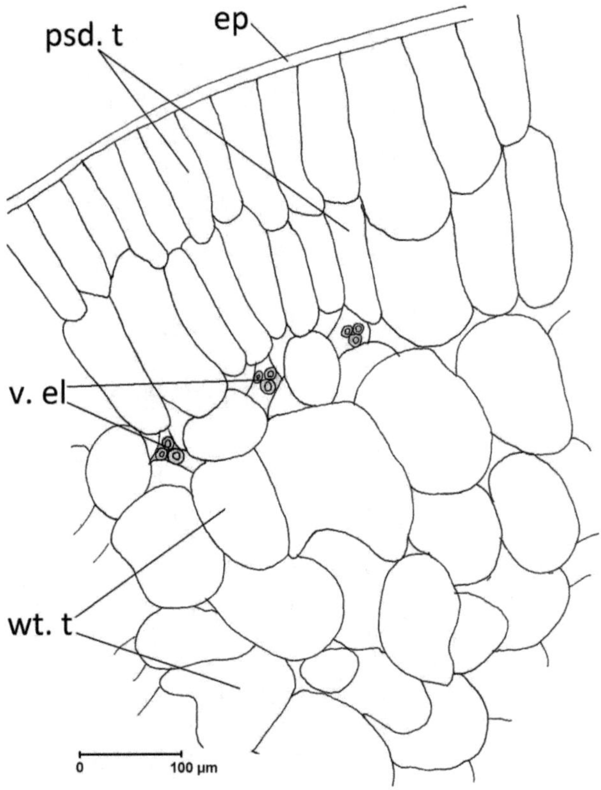

Fig. 3.90 Cross section through a fleshy segment of *Salicornia europaea* (*ep* epidermis, *psd t* palisade tissue, *v el* vascular elements, *wt t* water storage tissue) (original)

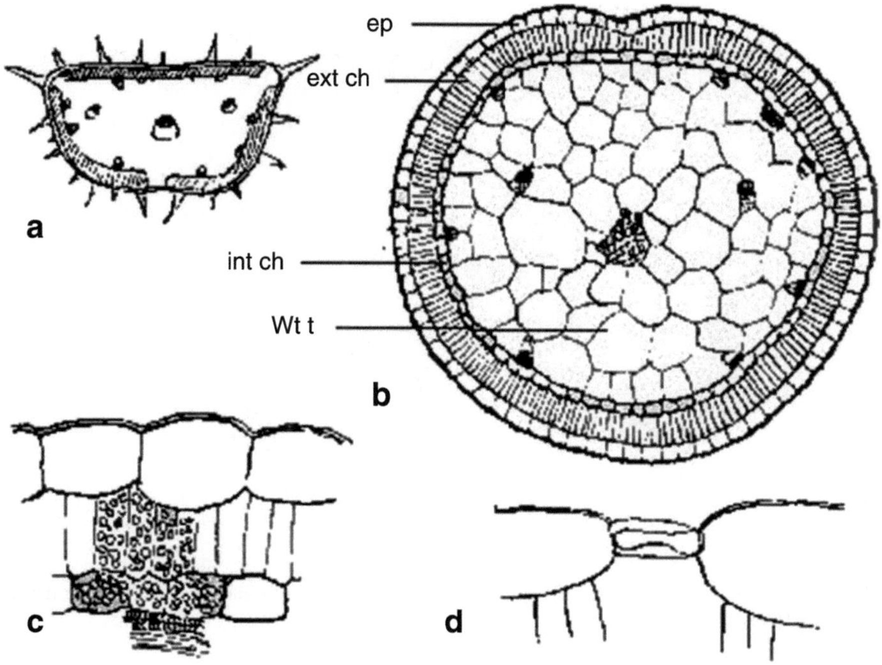

Fig. 3.91 Cross section through the lamina of *Salsola kali*: (**a**) general aspect, (**b**, **c**) details, (**d**) stomata (*ep* epidermis, *ext ch* external chlorenchyma, *int ch* internal chlorenchyma, *wt t* water storage tissue) (Warming 1906)

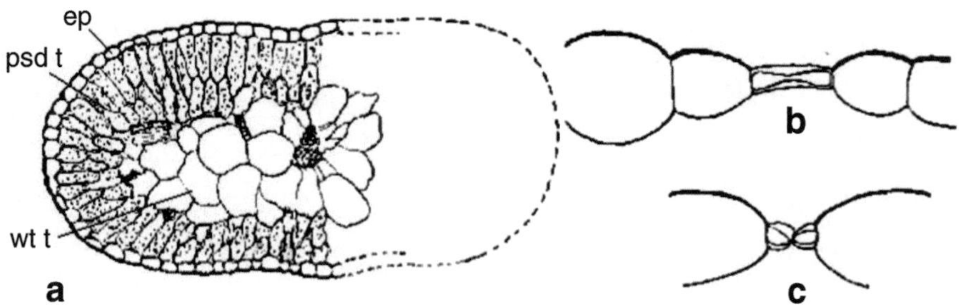

Fig. 3.92 Cross section through the lamina of *Kochia hirsuta*: (**a**) detail, (**b**, **c**) stomata (*ep* epidermis, *psd t* palisade tissue, *wt t* water storage tissue) (Warming 1906)

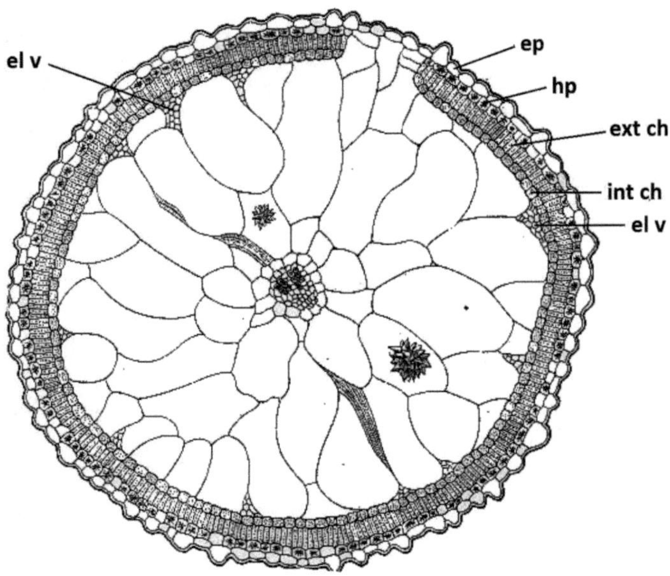

Fig. 3.93 Cross section through a fleshy segment of *Salsola longifolia* (*wt t* water storage tissue) (Volkens 1887)

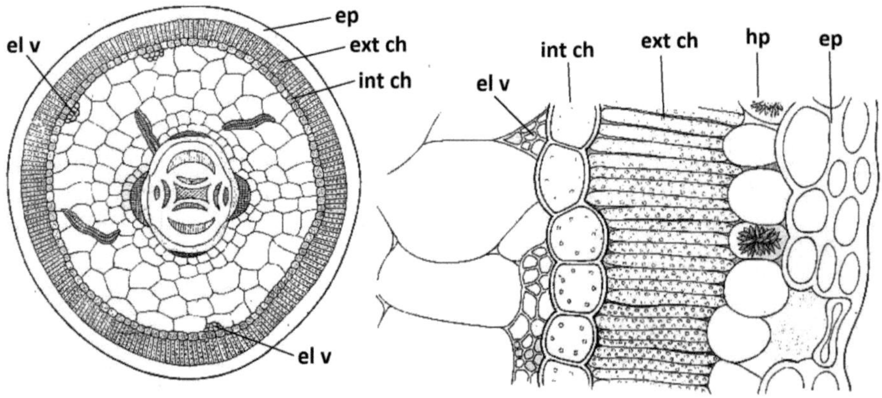

Fig. 3.94 Cross section through a fleshy segment of *Haloxylon schweinfurthii* (*wt t* water storage tissue, *left*—general view; *right*—detail) (Volkens 1887)

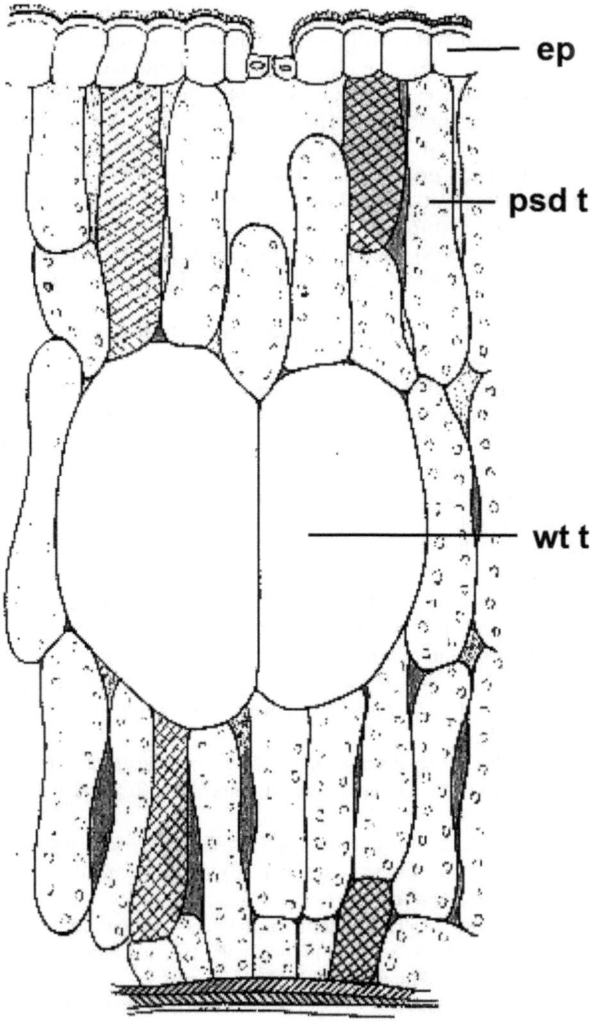

Fig. 3.95 Cross section through a leaf of *Nitraria retusa* (*ep* epidermis, *wt t* water storage tissue, *psd t* palisade tissue) (Volkens 1887)

References

Abd Elbar Ola H, Abd El-Maboud MM (2013) Anatomical and physiological responses of three species of *Suaeda* Forssk. ex Scop. under different habitat conditions. J Appl Sci Res 9 (8):5370–5379

Abrams L (1944) An illustrated flora of the Pacific States: Washington, Oregon, and California, vol 2. Stanford University, Stanford University Press, California

Adriani MJ (1956) Der Wasserhaushalt der Halophyten. In: Ruhland W (ed) Encyclopedia of plant physiology, Water relations of plants, vol 3. Springer, Berlin, pp 902–914

Ahmad I, Wainwright SJ (1976) Ecotype differences in leaf surface properties of *Agrostis stolonifera* from salt marsh, spray zone and inland habitats. New Phytol 76:361–366

Anderson CE (1974) A review of structure in several North Carolina salt marsh plants. In: Reimold RJ, Queen WH (eds) Ecology of halophytes. Academic, New York, London, pp 307–344

Arnold A (1955) Die Bedeutung der Chlorionen für die Pflanze, insbesondere deren physiologische Wirksamkeit; eine monographische Studie mit Ausblicken auf das Halophytenproblem, Botanische Studien, vol 2. Gustav Fischer, Jena

Batalin A (1886) Wirkung des Chlornatriums auf die Entwicklung von *Salicornia herbacea* L. Bull du Congr Intern de Bot et d'Hort de S Petersbourg

Baylis GTS (1940–1941) Leaf anatomy of the New Zealand mangroves. Trans Proc Royal Soc New Zeal 70:164–170

Bentham G (1858) Handbook of the British flora. Lowell Reeve, London

Bernstein L (1961) Osmotic adjustment of plants to saline media. I. Steady State. Am J Bot 48:908–918

Bernstein L (1963) Osmotic adjustment of plants to saline media. II. Dynamic phase. Am J Bot 50:360–370

Bickenbach K (1932) Zur Anatomie und Physiologie einiger Strand und Dünenpflanzen. Beitrage zum Halophytenproblem. Beitr Biol Pflanz 19:334–370

Biebl R, Kinzel H (1965) Blattbau und Salzhaushalt von *Laguncularia racemosa* (L.) Gaertn. f. und anderer Mangrovebäume auf Puerto Rico. Österr Bot Zeit 112(1-2):56–93

Black RF (1958) Effects on NaCl on the leaf succulence and area of *Atriplex hastata* L. Aust J Bot 6:306–321

Bowman HHM (1921) Histological variations in *Rhizophora mangle*. Michigan Acad Sci Rep 22:129–134

Boyce SG (1951) Salt hypertrophy in succulent dune plants. Science 114:544–545

Camilleri JC, Ribi G (1983) Leaf thickness of mangroves (*Rhizophora mangle*) growing in different salinities. Biotropica 15:139–141. (abstract)

Chapman VJ (1942) The new perspective in the halophytes. Quart Rev Biol 17(4):291–311

Chermezon H (1910) Recherches anatomiques sur les plantes littorales. Ann Sci Nat sér 9 Bot 12:117–313

Cooke FW (1911) Observations on *Salicornia australis*. Trans Proc New Zeal Inst 44:349–362

Cross BD (1909) Observations on some New Zealand halophytes. Trans Proc New Zeal Inst 42:545–574

Dangeard PA (1888) Note sur la gaine foliaire des *Salicornieae*. Bull Soc Bot France 35:157–160

de Bary A (1884) Comparative anatomy of the vegetative organs of the Phanerogams and Ferns. Clarendon, Oxford

de Fraine E (1912) The anatomy of the genus *Salicornia*. Linn J Bot 41:317–346

Duval-Jouve M (1868) Des *Salicornia* de l'Hérault. Observations anatomiques et morphologiques. Bull Soc Bot France 15:132–140

Elhalim ME, Abo-Alatta OK, Habib SA, Abd Elbar OH (2016) The anatomical features of the desert halophytes *Zygophyllum album* L.F. and *Nitraria retusa* (Forssk.) Asch. Ann of Agric Sci 61(1):97–104

Engler A, Drude O (1921) Die Vegetation der Erde, III.2. IX. Die Pflanzenwelt Afrikas insbesondere seiner tropischen Gebiete. Verlag von Wilhelm Engelmann, Leipzig

Evenari M (1938) The physiological anatomy of the transpiratory organs and the conducting stems of certain plants typical of the Wilderness of Judaea. J Linn Soc London 51:389–407

Fahn A (1963) The fleshy cortex of articulated *Chenopodiaceae*. J Indian Bot Soc 42(A):39–45

Fahn A, Arzee T (1959) Vascularization of articulated *Chenopodiaceae* and the nature of their fleshy cortex. Am J Bot 46:330–338

Flowers TJM, Troke PF, Yeo AR (1977) The mechanism of salt tolerance in halophytes. Annu Rev Plant Physiol 28:89–121

Flowers TJ, Hajibagheri MA, Clipson NJW (1986) Halophytes. Quart Rev Biol 61(3):313–337

Gale J, Poljakoff-Mayber A (1970) Interrelationships between growth and photosynthesis of salt bush (*Atriplex halimus* L.) grown in saline media. Aust J Biol Sci 23:937–945

Ganong WF (1903) The vegetation of the Bay of Fundy Salt and Diked marshes: an ecological study. Bot Gaz 36(3):161–186. 280–302, 349–367, 429–455

Greenway H (1968) Growth stimulation by high sodium chloride concentrations in halophytes. Isr J Bot 17:169–178. (abstract)

Grigore M-N (2008) Introducere în Halofitologie. Elemente de Anatomie Integrativă. Ed. PIM, Iaşi

Grigore M-N, Toma C (2007) Histo-anatomical strategies of *Chenopodiaceae* halophytes: adaptive, ecological and evolutionary implications. WSEAS Trans on Biol and Biomed 12 (4):204–218

Grigore M-N, Toma C (2008) Ecological anatomy of halophyte species from the *Chenopodiaceae* family. Advanced topics on mathematical biology and ecology (Proceedings of the 4th WSEAS International Conference on Mathematical Biology and Ecology—MABE '08, Acapulco, Mexico, January 25–27, 2008), p 62–67.

Grigore M-N, Toma C (2010a) Halofitele. Aspecte de anatomie ecologică. Edit. Univ. "Al. I. Cuza", Iaşi

Grigore M-N, Toma C (2010b) A proposal for a new halophytes classification, based on integrative anatomy observations. Muz. Olteniei Craiova. Studii şi Comunicări, Ştiinţele Naturii 26 (1):45–50

Grigore M-N, Toma C (2010c) Structuri secretoare de săruri la halofite. O abordare integrativă. Edit. Academiei Române, Bucureşti

Grigore M-N, Toma C (2011a) Halofitele, o categorie ecologică polimorfă. Î}\^{I}{ntre seceta fiziologică a solului şi stresul salin. Revista Botanică (Chişinău) 2(3):38–46

Grigore M-N, Toma C (2011b) Observaţii ecologice preliminare referitoare la speci de halofite de la rezervaţia naturală "Valea Ilenei" (Iaşi). Materialele Simpozionului Ştiinţific Internaţional "Rezervaţia Codrii, 40 de ani", pp 180–183

Grigore M-N, Toma C (2014) Integrative ecological notes on halophytes from "Valea Ilenei" (Iaşi) nature reserve. Memoirs of the Scientific Sections of the Romanian Academy 37:19–36

Grigore M-N, Boscaiu M, Vicente O (2011a) Assessment of the relevance of osmolyte biosynthesis for salt tolerance of halophytes under natural conditions. Eur J Plant Sci Biotechnol 5 (Special Issue 2):12–19

Grigore M-N, Toma C, Boscaiu M (2011b) Ecological notes on halophytes species from Mediterranean climate. Lucr Şt (Horticultură), USAMV "Ion Ionescu de la Brad". Iaşi 54(1):29–34

Grigore M-N, Toma C, Ivănescu L (2011c) Anatomical and ecological observations on Mediterranean halophytes: *Suaeda* Forssk. ex Scop. genus. Lucr Şt (Horticultură), USAMV "Ion Ionescu de la Brad". Iaşi 54(1):23–28

Grigore M-N, Toma C, Boscaiu M, Zamfirache M-M, Ivănescu L (2012a) Anatomical and ecological observations on psammo-halophytes species (Eastern part of Spain). Lucr Şt (Horticultură), USAMV "Ion Ionescu de la Brad". Iaşi 55(2):19–24

Grigore M-N, Toma C, Zamfirache M-M, Ivănescu L (2012b) A survey of anatomical adaptations in Romanian halophytes. Towards an ecological interpretation. Fresen Environ Bull 21 (11b):3370–3375

Grigore M-N, Villanueva M, Boscaiu M, Vicente O (2012c) Do halophytes really require salt for their growth and development? An experimental approach. Not Sci Biol 4(2):23–29

Grigore M-N, Toma C, Zamfirache M-M, Ivănescu L, Daraban I (2013) Anatomical and ecological observations in succulent (articulated) halophytes from *Chenopodiaceae*. Lucr Şt (Horticultură), USAMV "Ion Ionescu de la Brad". Iaşi 56(2):19–24

Grigore M-N, Ivănescu L, Toma C (2014) Halophytes. An integrative anatomical study. Springer, Cham

Hajibagheri MA, Hall JL, Flowers TJ (1983) The structure of the cuticle in relation to cuticular transpiration in leaves of the halophyte *Suaeda maritima* (L.) Dum. New Phytol 94:125–131

Halket AC (1928) The morphology of *Salicornia*—an abnormal plant. Ann Bot (London) 42:523–530

Handley JF, Jennings DH (1977) The effect of ions on growth and leaf succulence of *Atriplex hortensis* var. *cupreata*. Ann Bot 41:1109–1112

Hayward HE, Long EM (1941) Anatomical and physiological response of the tomato to varying concentrations of sodium chlorides, sodium sulphate and nutrient solution. Bot Gaz 102:437–462

Holtermann C (1907) Der Einfluss des Klimas auf den Bau der Pflanzengewebe. Verlag Von Wilhelm Engelmann, Leipzig

Hooker JD (1884) The student's flora of the British Islands. MacMillan & Co, London

James L, Kyhos DW (1961) The nature of fleshy shoot of *Allenrolfea* and allied genera. Am J Bot 48:101–108

Jennings DH (1968) Halophytes, succulence and sodium in plants—a unified theory. New Phytol 67:899–911

Jennings DH (1976) The effects of sodium chloride on higher plants. Biol Rev 51:453–486

Jepson WL (1923) A manual of the flowering plants of California. University of California Press, Berkeley

Joshi GV, Karekar MD, Gowda CA, Bhosale L (1974) Photosynthetic carbon metabolism and carboxylating enzymes in algae and mangrove under saline conditions. Photosynthetica 8:51–52

Jussieu D (1717) Histoire du Kali d'Alicante. Mém Math Phys Acad Royale Sci Paris:73–78

Keller B (1925) Halophyten und xerophyten studien. J Ecol 13:224–261

Keller B (1951) Extreme salt resistance of higher plants in nature and the problem of adaptation. In: Selected works: 212–236, Akad. Nauk SSSR, Moskwa. (First published in 1940 in Plant and Environment—Rastyenye y sreda), Akad. Nauk SSSR (in Russian)

Lacerda CF, Assis Júnior JO, Filho LCAL, de Oliveira TS, Guimarães VA, Gomes-Filho E, Prisco JT, Bezerra M (2006) Morpho-physiological responses of cowpea leaves to salt stress. Braz J Plant Physiol 18(4):455–465

Lagerwerff JV, Eagle HE (1961) Osmotic and specific effects of excess salt on beans. Plant Physiol 36:472–477

Lesage PM (1890) Recherches expérimentales sur les modifications des feuilles chez les plantes maritimes. Rév Gén Bot 2:55–65. 106–121,163–173

Leisle FF (1949) K ekologhii i anatomii galofitov i kserofitov s redutirovanii listiami (Ecology and anatomy of halophytes and xerophytes with reduced leaves). Bot J SSSR 34(3):253–266

Longstreth DJ, Nobel PS (1979) Salinity effects on leaf anatomy: Consequences for photosynthesis. Plant Physiol 63:700–703

Meiri A, Poljakoff-Mayber A (1970) Effect of various salinity regimes on growth, leaf expansion and transpiration rate of bean plants. Soil Sci 109(1):26–32

Mendoza MM (1971) The effects of NaCl on anatomical and physiological processes in *Atriplex hastata* L. M.S. Thesis. Univ. Utah, Salt Lake City

Monteil P (1906) Anatomie comparée de la feuille des Chénopodiacées, Thèse, Ecole Supérieure de Pharmacie, no. 9, Université de Paris.

Moquin AA (1831) Premier mémoire sur la famille des *Chénopodées*. Essai monographique sur le genre *Suaeda* et sur les *Chénopodées* les plus voisines. Ann Sci Nat 23:278–325

Mullan DP (1931) Observations on the water-storing devices in the leaves of some Indian halophytes. J Indian Bot Soc 10:126–133

Muntz PA (1959) A California flora. University of California Press, Berkeley and Los Angeles

Norkrans B, Kylin A (1969) Regulation of the potassium to sodium ratio and of the osmotic potential in relation to salt tolerance in yeasts. J Bacteriol 100(2):836–845

Osterhout WJ (1906) On the importance of physiologically balanced solutions for plants. Bot Gaz 42:127–134. (abstract)

Pax F (1897) Myrsinaceae, In: Engler A, Prantl K (eds) Die natürlichen Pflanzenfamilien 4(1-2). Leipzig, von Wilhelm Engelmann Verlag, p 84–97

Peck ME (1941) A manual of the higher plants of Oregon. Binfords and Mort Publishers, Portland

Poljakoff-Mayber A (1975) Morphological and anatomical changes in plants as a response to salinity stress. In: Poljakoff-Mayber A, Gale J (eds) Plants in saline environments. Springer, New York, pp 97–117

Qiu D-L, Lin P, Guo SZ (2007) Effects of salinity on leaf characteristics and CO_2/H_2O exchange of *Kandelia candel* (L.) Druce seedlings. J Forest Sci 53(1):13–19

Rao GC, Basha SKM, Rao GR (1981) Effect of sodium chloride salinity on amount and composition of epicuticular wax and cuticular transpiration rate in peanut *Arachis hypogaea*. Indian J Exp Biol 19:880–881

Reinders-Gouwentak CA (1953) *Sonneratiaceae* and other mangrove—swamp families, anatomical structure and water relations. Fl Males 1(4):513–515

Repp G (1939) Ökologische Untersuchungen im Halophytengebiet am Neusiedler See. Jahrb f wiss Bot 88(4):554–632

Sabnis TA (1920) The physiological anatomy of the plants of the Indian Desert. J Indian Bot 1 (6-7):183–205

Sabnis TA (1921) The physiological anatomy of the plants of the Indian Desert. J Indian Bot 2 (4–5):93–115

Schimper AFW (1891) Die Indo-Malayische Strandflora. Bot Mit Trop 3:1–204

Schimper AFW (1898) Rhizophoraceae. In: Engler A, Prantl K (ed by) Die natürlichen Pflanzenfamilien, Leipzig, Verlag von Wilhelm Engelmann, III(7–8):42–56

Schimper AFW (1903) Plant Geography upon a physiological basis. Clarendon, Oxford

Schischkin BK (1936) *Chenopodiaceae*. In: Komarov L (ed) *Flora of the U.R.S.S*, vol 6. Izdatel'stvo Akademii Nauk SSSR, Moskva, Leningrad, pp 1–354

Schratz E (1934) Beitrage zur Biologie der Halophyten. Jahrb F wiss Bot 80:112–142

Schulze E-D, Beck E, Müller-Hohenstein K (2005) Plant Ecology. Springer, Berlin, Heidelberg

Shennan R, Macrobbie AC (1987) Salt tolerance in *Aster tripolium*. I. The effect of salinity on growth. Plant Cell Environ 10:59–65

Shmueli E (1948) The water balance of some plants of the Dead Sea salines. Palest J Bot (Jerusalem ser) 4:117–142

Smith JAC, Popp M, Lüttge U, Cram WJ, Diaz M, Griffiths H, Lee HSJ, Medina E, Schäfer C, Stimmel K-H, Thonke B (1989) Ecophysiology of xerophytic and halophytic vegetation of a coastal alluvial plain in northern Venezuela. VI. Water relations and gas exchange of mangroves. New Phytol 111:293–307

St Omer L, Schlesinger WH (1980a) Regulation of NaCl in *Jaumea carnosa* (*Asteraceae*), a salt marsh species, and its effect on leaf succulence. Am J Bot 67:1445–1454

St Omer L, Schlesinger WH (1980b) Field and greenhouse investigations of the effect of increasing salt stress on the anatomy of *Jaumea carnosa*, a salt marsh species. Am J Bot 67:1455–1465

Stace CA (1966) The use of epidermal characters in phylogenetic considerations. New Phytol 65:304–318

Stocker O (1928) Das Halophytenproblem. Ergeb Biol 3:265–353

Stocker O (1933) Salzpflanzen. Handb Naturwiss 8:699–712

Storey R, Wyn Jones RG (1979) Response of *Atriplex spongiosa* and *Suaeda monoica* to salinity. Plant Physiol 63:156–162

Strogonov BP (1962) Fiziologhiceskie osnovy soleustocivosti rastenij (Physiological basis of salt tolerance of plants). Akademia Nauk SSSR, Moskva

Strogonov BP (1964) Physiological basis of salt tolerance of plants (as affected by various types of salinity). Akad. Nauk. SSSR. Transalted from Russian, Israel Progr. Sci. Transl., Jerusalem.

Toma C, Flenchea-Teodorescu G, Răşcanu S, Zaharia M (1979) Trăsăturile anatomo-ecologice ale unor plante litorale (*Cakile maritima* Scop. şi *Eryngium maritium* L.). Culegere de Stud. şi artic. de Biologie, Univ. "Al.I.Cuza" Iaşi (Grăd. Bot.) (Lucrările simpozionului "120 de ani de la înfiinţarea la Iaşi a primei grădini botanice din România") 1:273–287

Tullin V (1954) Response of sugar beet to common salt. Physiol Plant 7:810–834

Udovenko GV, Gradchaninova OD, Semushina LA (1970) Morphological and anatomical changes in wheat leaves and roots with increasing soil salinity. Bot J 55:931–937
Van Eijk M (1939) Analyse der Wirkung des NaCl auf die Entwicklung Sukkulenze und Transpiration bei *Salicornia herbacea*, sowie Untersuchungen über den Einfluss der Salzaufnahme auf die Wurzelatmung bei *Aster tripolium*. Rec Trav Bot Neerl 36:559–657
Volkens G (1887) Die Flora der aegyptisch-arabischen Wueste auf Grundlage anatomisch-physiologischer Forschungen. Gebrüder, Borntraeger, Berlin
Volkens G (1893) Chenopodiaceae. In: Engler A, Prantl K (ed by) Die natürlichen Pflanzenfamilien, Leipzig, Verlag von Wilhelm Engelmann., III. Teil. 1a: 36–91
Waisel Y (1972) Biology of halophytes. Academic, New York, London
Walter H (1937) Die Ökologishen Verhältnisse in der Namib Nebelwüste (Südwestafrika). Jahrb f wiss Bot 84:58–219
Walter H, Steiner M (1936) Die Őkologie der Ostafrikanischen Mangroven. Zeitschrift Bot 30:65–193
Warming E (1890) Botaniske exkursioner. 1. Fra Vesterhavskystens Marskegne. Vidensk Meddel Fra D naturh Foren Kjøben V(1):206–239
Warming E (1891) Botaniske exkursioner. 2. De psammophile Formationer i Danmark. Vidensk Meddel Fra D naturh Foren Kjøben V(3):153–202
Warming E (1897) Halophyt-studier. D Kgl Danske Vidensk Selsk. Skr, 6, Raekke, naturvidenskabeling og mathematisk Afd. VIII 4:173–272
Warming E (1906) Dansk Plantevaekst. 1. Strandvegetation, Gyldendalske Boghandel Nordisk Forlag, København Kristiania.
Warming E (1909) Oecology of Plants. An introduction to the study of plant-communities. Clarendon, Oxford
Werner A, Stelzer R (1990) Physiological responses of the mangrove *Rhizophora mangle* grown in the absence and presence of NaCl. Plant Cell Environ 13:243–255
Williams MC (1960) Effect of sodium and potassium salts on growth and oxalate content of *Halogeton*. Plant Physiol 35:500–505
Yeo AR, Flowers TJ (1980) Salt tolerance in the halophyte *Suaeda maritima* L. Dum.: Evaluation of the effect of salinity upon growth. J Exp Bot 31:1171–1183. (abstract)

Chapter 4
Tracheoidioblasts (Spiral Cells) and Stereides (Spicular Cells)

These intriguing structures do deserve a separate chapter despite they are found only in a few halophytic articulated species from *Chenopodiaceae* (*Salicornia, Arthrocnemum*—Fig. 4.1, *Sarcocornia*—Fig. 4.2). Actually, the fact that they are restricted in succulent, stem-articulated halophytic chenopods would further question their ecological and adaptive value within these "extreme" halophytes. Going deeper, it is also striking that both spiral cells and stereides (spicular cells) seem to not occur in the same species; however, in the rare situations when they are reported in the same species, they are located in different anatomical areas, with spiral cells occurring always in the palisade region (de Fraine 1912). In this way, a sort of "specificity" may be suggested and proposed; it can be used for taxonomical purposes, as the difficulties related to correct identification among these species are well known.

Their terminology was confusing at that time and even a clear distinction between them was not made until de Fraine (1912) accurately described them and established proper terms, as are being also used by us: spiral cells (also known as tracheoidioblasts) and spicular cells or stereides.

French botanists Duval-Jouve (1868), Mangin (1882), and Dangeard (1888) first evidenced and described them accordingly; they also tried to attribute different roles in the plants' life. Botanists who followed also adopted the previous explanations and added new clarifications and assumed functions to spiral and spicular cells (Volkens 1884; Monteil 1906; Warming 1909; Chermezon 1910; Cooke 1911; Baumgärtel 1917; Ganong 1903; de Fraine 1912; Grigore et al. 2013, 2014). The spiral-like appearance of spiral cells (see Figs. 4.7, 4.9, and 4.16) actually suggested to all researchers to keep the same nomenclature: (*les grandes*) *cellules spiralées* (French botanists—Duval-Jouve 1868; Mangin 1882; Dangeard 1888), *spiral cells* (English—de Fraine 1912), *spiral tracheids, spicular cells* (translated in English from German—Solereder 1908), *spiraltracheiden*, *spikularzellen* (German—Baumgärtel 1917; Holterman 1907), *stereïden* (German—Volkens 1884), and *spiralceller* (Danish—Warming 1890).

M.-N. Grigore, C. Toma, *Anatomical Adaptations of Halophytes*,
DOI 10.1007/978-3-319-66480-4_4

Fig. 4.1 *Arthrocnemum macrostachyum*—General view (1) and details (2–8) (Reichenbach and Reichenbach 1909)

Duval-Jouve (1868) evidenced spiral cells (tracheoidioblasts) (*les grandes cellules spiralées*) in fleshy segments of *Salicornia fruticosa* (Figs. 4.3 and 4.4) and spicular cells (stereides) in *Salicornia macrostachya* (Figs. 4.5 and 4.6). He uses the generic term—*les cellules aérifères*—thus proposing an air-storing function for both of them. Spiral cells are different from the palisade cells by their length and conformation, being always in a much smaller number than the cells of the palisade parenchyma, in the mass of which they are located. They are almost always situated

Fig. 4.2 *Sarcocornia fruticosa*—General view (1) and details (2–7) (Reichenbach and Reichenbach 1909)

on the direction and under the stomata, *without being in contact* (our emphasis) with any of these. On his drawings, spiral cells lack typical spiral-like thickenings. On the material analyzed by Duval-Jouve, all these tracheoidioblasts are full of air; the author has reached this conclusion noticing that, by simply pressing the tracheids from histological slides, the air is released outside, in the aqueous media. Duval-Jouve stated that *S. patula* and *S. sarmentosa* have also tracheoidioblasts in their succulent segments, without giving any drawing of them.

In *Salicornia macrostachya*, spiral cells are replaced by spicular cells (stereides) (Figs. 4.5 and 4.6); according to Duval-Jouve (1868), they occur in the same

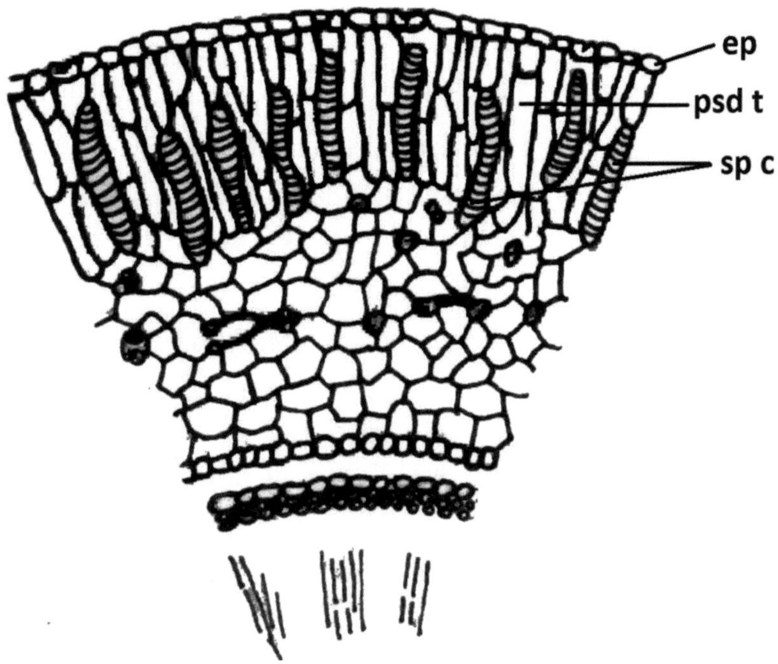

Fig. 4.3 Spiral cells (tracheoidioblasts) in the fleshy segment of *Salicornia fruticosa*, general view (*ep* epidermis, *psd t* palisade tissue, *sp c* spiral cells, filled in *black* color) [adapted and slightly modified from Duval-Jouve (1868)]

position as spiral cells—a very important anatomical observation in the frame of a further classification of halophytic chenopods based on the presence of these structures. Their cells walls are very thick.

Mangin (1882) believed that the role of spiral cells is supporting (mechanical) and reinforced Duval-Jouve's idea (1868) that they are not in contact with the stomata, although he observed them up at the level of the stomata, at a small distance from the stomatal crypt. He refers to spicular cells as *cellules scléreuses*, in contrast to spiral cells.

Van Tieghem (1884) also attributed to spiral cells a mechanical (supporting) role construing them as elements of the cortical stereome.

De Bary (1884) includes the tracheoidioblasts of *Salicornia* in the category of the isolated tracheids, located outside the vascular bundles; he does not explicitly grant them a particular function, but the fact that they are mentioned in this chapter could suggest that their function is a vascular one.

Dangéard (1888) observed spiral cells in foliar sheaths of *Salicornia peruviana*, *S. virginica*, and *Arthrocnemum ambiguum*; according to him, the transversal spiral cells would belong to the cortical stereome.

Ganong (1903) refers to *Salicornia herbacea* air storage system, talking about "(...) *certain air-storing tracheids near the stomata*." As easily noticeable, he attributed an air-storing role of these structures, suggesting that they are not in direct contact with stomata.

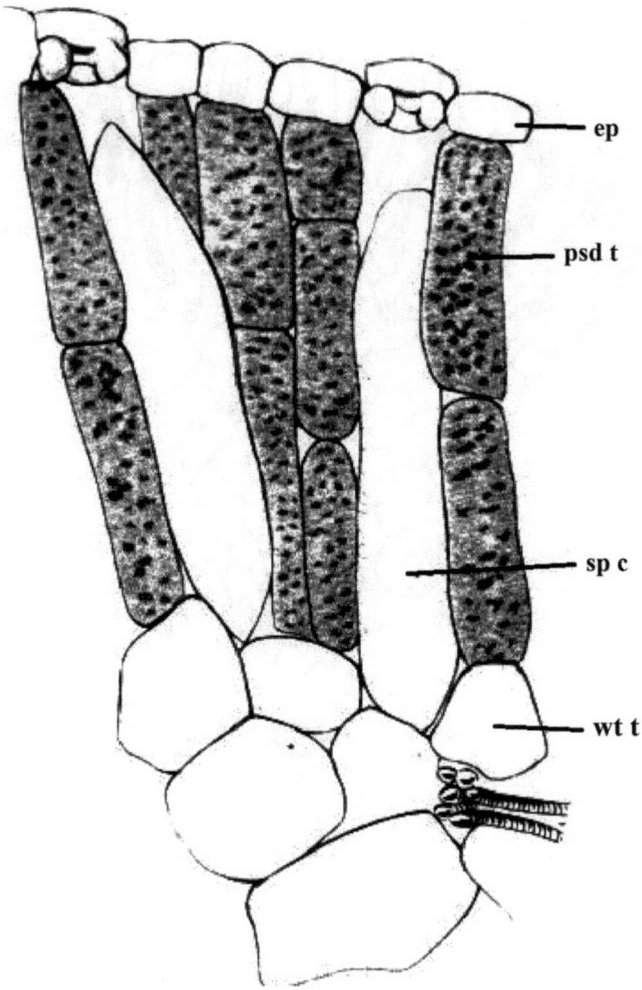

Fig. 4.4 Spiral cells (tracheoidioblasts) in the fleshy segment of *Salicornia fruticosa*, detail (*ep* epidermis, *psd t* palisade tissue, *sp c* spiral cells, *wt t* water storage tissue) (Duval-Jouve 1868)

Holterman (1907) mentions spiral cells (*spiraltracheiden*) in *Arthrocnemum indicum* stating that they are located in the water storage tissue surrounding central cylinder; he believes that tracheoidioblasts are involved in conducting water toward peripheral regions of the plant.

Monteil (1906) evidenced these structures in *Salicornia patula* (Fig. 4.7), *S. sarmentosa*, and *S. fruticosa,* calling them air-storing cells (*cellules aérifères*).

Monteil also delivered a drawing with cross section through the fleshy segment of *Salicornia macrostachya* (Fig. 4.8), where spiral cells (air-storing cells) are depicted (p. 130, op. cit.), but he became rather confused when specifying in the text description that [as compared to *S. fruticosa*]: "*there are no air-storing cells*"

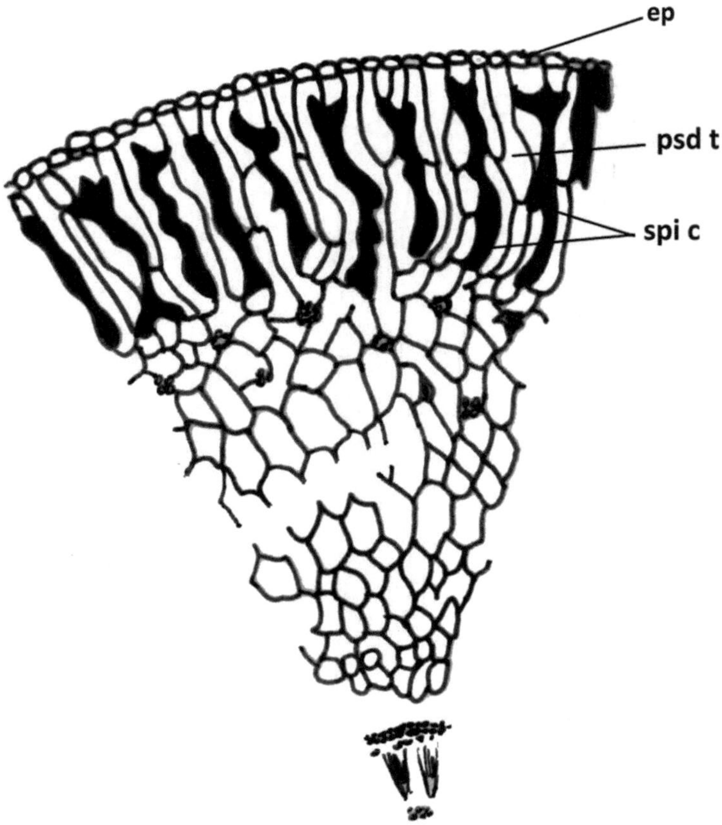

Fig. 4.5 Spicular cells (stereides) in the fleshy segment of *Salicornia macrostachya*, general view (*ep* epidermis, *psd t* palisade tissue, *spi c* spicular cells, filled in *black* color) (adapted and slightly modified from Duval-Jouve 1868)

(*mais sans cellules aérifères*); instead, he clearly (and correctly) captures spicular cells—called *sclérites*, thus suggesting their thickened aspect. The same inconsistency to Monteil's work has also been discussed by de Fraine (1912).

Warming (1909) names them the *water-storing tracheids*, as he thinks that they play the same role as the xylem vessels from the vascular bundles since they are full of water.

Chermezon (1910) evidenced spiral cells in the fleshy segments of *Salicornia fruticosa* (Fig. 4.9) and delivered a nice picture of them. As his precedent French botanists, he also pointed out that these elements have a water storage function.

Cooke (1911) believed that spiral cells (*scattered tracheides*) have an air-storing role in *Salicornia australis* (Fig. 4.10). Keshavarzi and Zare (2006) believe that they interfere with water transport toward peripheral tissues. Anderson (1974) also underlines their role in the water balance.

De Fraine (1912) in his excellent study on the anatomy of the *Salicornia* genus delivered a very detailed description of spiral (tracheoidioblasts) and spicular cells

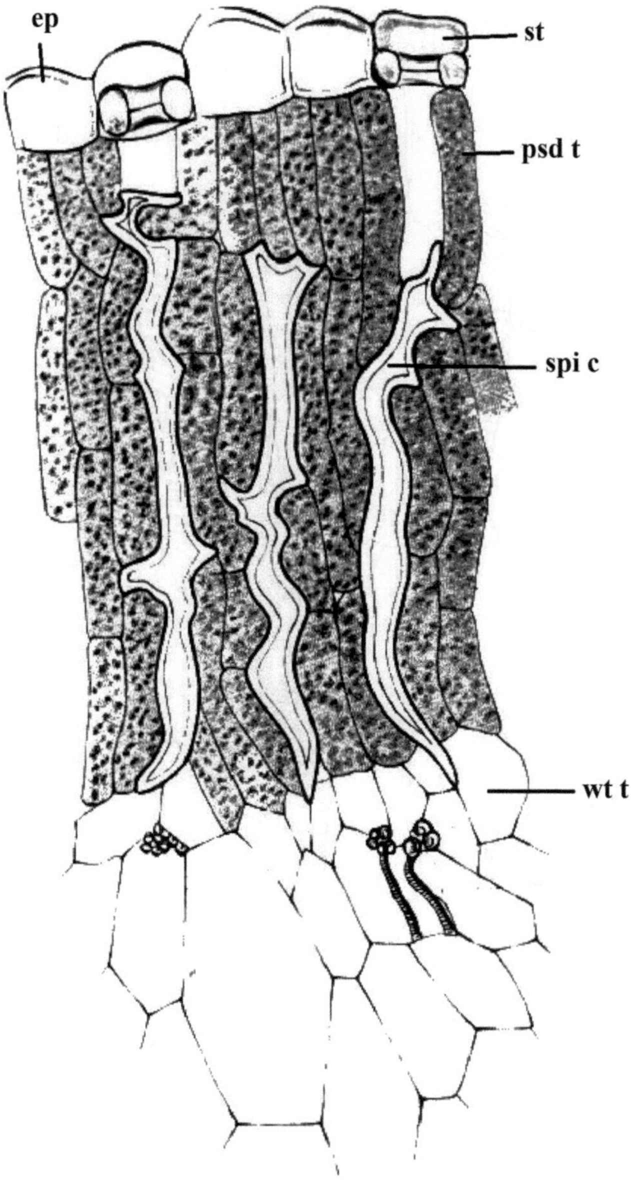

Fig. 4.6 Spicular cells (stereides) in the fleshy segment of *Salicornia macrostachya*, detail (*ep* epidermis, *psd t* palisade tissue, *spi c* spicular cells, *st* stomata, *wt t* water storage tissue) (Duval-Jouve 1868)

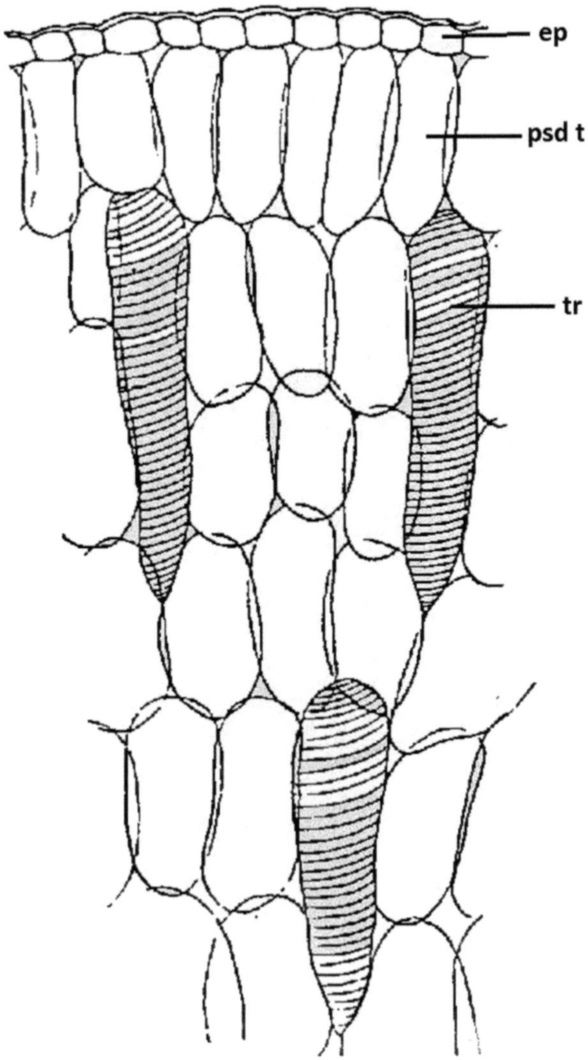

Fig. 4.7 Tracheoidioblasts (spiral cells) in the fleshy segment of *Salicornia patula* (*psd t* palisade tissue, *ep* epidermis, *tr* tracheoidioblasts) (Montéil 1906)

(stereides). He is, likely, the first botanist who clearly delineated them and used correct related terminology. De Fraine underlined that a characteristic feature in the anatomy of many species of *Salicornia* is the occurrence of spiral, or spicular cells, or both.

According to him, spiral cells are large, sac-like, colorless elements which occur wedged in between the palisade cells of the assimilating tissue of the foliar organs, their long axis parallel to the palisade cells (Fig. 4.11). The cell walls of spiral cells are thin and composed of cellulose, despite in very few cases of *Salicornia fruticosa*, a slight trace of lignifications has been noticed.

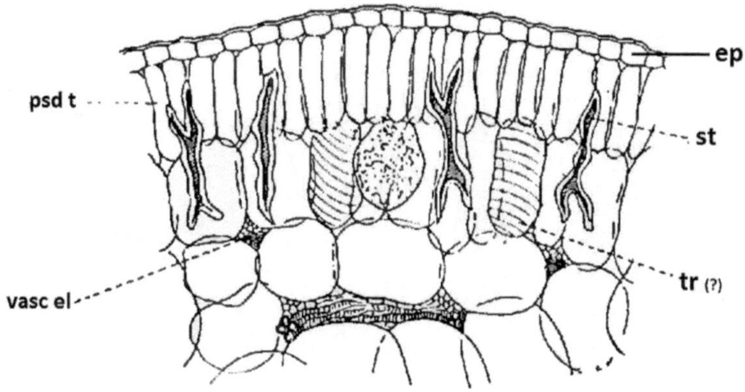

Fig. 4.8 Spicular cells (stereides) in the fleshy segment of *Salicornia macrostachya* (*psd t* palisade tissue, *ep* epidermis, *st* stereides, *vasc el* vascular elements, *tr (?)* tracheoidioblasts) (Monteil 1906)

De Fraine (1912) indicates that the distribution of spiral cells in the different parts of the plant seems to be similar in all species in which they occur. In the reproductive shoots, the spiral cells reach their maximum development, in both size and number. They were reported in flowering spikes of *Salicornia prostrata, S. herbacea,* and *S. pusilla*, while in vegetative shoots they generally lack.

Spicular cells (stereides) were observed by de Fraine (1912) in *Salicornia glauca* (Fig. 4.12), in which spiral cells were replaced, as far as position is concerned, by them. These elements are long, slightly branched with thick lignified walls, freely perforated by simple pits. They occur in the palisade region and usually do not extend deep in the water storage tissue (Fahn and Arzee 1959).

De Fraine believes that spiral and spicular cells are homologous structures; he suggests that the function of spiral cells is water storage, while that of stereides is clear—part of the mechanical support of the plant.

Summarizing his results, de Fraine (1912) proposed a classification of *Salicornia* species, based on the presence of spiral and spicular cells:

1. Stereides only: *S. glauca;*
2. Stereides and spiral cells—the latter only in the palisade region:
   a. Stereides in both vegetative and reproductive shoots: *S. fruticosa.*
   b. Stereides in reproductive shoots only: *S. perennis, S. disarticulata,* and *S. gracillima.*
3. Spiral cells only:
   a. Always very few. Very often entirely absent in the vegetative shoots: *S. pusilla, S. ramosissima, S. appressa, S. herbacea,* and *S. prostrata.*
   b. Absent in the vegetative shoots. Few in reproductive shoots: *S. prostrata* var. *smithiana*

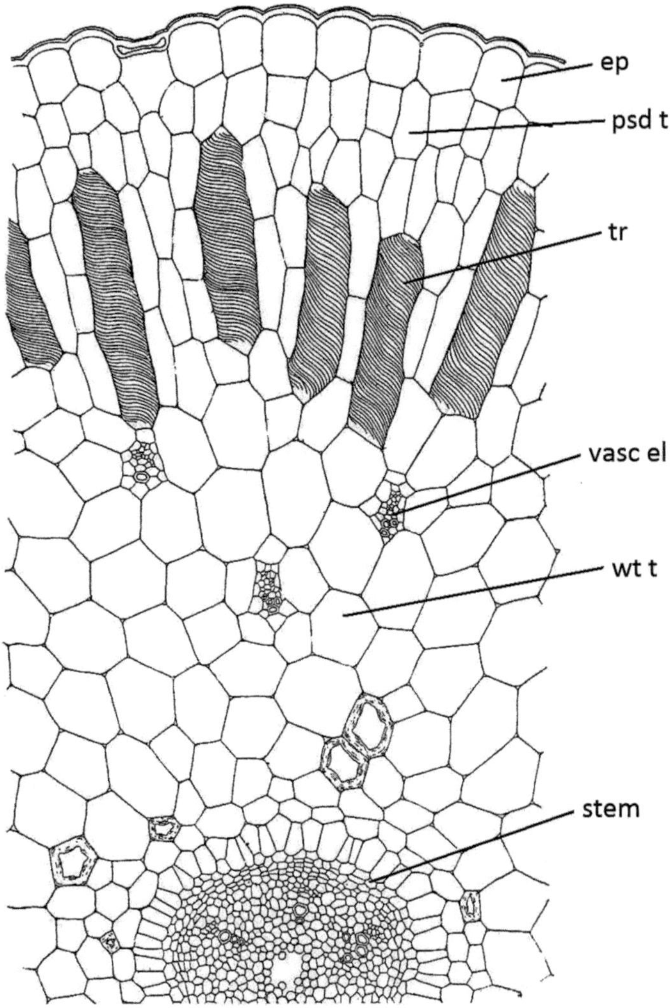

Fig. 4.9 Spiral cells (tracheoidioblasts) in the fleshy segment of *Salicornia fruticosa* (*psd t* palisade tissue, *ep* epidermis, *vasc el* vascular elements, *tr* tracheoidioblasts, *wt t* water storage tissue) (Chermezon 1910)

4. Stereides and spiral cells absent: *S. oliveri, S. dolichostachya.*

Baumgärtel (1917) in his monograph about the anatomy of the *Arthrocnemum* genus mentioned spiral cells (*spiraltracheiden*) and spicular cells (*spikularzellen*); the latter are accurately depicted (Fig. 4.13) in *Arthrocnemum*, albeit the species is not clearly mentioned.

Anderson (1974) evidenced spiral cells in fleshy segments of *Salicornia virginica*, and he believes that these large spirally thickened cells may be involved

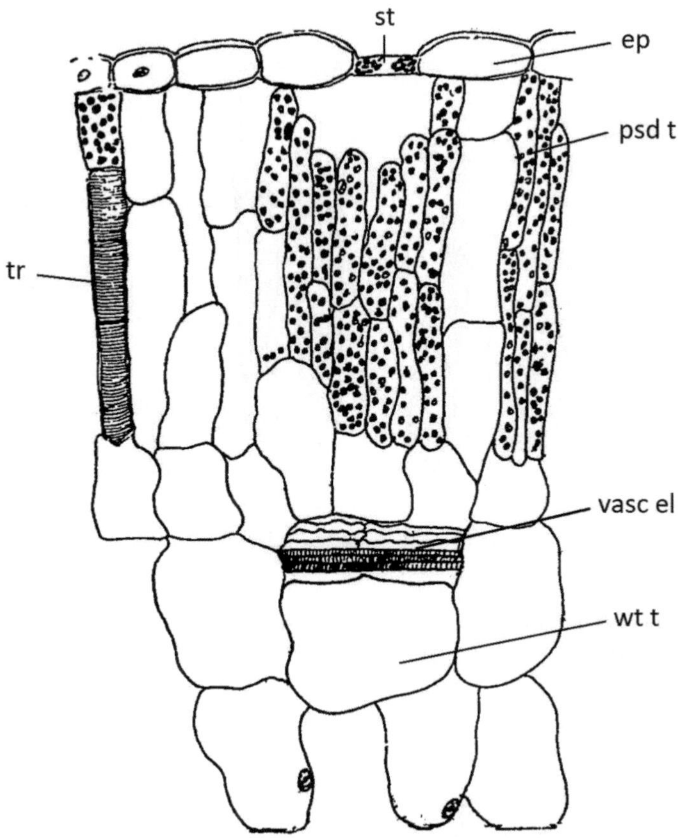

Fig. 4.10 Tracheoidioblasts (spiral cells) in the fleshy segment of *Salicornia australis* (*psd t* palisade tissue, *ep* epidermis, *tr* tracheoidioblasts, *vasc el* vascular elements, *wt t* water storage tissue) (Cooke 1911)

in the maintenance of water balance; actually, he is the only author who uses (properly) the term *tracheoidioblasts*.

Our research on halophytes (Grigore and Toma 2010; Grigore et al. 2014 and references therein) reveals that spiral and spicular cells are found only in articulated chenopods (Grigore and Toma 2007, 2008; Grigore et al. 2013, 2014) (Figs. 4.14, 4.15, 4.16, 4.17, 4.18, 4.19, 4.20, 4.21, 4.22, 4.23, and 4.24); this may suggest a specific structural coevolution within this restricted group of halophytes from *Chenopodiaceae*. What do they have in common? Why they are not found in other (succulent) chenopods?

First of all, let's try to cluster. They are succulent, segmented plants; generally, without any exception, they are euhalophytes—thus vegetating only in high saline environments (Grigore 2012; Grigore et al. 2014). Usually, they must also withstand long periods of waterlogging exposure. Their mechanical tissues—at least at the levels of segments—are less developed; on the other hand, their water storage

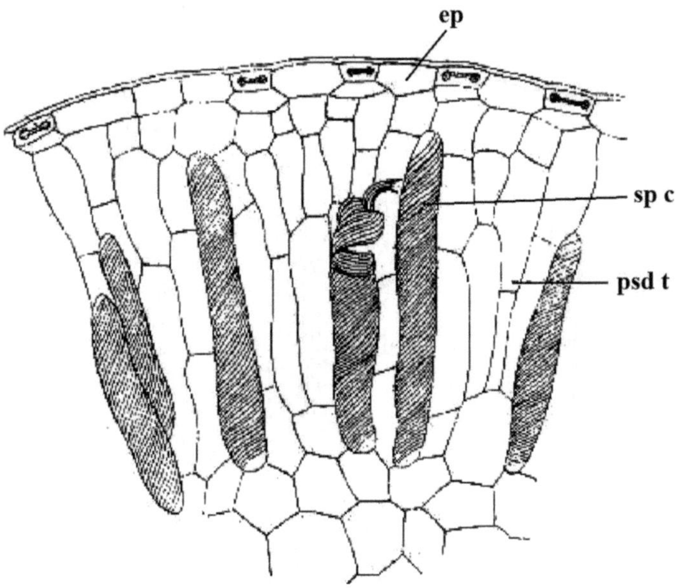

Fig. 4.11 Tracheoidioblasts (spiral cells) in the fleshy segment of *Salicornia fruticosa* (*psd t* palisade tissue, *ep* epidermis, *sp c* spiral cells) (de Fraine 1912)

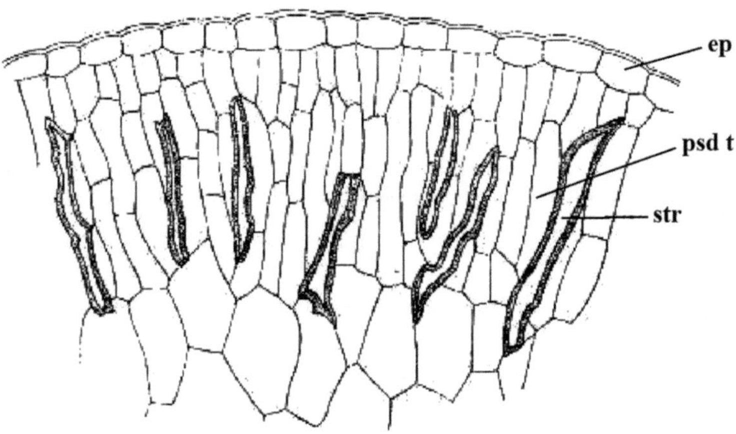

Fig. 4.12 Stereides (spicular cells) in the fleshy segment of *Salicornia glauca* (*psd t* palisade tissue, *ep* epidermis, *str* stereides) (de Fraine 1912)

tissue is strongly developed, usually surrounding the central cylinder. Therefore, this architectural conformation would manage to ensure the water supply—involved mainly in the salt dilution and, to some extent, in providing a strong cell turgor, needed for plants living in such ecological conditions (Grigore 2008; Grigore et al. 2014). Therefore, in this context, perhaps the water-storing function of spiral cells is not strongly supported.

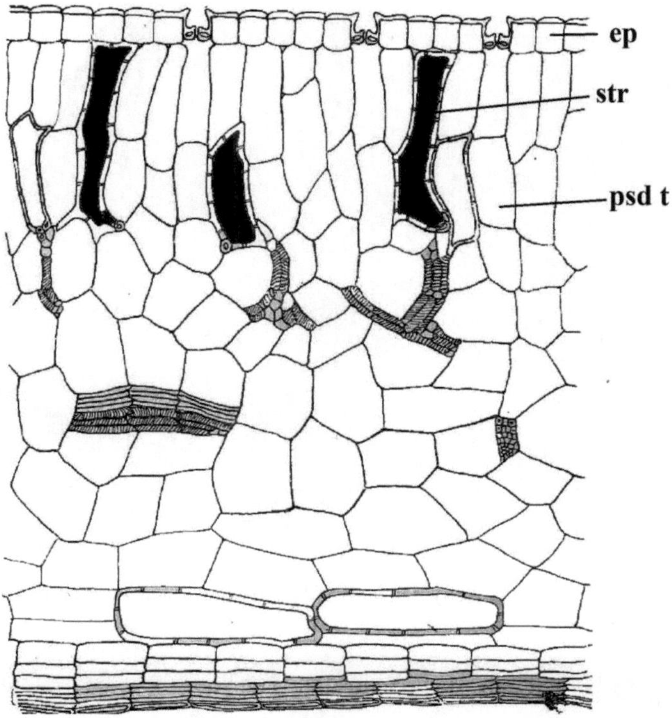

Fig. 4.13 Stereides (spicular cells) in the fleshy segment of *Arthrocnemum ssp* (*psd t* palisade tissue, *ep* epidermis, *str* stereides) (adapted and slightly modified from Baumgärtel 1917)

In contrast, mechanical tissues (sclerenchyma and collenchyma) are poorly expressed in these fleshy segments; they surround completely the central cylinder (for extended discussions about its origin in the frame of leafless like aspect of these plans, see Grigore et al. 2014). Considering the segments separated by the central cylinder of the plant axis—as independent units, it would imply that they lack absolutely mechanical tissues. For this reason, they would require a certain degree of mechanical support and spiral cells (and especially spicular cells, depending on the species) would act as a mechanical anchor, reinforcing this structural and functional unit (the segment), where the high internal turgor would represent a pressure factor itself. In addition, the intensely lignified and thickened cell walls of these elements may suggest their mechanical role.

The air-storing proposed function of spiral cells may be explained by the fact that segmented chenopods face periodic flooding, and conditions of hypoxia do occur in such ecological conditions; however, they seem not to be connected to stomata or other structures involved in the air-circulating or storing system of plant.

Unlike de Fraine (1912), in all material analyzed by us, no spiral and spicular cells have been found in the same species.

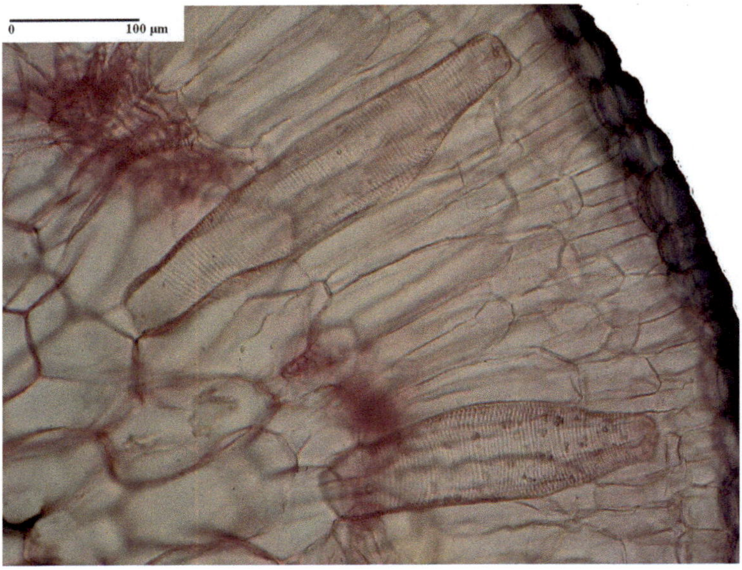

Fig. 4.14 Tracheoidioblasts in fleshy segments of *Salicornia europaea* (Grigore et al. 2014)

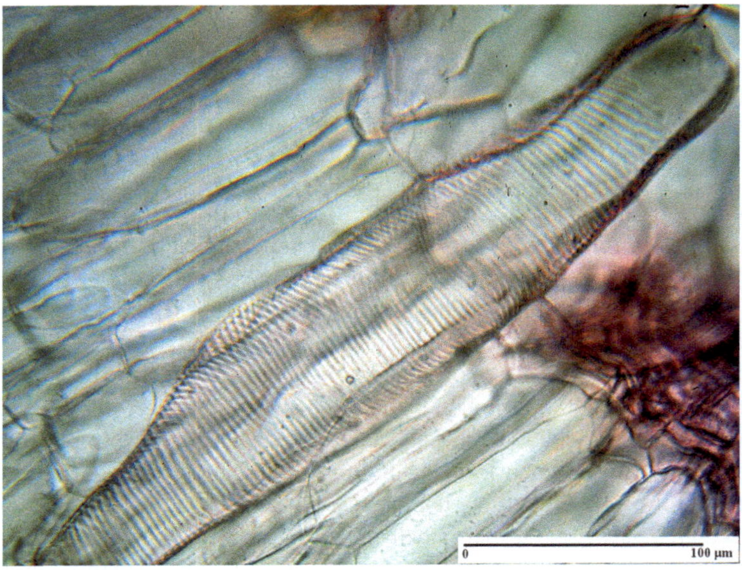

Fig. 4.15 Tracheoidioblasts in fleshy segments of *Salicornia europaea* (Grigore et al. 2014)

Keshavarzi and Zare (2006) describe these elements as large, spiral cells, occurring between the palisades of the assimilating layer of the foliar organs. The epidermis is rarely in contact with these tracheoidioblasts, and there is no connection between them and the vascular system. Their shape is cylindrical and their ends are sometimes oblique.

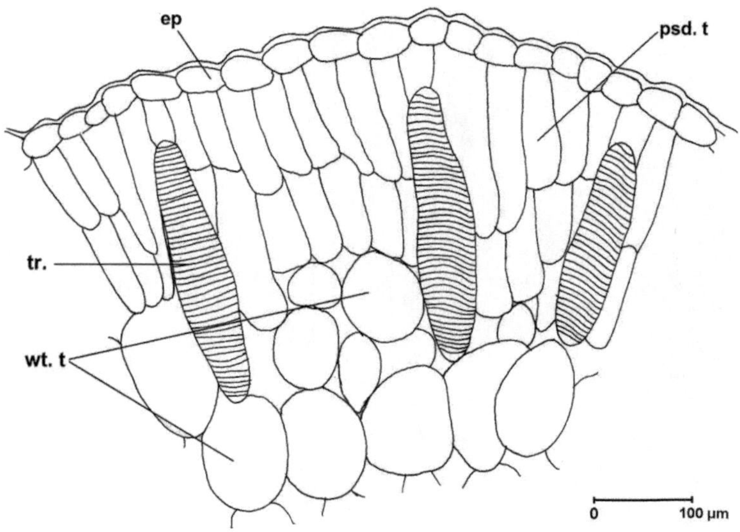

Fig. 4.16 Tracheoidioblasts in fleshy segments of *Salicornia europaea* (*ep* epidermis, *psd t* palisade tissue, *tr* tracheoidioblasts, *wt t* water storage tissue) (Grigore et al. 2014)

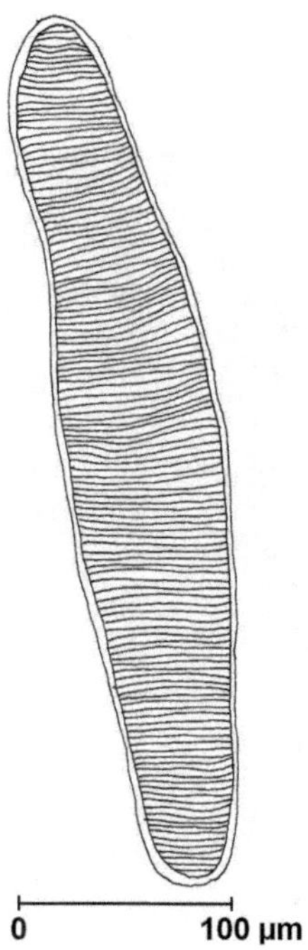

Fig. 4.17 Isolated tracheoidioblast from the segment of *Salicornia europaea* (Grigore et al. 2014)

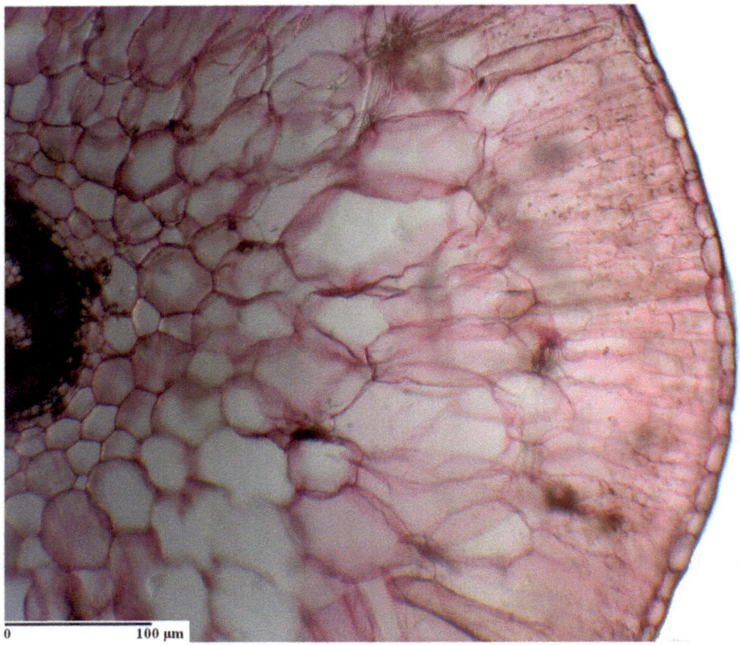

Fig. 4.18 Tracheoidioblasts in fleshy segments of *Salicornia ramosissima* (Grigore et al. 2014)

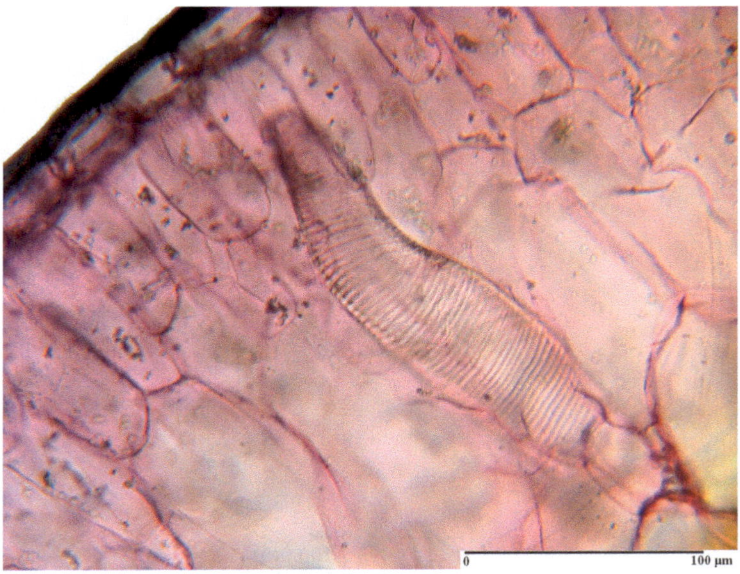

Fig. 4.19 Tracheoidioblasts in fleshy segments of *Salicornia ramosissima* (Grigore et al. 2014)

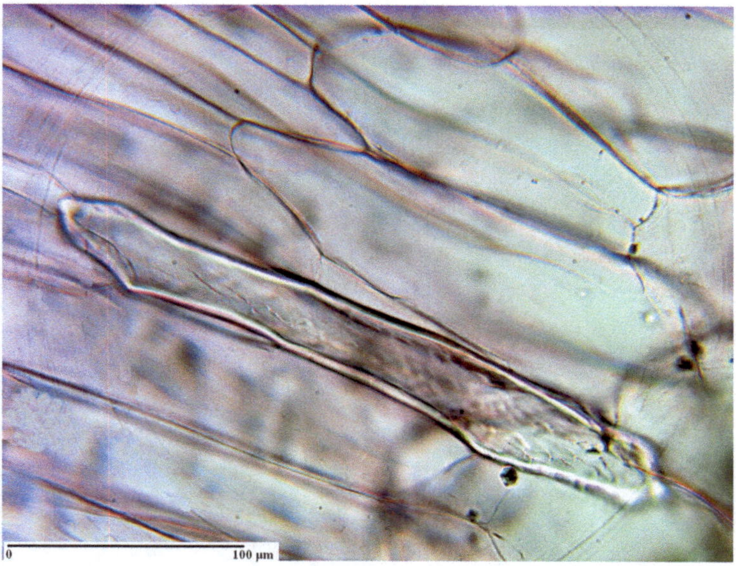

Fig. 4.20 Spicular cells (stereides) in fleshy segments of *Sarcocornia fruticosa* (Grigore et al. 2014)

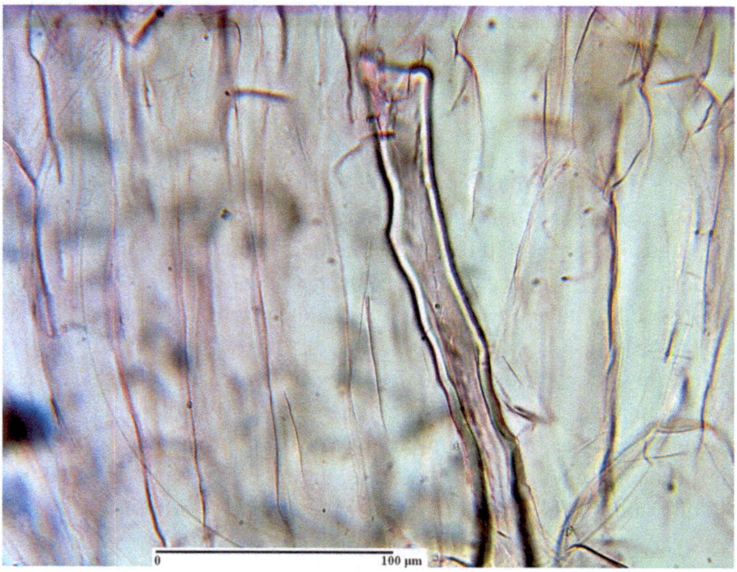

Fig. 4.21 Spicular cells (stereides) in fleshy segments of *Sarcocornia fruticosa* (Grigore et al. 2014)

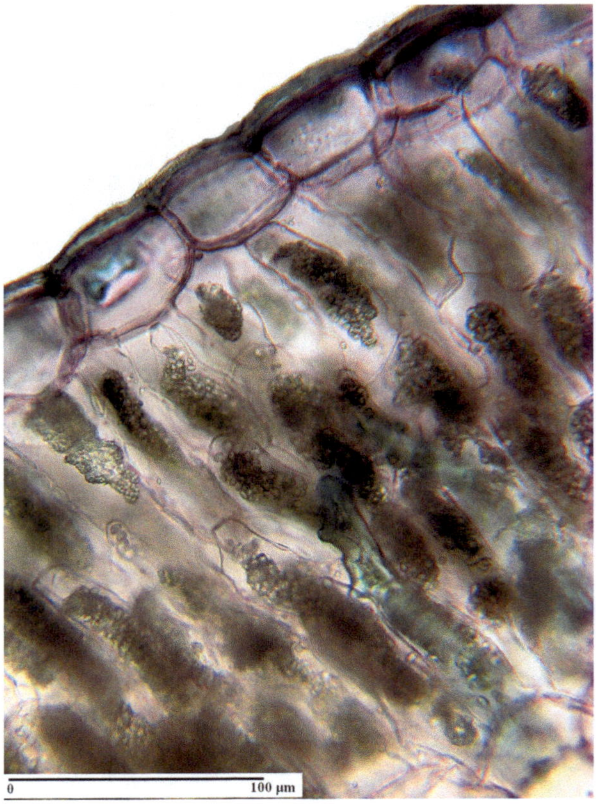

Fig. 4.22 Spicular cells (stereides) in fleshy segments of *Arthrocnemum macrostachyum* (Grigore et al. 2014)

In very rare situations, these spiral cells have been erroneously described as salt glands in *Salicornia europaea* species (Bercu and Bavaru 2005).

Indeed, it has been reported that these tracheids connect the water storage tissue to the epidermis, as it is the case of the *Arthrocnemum fruticosum* species, where they are supposed to have a role in the uptake of dew by the epidermis (Saadeddin and Doddema 1986).

However, these "sclereids" (actually, they are, according to the nomenclature used throughout this chapter—spiral cells; see also Waisel's images), as Waisel (1972) defines them, seem to attain larger sizes in plants exposed to higher saline conditions, especially with NaCl; therefore, they seem to be responsive to salinity.

Saadeddin and Doddema (1986) show that *leaves contain an assimilatory palisade layer and a water storage tissue,* ***connected****(our emphasis) with the epidermis by a number of vessel-like tracheoidioblasts.*

Pirwitz (1931) demonstrated that these cells are water filled, are alive, and contain a nucleus, subjected to plasmolysis. These observations have been subsequently confirmed by electronic microscopy techniques (Hess et al. 1975).

Weber et al. (1977) agree with this but elaborate on it by suggesting that they may function as a reservoir of low–salt water in an otherwise high-salt

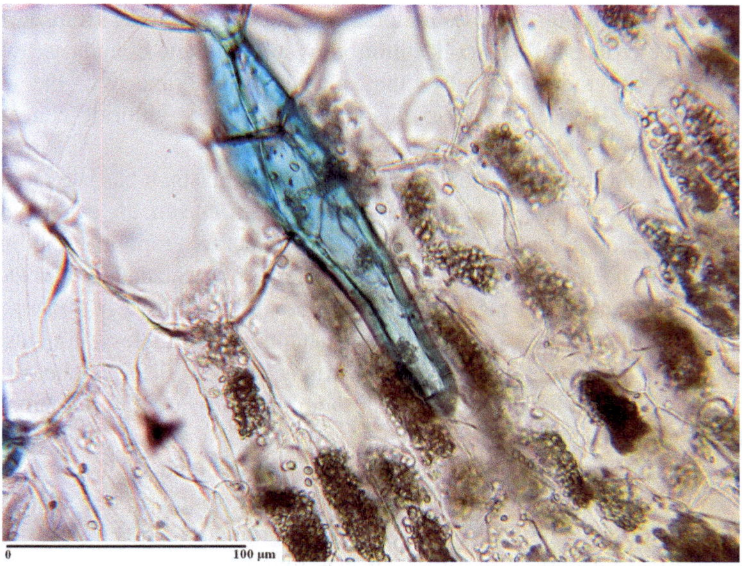

Fig. 4.23 Spicular cells (stereides) in fleshy segments of *Arthrocnemum macrostachyum* (Grigore et al. 2014)

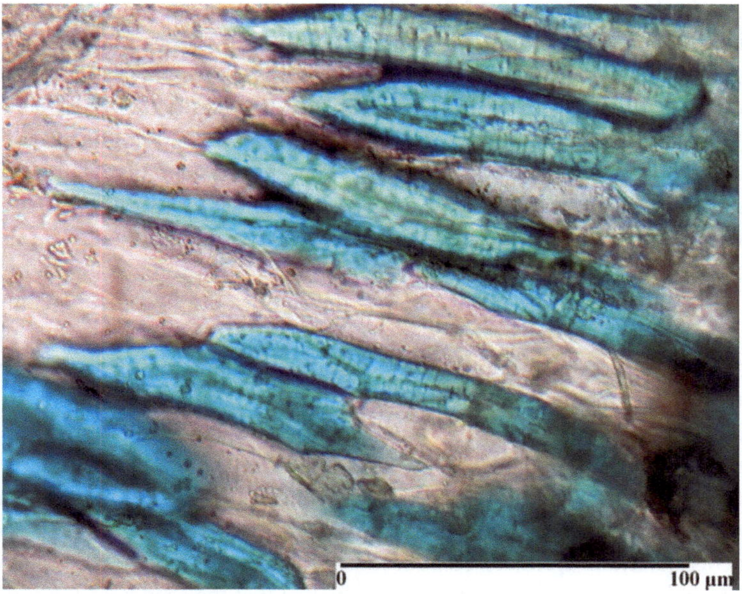

Fig. 4.24 Spicular cells (stereides) in fleshy segments of *Halocnemum strobilaceum* (Grigore et al. 2014)

environment. According to these authors, the salt water moving from the spongy tissue could be reduced in salt concentration by a membrane boundary, by reverse ion-transport enzymes, and by the plasma membrane of the isolated tracheoidioblasts. Hess et al. (1975) also suggest that the chloride concentration inside the tracheoidioblasts is indeed low.

Another role has been assigned to these tracheoidioblasts. It has been noticed for a long time that many halophyte species, including *Salicornia*, are capable of water uptake through their aerial parts when they are subject to temporary flooding. This uptake seems to be able to replace previous water losses. Saadeddin and Doddema (1986) suggest that such a mechanism could be also operative in *Arthrocnemum fruticosum*, by the uptake of dew by the shoots. Nevertheless, the discussion remains open whether there are any connections between these tracheids and the epidermis, on the one hand, and between them and the vascular elements, on the other hand.

As already mentioned, perhaps the presence of spiral and spicular cells may be correlated with ecology of the species where they are found. In articulated chenopod species whose ecology has been studied (Grigore et al. 2011a, 2011b, Grigore and Toma 2014; Grigore et al. 2014), it was found, for instance, that *Salicornia europaea* is a hygro-halophyte which vegetates, just like *Suaeda maritima* (euhalophyte, but not articulated), from a wet, sandy, often waterlogged salt area. The presence of succulence, besides the previously mentioned dilutive effect, also interferes in the maintenance of the cell turgor, which means another way in order to maintain the erect position of the plant, as it is well known that the mechanical tissues are rudimentary and poorly developed in this species. "As soon as the turgor is no longer assured, due to the lack of water, the plant fatally dies out," says Prodan (1922), with reference to *Salicornia*. Therefore, the species is confined to intensely saline, chloride, but humid environments. The chloride salinity of the soil often induces succulence, playing the role of both diluting the toxic ions and maintaining the osmotic pressure, the turgor, which allows the plant to uptake saline soil solution and to keep the erect position of the plant. Therefore, as the soil is scarce in water, the plant may die either because the salts would become concentrated in cells (that would be lethal) or because of the loss, reversible or not, of the erect position. Therefore, the water balance, besides the balance of salts, is of great importance in segmented succulent halophytes.

Thus, *Salicornia europaea* is a hygrophilous species, from moderately to intensely halophilous, being developed on salinized water meadows wet in the depth and less wet toward the surface (Bucur et al. 1960), having basically the same ecological requirements as *Suaeda maritima*. They are both species that vegetate in chloride associations (Şerbănescu 1965). Moreover, speaking about the importance of water for *Salicornia*, one must mention the fact that the seed germination begins, in most cases, under the precipitation water, when the salts are much diluted.

References

Anderson CE (1974) A review of structure in several North Carolina salt marsh plants. In: Reimold RJ, Queen WH (eds) Ecology of halophytes. Academic, New York, London, pp 307–344

Baumgärtel O (1917) Die anatomie der gattung *Arthrocnemum* Moqu. Sitz. Kaiser. Akad. Wissensch. in Wien. Matematisch-naturwissen kl 126:41–74

Bercu R, Bavaru E (2005) Contribuţii la cunoaşterea anatomiei speciei *Salicornia europaea* L. (*Chenopodiaceae*). Lucr Şt Univ Şt Agr Med Vet "Ion Ionescu de la Brad", Iaşi, ser Hort 1(48):625–630

Bucur N, Dobrescu C, Turcu G, Lixandru G, Teşu C (1960) Contribuţii la studiul halofiliei plantelor din păşuni şi fâneţe de sărătură din Depresiunea Jijia-Bahlui (partea a II-a). Stud şi Cerc (Biol şi Şt Agricole) Acad R.P.R., filiala, Iaşi 11(2):333–347

Chermezon H (1910) Recherches anatomiques sur les plantes littorales. Ann Sci Nat sér 9, Bot 12:117–313

Cooke FW (1911) Observations on *Salicornia australis*. Trans Proc New Zeal Inst 44:349–362

Dangeard PA (1888) Note sur la gaine foliaire des *Salicornieae*. Bull Soc Bot France 35:157–160

de Bary A (1884) Comparative anatomy of the vegetative organs of the Phanerogams and Ferns. Clarendon, Oxford

de Fraine E (1912) The anatomy of the genus *Salicornia*. Linn J Bot 41:317–346

Duval-Jouve M (1868) Des *Salicornia* de l'Hérault. Observations anatomiques et morphologiques. Bull Soc Bot France 15:132–140

Fahn A, Arzee T (1959) Vascularization of articulated *Chenopodiaceae* and the nature of their fleshy cortex. Am J Bot 46:330–338

Ganong WF (1903) The vegetation of the Bay of Fundy Salt and Diked marshes: an ecological study. Bot Gaz 36(3):161–186. 280–302, 349–367, 429–455

Grigore M-N (2008) Introducere în halofitologie. Elemente de anatomie integrativă. Pim, Iaşi

Grigore M-N (2012) Romanian salt tolerant plants. Taxonomy and ecology. Tehnopress, Iasi

Grigore M-N, Toma C (2007) Histo-anatomical strategies of *Chenopodiaceae* halophytes: adaptive, ecological and evolutionary implications. WSEAS Trans Biol Biomed 12(4):204–218

Grigore M-N, Toma C (2008) Ecological anatomy of halophyte species from the *Chenopodiaceae* family. Advanced topics on mathematical biology and ecology (Proceedings of the 4th WSEAS International Conference on Mathematical Biology and Ecology—MABE '08, Acapulco, Mexico, January 25–27, 2008), pp 62–67

Grigore M-N, Toma C (2010) Halofitele. Aspecte de anatomie ecologică. Edit. Univ. "Al. I. Cuza", Iaşi

Grigore M-N, Toma C, Boscaiu M (2011a) Ecological notes on halophytes species from Mediterranean climate. Lucr Şt (Horticultură), USAMV "Ion Ionescu de la Brad", Iaşi 54(1):29–34.

Grigore M-N, Toma C, Ivănescu L (2011b) Anatomical and ecological observations on Mediterranean halophytes: Suaeda Forssk. ex Scop. genus. Lucr Şt (Horticultură), USAMV "Ion Ionescu de la Brad", Iaşi 54(1):23–28

Grigore M-N, Toma C (2014) Integrative ecological notes on halophytes from "Valea Ilenei" (Iaşi) nature reserve. Mem Sci Sect Rom Acad 37:19-36

Grigore M-N, Toma C, Zamfirache M-M, Ivănescu L, Daraban I (2013) Anatomical and ecological observations in succulent (articulated) halophytes from *Chenopodiaceae*. Lucr Şt (Horticultură), USAMV "Ion Ionescu de la Brad", Iaşi 56(2):19–24

Grigore M-N, Ivănescu L, Toma C (2014) Halophytes. An integrative anatomical study. Springer, Cham

Hess WM, Hansen DJ, Weber DJ (1975) Light and electron microscopy localization of chloride ions in cells of *Salicornia pacifica var. utahensis*. Can J Bot 53:1176–1187

Holterman C (1907) Die einfluss der klimas auf den bau der pflanzengewebe. Anatomisch-physiologische untersuchungen in den tropen. Verlag von Wilhelm Engelman, Leipzig

Keshavarzi M, Zare G (2006) Anatomical study of *Salicornieae* Dumort. (*Chenopodiaceae* Vent.) native to Iran. Int J Bot 2(3):278–285

Mangin L (1882) Sur le development des cellules spiralées. Bull Soc Bot France 29:14–17
Monteil P (1906) Anatomie comparée de la feuille des Chénopodiacées, Thèse, Ecole Supérieure de Pharmacie, no. 9, Université de Paris
Pirwitz K (1931) Physiologische und anatomische Untersuchungen an Speichertracheiden und Velamina. Planta 14:19–76
Prodan I (1922) Oecologia plantelor halofile din România (comparate cu cele din Ungaria şi Şesul Tisei din regatul SHS). Bul Inf Grăd Bot şi Muz Bot din Cluj 2(1):1–17. 2(2):37–52; 2(3): 69–84, 101–114
Reichenbach L, Reichenbach HG (1909) Icones florae Germanicae et Helveticae simul terrarum adjacentium ergo Media Europae, vol 24. Lipsia et Gerae, Sumptibus Friederici de Zezschwitz
Saadeddin R, Doddema H (1986) Anatomy of the 'Extreme' halophyte *Arthrocnemum fruticosum* (L.) Moq. in relation to its physiology. Ann Bot 57:531–544
Şerbănescu I (1965) Asociaţiile halofite din Câmpia Română. Com Geol Instit Geol St Tehn şi Econ seria C Pedologie 15:1–148
Solereder H (1908) Systematic anatomy of the Dicotyledons. A handbook for laboratories of pure and applied Botany, vol 2. Clarendon, Oxford
van Tieghem P (1884) Traité de botanique. Librairie F, Savy, Paris
Volkens G (1884) Zur kenntniss der beziehungen zwischen standort und anatomischem bau der vegetationsorganen. Jahr des Königl Botanisch Gartens und des Botanisch. Museums zu Berlin 3:1–46
Waisel Y (1972) Biology of halophytes. Academic Press, New York, London
Weber DJ, Rasmussen HP, Hess WM (1977) Electron microprobe analyses of salt distribution in the halophyte *Salicornia pacifica* var. *utahensis*. Can J Bot 55:1516–1523
Warming E (1890) Botaniske Exkursioner. 1. Fra Vesterhavskystens Marskegne. Vidensk Meddel Fra D naturh Foren Kjøben 52:206–239
Warming E (1909) Oecology of plants: an introduction to the study of plant-communities. Clarendon, Oxford

Chapter 5
Salt Secretion

It is known that salts are continuously transported toward aerial parts of the plant due to the uninterrupted flow of water, via xylem stream. In plants growing in saline habitats, accumulation of salts may reach at certain times high (toxic) levels that require, for the survival of these species, the reduction of salt content in plant' shoots (Grigore 2008a, b).

In such circumstances, it is compulsory for the plant to secrete excess of ions from its organs. The well-documented mechanism in this respect is salt secretion via salt glands, which therefore represent effective devices for adjusting the mineral content at the level of stems and leaves. This is, however, only one of the mechanisms by which salts are removed from plant's organs. Salts can be also removed through the cuticle or by the process of guttation. They may also be retranslocated via phloem back to the roots and soil or may be concentrated in hairs located at leaf level.

Salt secretion is defined as a very important adaptive-ecological strategy, whose efficiency may depend on the ability of the plant to survive in a given habitat, to the detriment of other species. As known, salt-secreting structures may be found especially in non-succulent halophytes. Usually, it has been suggested (Grigore et al. 2014) that succulence and salt secretion are not to be found in the same halophyte species, as a mechanism to regulate the salt content; seemingly, they are well-developed mechanisms built up during the plant evolution.

Salt glands were identified and described as early as the mid-nineteenth century. Initially considered hydathodes or even limestone (chalk)-secreting glands, they were regarded as rarities in the plant world (Grigore and Toma 2010). It seems that the first researcher who revealed and characterized these glands was Licopoli (1866). Later, authors like Marloth (1887), Volkens (1884), or Vuillemin (1887) investigated these glands, the first handling *Tamaricaceae*, the second *Plumbaginaceae*, and the third *Plumbaginaceae, Frankeniaceae,* and *Tamaricaceae*. They were the nineteenth-century pioneers of this field. Subsequently, investigations from the twentieth century, like the ones carried out by Schtscherback (1910) or Ruhland (1915), enrich the knowledge related to this field

M.-N. Grigore, C. Toma, *Anatomical Adaptations of Halophytes*,
DOI 10.1007/978-3-319-66480-4_5

and progressively started to emphasize their more detailed structure, function, and environmental significance.

Regarding the mechanism of production and elimination of substances to the exterior, Frey-Wyssling (1935a, b) distinguished three distinct situations, for which he recommended three different words:

a. *Excretion (exkrete)*, for the products of the catabolic (dissimilation) phase of the plant metabolism;
b. *Secretion (sekrete)*, for the substances formed during the anabolic (assimilation) phase of the plant metabolism;
c. *Recretion (rekrete)*, for the substances, eliminated in a state similar to the one they were absorbed in an unaltered state.

Among these words, recretion is the most adequate to describe the activity of the salt-bearing glands; Fahn (1988) also thought that these structures (just like hydathodes) eliminate substances that are metabolically unaltered or only slightly altered.

Nevertheless, as Stenlid (1958) also stated, these subtle differences were somewhat ignored by the authors following Frey-Wyssling; Stenlid seems to use the term secretion to refer rather to substances eliminated by special mechanisms (glands) and not necessarily to a particular class of substances.

Salt secretion is a phenomenon common to several halophyte genera (Table 5.1), such as *Cressa (Convolvulaceae), Frankenia (Frankeniaceae), Spartina, Chloris, Aeluropus (Poaceae), Limonium, Plumbago, Armeria (Plumbaginaceae), Glaux (Primulaceae), Tamarix, Reaumuria (Tamaricaceae),* as well as many species of mangrove forests: *Acanthus (Acanthaceae), Avicennia (Avicenniaceae), Laguncularia (Combretaceae), Aegiceras (Myrsinaceae), Ceriops, Bruguiera (Rhizophoraceae),* and *Sonneratia (Sonneratiaceae).* The salt-secreting formations of these taxa are usually involved in removing excess salt (Haberlandt 1914; Helder 1956; Scholander 1968; Scholander et al. 1962, 1965, 1966). However, there are many other plants with trichomes, glands, and glandular structures; yet, only further investigations would determine the exact nature of secreted products.

In addition, in the botanical literature, some inconsistency and the lack of a common point of view were present, as regards the definition of salt glands. Fahn (1988) clearly defined salt glands: "*Salt glands are specialized epidermal cells or trichomes, which play an active part in the secretion of solutions of mineral salts and often also contain organic substances.*" On the other hand, confusion between hydathodes and salt glands was also present.

Hydathodes are generally defined as eliminating structures for water; they are found on the plant surface. Haberlandt (1914) divided the hydathodes into two functional types:

1. Passive hydathodes, which are directly connected to the conducting system, in which secretion is a process of filtration under pressured conditions;
2. Active hydathodes, which have no connection with the conducting system and are active in the secretion process.

Table 5.1 Distribution of salt-secreting structures in halophytes

Family	Taxa	Observations
Salt glands (*stricto sensu*)		
Plumbaginaceae°	*Limonium gmelinii*°, *Limonium furfuraceum*°, *Limonium girardianum*°, *Limonium narborense*°, *Limonium latifolium, Plumbago capensis, P. europaea, Armeria maritima, Aegialitis**, *Limoniastrum*	*Mangrove species °Several Limonium species investigated by us
Tamaricaceae	*Tamarix sp., Reaumuria palaestina, Myricaria**	*Rare on saline soils
Avicenniaceae	*Avicennia marina**	*Mangrove species
Rhizophoraceae	*Ceriops sp., Bruguiera sp**	*Mangrove species, where the presence of salt glands is still questionable
Primulaceae	*Glaux maritima**, *Samolus littoralis, Samolus repens*	*Also investigated by us
Frankeniaceae	*Frankenia laevis**, *Frankenia pulverulenta, Frankenia hirsuta, Frankenia reuteri, Frankenia grandifolia, Frankenia pauciflora, Hypericopsis*	*Also investigated by us
Combretaceae	*Laguncularia**	*Mangrove species
Acanthaceae	*Acanthus, Neuracanthus*	
Convolvulaceae	*Cressa, Ipomoea**	*Only in species from saline soils
Myrsinaceae	*Aegiceras**	*Mangrove species
Poaceae	*Aeluropus**, *Distichlis, Spartina, Bouteloua, Buchloe, Cynodon, Coelachyrum, Crypsis, Dactyloctenium, Dinebra, Eleusine indica, Enteropogon, Sporobolus, Tetranche, Tetrapogon, Andropogon, Brachiaria, Cenchrus, Chrysopogon, Coix, Dichanthium, Digitaria, Echinochloa, Erianthus, Hyparrhenia, Panicum, Paspalum, Paspalidium, Saccharum, Setaria, Sorghum, Tricholaena, Porteresia coarctata, Zoysia*	*Several species also investigated by us
Salt hairs (vesicular hairs, bladders, salt bladders)		
Chenopodiaceae	*Atriplex**, *Chenopodium, Halimione**, *Salsola*	* Several species also investigated by us
Epidermal bladder cells		
Aizoaceae	*Dorotheanthus**, *Mesembryanthemum**, *Psilocaulon**°	*Only in species from saline soils *°Rare on saline soils

Based on Breckle (1995), Gorham (1995), Breckle (2002), Grigore and Toma (2010), Grigore et al. (2014)

For detailed clarifications regarding the nomenclature of salt-secreting structures, see Grigore and Toma's (2010) monograph

Stocking (1956) suggested active hydathodes to be assimilated to salt glands; however, in some instances, there is no clear-cut distinction between salt glands and limestone (chalk)-secreting glands (Metcalfe and Chalk 1972).

When describing the secreting glands of the *Plumbaginaceae* species, Metcalfe and Chalk (1972) classify them into two categories:

1. Chalk (chalk-secreting) glands, also known as Mettenius glands or Licopoli glands, which generally occur on or in the cavities on the inner side of the leave and stem; they are sometimes surrounded by groups of elongated epidermal cells or by simple hairs. Individual glands of this sort are made up of four or eight epidermal cells arranged in palisade surrounded by one or two layers, each made up of four "accessory" cells. The walls between the secreting cells of the gland and the surrounding ("accessory") cells are cutinized. The secreting "organs" of this sort have been generally described as chalk glands because they exude calcium salt and water; calcium salts are sometimes scattered on the leaf or stem surface by rain drops. The amount of secreted calcium salts depends on the type of soil, although, for instance, the British *Limonium* species analyzed by Fraine (1916) do not secrete limestone-containing substances.
2. Mucilage glands occur in some representatives of the *Plumbaginaceae* family; those in the axils of the upper side of basal leaf from *Limonium bellidifolium* and *L. binervosum*, described by Fraine (1916), comprise a head resting on a head borne on a base consisting of few cells with very thick cuticle-lined walls.

Grigore and Toma (2010) have adapted, modified, and completed a previous definition of Fahn's (1988) regarding secretory tissues in vascular plants, using their anatomical vision in the salt-secreting structures of halophytes.

In this sense, the salt-secreting structures comprise:

1. Structures eliminating salts into the vacuole, the situation of *Atriplex* and *Chenopodium* species. Salt is eliminated into a central vacuole of the bladder cell of the leaf trichomes. These cells are situated on top of narrow, 1- to 3-celled stalks. Growth of bladder cell is accompanied by the formation and expansion of a central vacuole;
2. Structures (glands, stricto sensu) eliminating salts outside of the cells. Here, two types of glands may be included:
    a. Bicellular and monocellular glands, mainly on *Poaceae* species;
    b. Multicellular glands, such as those of *Limonium, Tamarix, Avicennia, Frankenia, Cressa cretica,* and *Lavatera arborea*.
3. Epidermal bladder cells of *Mesembryanthemum* species, which occupy an intermediary and special position between salt secretory structures. Despite the fact that some authors have included these bladder cells in the group of secretory trichomes, our opinion is that these cells must be regarded, from the histo-anatomical point of view, as special epidermal cells, which accumulate and eliminate salts outside of plant body.

5.1 Plumbaginaceae

When referring to the ***Plumbaginaceae*** family, it should be emphasized that one of the most striking traits of its representatives is the presence of these epidermal glands (chalk glands—Mettenius or Licopoli glands and mucilage glands) located on leaves and stems (Grigore and Toma 2010). The structure of these glands was frequently differently interpreted by some authors, although these controversies are related rather to details than their basic structure. These formations drew the botanists' attention as early as the second half of the nineteenth century, as it will be detailed herein.

Recently, Grigore and Toma (2016) reviewed the structure of these mucilage glands, as reflected by early botanists from the nineteenth century; the "paternity" of gland description is being discussed, addressing to results reported by Mettenius (1856) and Italian Licopoli (1866, 1879), the two botanists who gave the name of mucilage glands.

As is well known, *Plumbaginaceae* constitute a well-represented cosmopolitan family in the temperate zones of the Northern Hemisphere and show preferences for arid or saline, often coastal, environments (Kubitzki 1993). The Angiosperm Phylogeny Group classification of flowering plants (APG 2003) included this family in the *Caryophyllales* order, together with other families adapted to extreme environments including oligotrophic soils, arid zones, and soils with high salt content. The taxonomy and taxonomical affinities of this striking family are still very problematic and controversial (Cronquist 1981; Lledo et al. 1998, 2001, 2005; Reyes 1997; Short and Wightman 2011; Takhtajan 2009). For this reason, the number of genera and species included in the *Plumbaginaceae* differs greatly from one author to another: from about 12 genera and 400–500 species (Reyes 1997) to 10–27 genera and about 1000 species (Short and Wightman 2011). *Plumbaginaceae* is a well-known halophytic family (Grigore 2008a, b, 2012; Grigore and Toma 2010; Grigore et al. 2014), a reality since long recognized in botanical research (Endlicher 1836–1840; Lincevskii and Cerniakovskoi 1952; Bentham and Hooker 1876; Volkens 1884; Pax 1897; Strasburger et al. 1894; Lindley 1846; Răvăruț 1960; Moore 1972; Takhtajan 2009).

The chalk secretion and deposit on the surface of these organs have been noted long before the detection and description of these glands. Thus, Braconnot (1836) (not 1830, as quoted by Maury 1886) was the first who tried to analyze this mineral substance secreted by glands and also detected the existence of these special secreting "formations."

Braconnot (1836) analyzed the mineral substance secreted by glands of different species of *Statice*, *S. monopetala*, *S. pruinosa*, *S. aphylla*, and others, and of *Plumbago*, *P. zeylanica*, *P. auriculata*, *P. scandens*, and *P. rosea*. He investigated the "*inorganic scales* (*écailles de nature inorganique*) *produced by species of Plumbaginaceae family*"; when examined with a magnifier glass, these white deposits on the surface of leaves appeared to Braconnot as a "*small parasitic fungus embedded in the tissues of host plant.*" He has also anticipated the existence of special secreting formations involved in the occurrence of these deposits, but he did not use a specific term to nominate them. However, he made an interesting anatomical-like observation: after washing the leaves of *Statice* species with acids, he observed on their surface "*visible cavities indicating the places where the stalks of these small scales were embedded.*" After having treated the leaves of several *Statice* species with hydrochloric acid, he performed the dissolution of the secreted substance, which he identified as calcium carbonate and which contained suspended transparent formations, which he assumed to be the "*organs*" considered to have secreted this carbon-containing substance. Unfortunately, his findings seemed to remain unknown to many future botanists for a long time, since no record of this published study has been registered until Mettenius (1856), who does mention Braconnot's work.

Maury (1886) and Vuillemin (1887) believed that Italian botanist Licopoli (1866) was the first researcher who made a histological description of these calcium carbonate-secreting "organs"; likely, it seems that they were not aware of Mettenius' work. In this respect, Grigore and Toma (2016) show that even nowadays, the terms "Licopoli" and "Mettenius" organs are being used in parallel in botanical literature. The reason for this perception is perhaps explained by the fact that some authors knew only Mettenius's or only Licopoli's paper and not both of them, so that they could not have an accurate historical picture. For instance, neither Maury nor Vuillemin do mention Mettenius's work, whereas, out of the two French botanists, only Maury (1886) mentions Braconnot's earliest paper. One may assume that Mettenius's paper, published in German, was inaccessible to French botanists, and thus, it has not been consulted; however, Mettenius (1856) does mention Braconnot's findings.

Mettenius (1856) described chalk glands in a very succinct but quite precise manner, in the way that he did not hesitate at all in using correct terms related to the chalk-secreting function of these glands: *Kalksecretion* (chalk secretion) and *Kalkschüppchen* (chalk scales). He described the chalk glands of *Goniolimon tataricum* (Fig. 5.1), *Limoniastrum monopetalum* (Fig. 5.2), *Plumbago europaea* (Fig. 5.3), and *P. zeylanica* (Fig. 5.4).

Nevertheless, Mettenius's work (1856) represents a significant progress in the research of chalk glands as compared to earlier Braconnot's (1836) paper, assumed to be the first in signaling chalk-secreting process. In his brief considerations, Mettenius underlined several important aspects. For instance, he has correctly shown that chalk glands belong to the epidermal complex, that they are derived from epidermal cells divisions, and—most important—that they are not connected with stomata or the vascular system. However, the structure of gland was incorrectly described by Mettenius as consisting of a group of four cells; his mistake was maintained subsequently by Licopoli (1866, only in part, as it will be shown) and Maury (1886).

Before and after Mettenius' findings, several botanists linked species from *Plumbaginaceae* to a secretion produced by leaves and stems; albeit they were not able to specify if this fluid secretion is of chalk (salt) origin, it may be assumed that in several species a secretion produced by Mettenius glands could be localized.

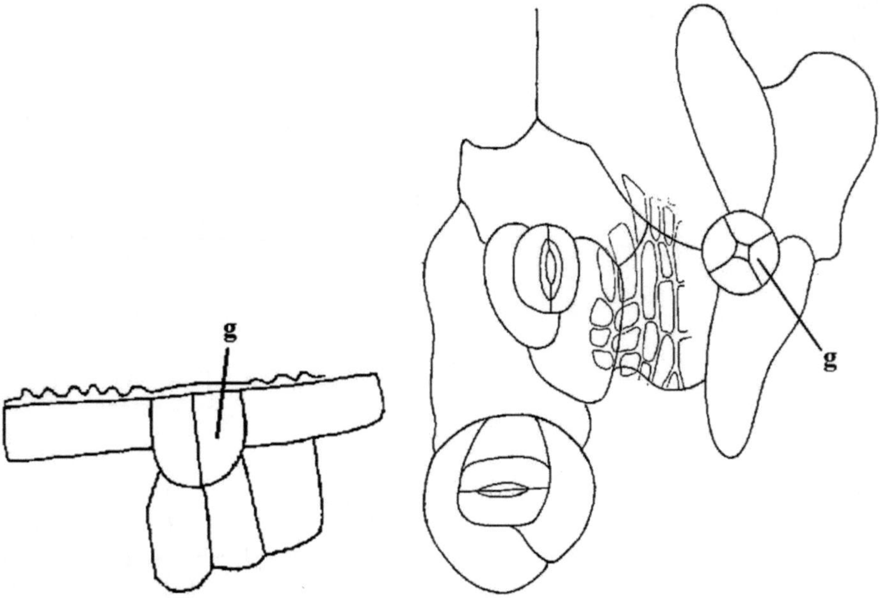

Fig. 5.1 Chalk glands (*g*) in *Goniolimon tataricum* (*left*—cross section; right—surface view) (Mettenius 1856)

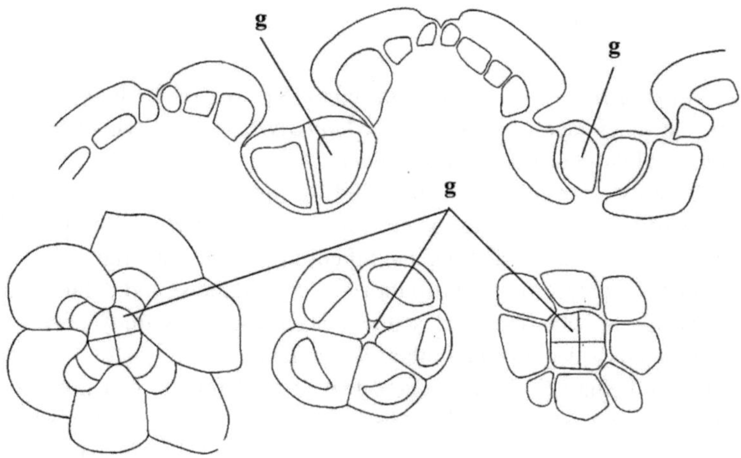

Fig. 5.2 Chalk glands (*g*) in *Limoniastrum monopetalum* (Mettenius 1856)

Thus, Griffith (1854) describes leaves of *Aegialitis rotundifolia* as being viscous or watery: "*[...] fluida viscosa-aquosa*." Bunge (1872) in his monograph on *Acantholimon* genus gave a general correct observation, stating that: *"in most species, leaves and aerial organs have small piths, where chalk is being excreted as rounded scales; these lack in several species, vary in shape and density and may*

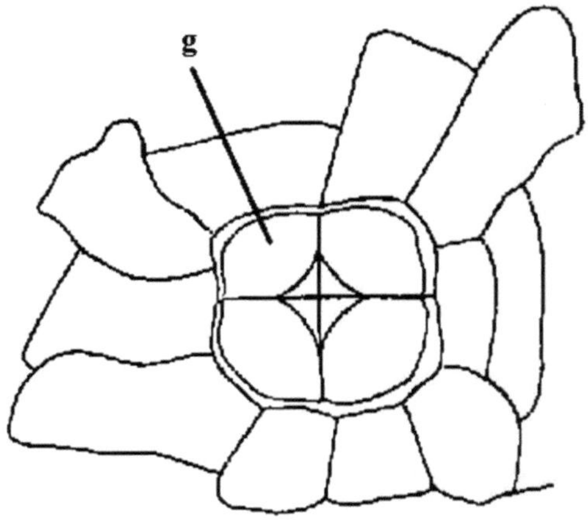

Fig. 5.3 Chalk glands (*g*) in *Plumbago europaea* (Mettenius, 1856)

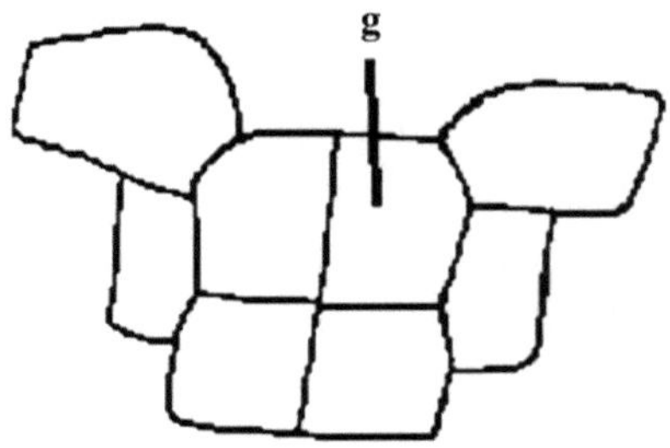

Fig. 5.4 Chalk glands (*g*) in *P. zeylanica* (Mettenius 1856)

affect the color of the whole plant." Martinet (1872) refers to the external glands located in the inner face of calyx from *Plumbago capensis*, where he observed: "*small spherical masses produced by epidermis*."

Italian Botanist Licopoli still considered the first who made a description of chalk glands of *Plumbaginaceae* species (*Statice monopetala*, 1866) gave detailed descriptions of these glands and depicted them in several drawings (Figs. 5.5, 5.6, 5.7, 5.8, and 5.9). Indeed, his contribution is very extended and detailed; unfortunately, it has no references included, so that it is almost impossible to assert whether he knew Mettenius's paper or had other data in hand.

Grigore and Toma (2016) underlined that, except for the fact that he did not nominate the exact types of gland-consisting cells, he was however able to distinguish them from an anatomical point of view and finally to deliver an accurate description of glands (known, as shown as "Licopoli organs"). In addition, he pointed out several important details with respect to these "organs"; he correctly concluded that these glands are not connected with the vascular system, nor with the stomata of a plant leaf. Another important observation is that calcium carbonate is the excreted material of these glands; on his microscopic observations, he identified the deposit of the chalk at the top of the glands—clearly nominated as "*glandole*."

Licopoli resumed his observations in a paper from 1879, where he used the term "*glandole calcifere*." He states that "*these glands have an organization (structure)*

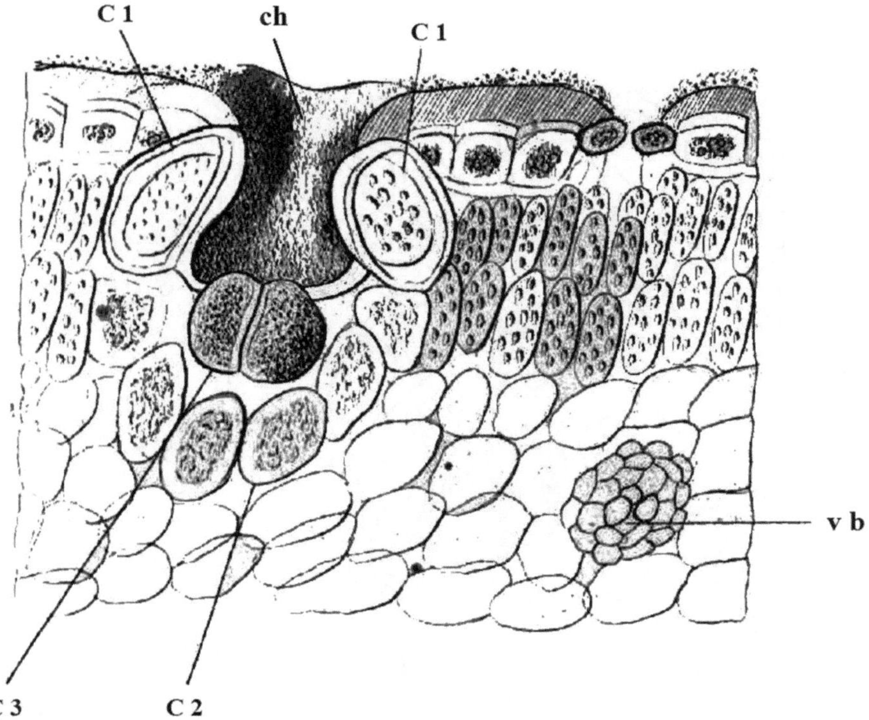

Fig. 5.5 Licopoli "organs" in the lamina of *Statice monopetala*. C 1, C 2—different types of cells; C 3—a complex of cells—*borsetta*, forming the bottom of the gland; *ch* chalk deposit (Licopoli 1866)

based on the type discovered and described in Statice monopetala in my previous work"—that from 1866. Apart from figures with glands of *Statice monopetala,* he included in the paper from 1879 new additional data and drawings referring on *Statice duriaei* (Fig. 5.10), *S. splendens* (Fig. 5.11), and *S. limonium* (Fig. 5.12). Despite very detailed descriptions of these glands, he was not able to explicitly specify the eight-cell structure of these glands (1866, 1879). However, on a deeper text analysis, it could be foreseen that Licopoli may refer in Licopoli 1879 to an eight-cell structure of these glands; for instance, when describing glands from *Statice splendens* (Fig. 5.11), he referred to two distinct groups of four cells and even clearly depicted them in a surface view drawing (thus, eight cells).

After Licopoli's findings—already known and commented by the botanists to come—the interest for the study of chalk glands was intensified toward the end of the nineteenth century; the great majority of botanists recognize these glands as "Licopoli" or rather as "Mettenius" glands.

Maury (1886) evidenced and described Licopoli "organs" in *Plumbago europaea* (Figs. 5.13 and 5.14), *P. larpentae* (Fig. 5.15), *Armeria plantaginea* (Fig. 5.16), *Statice limonium* (Figs. 5.17 and 5.18), *S. elata* (Fig. 5.19), and *S. lychnidifolia* (Fig. 5.20).

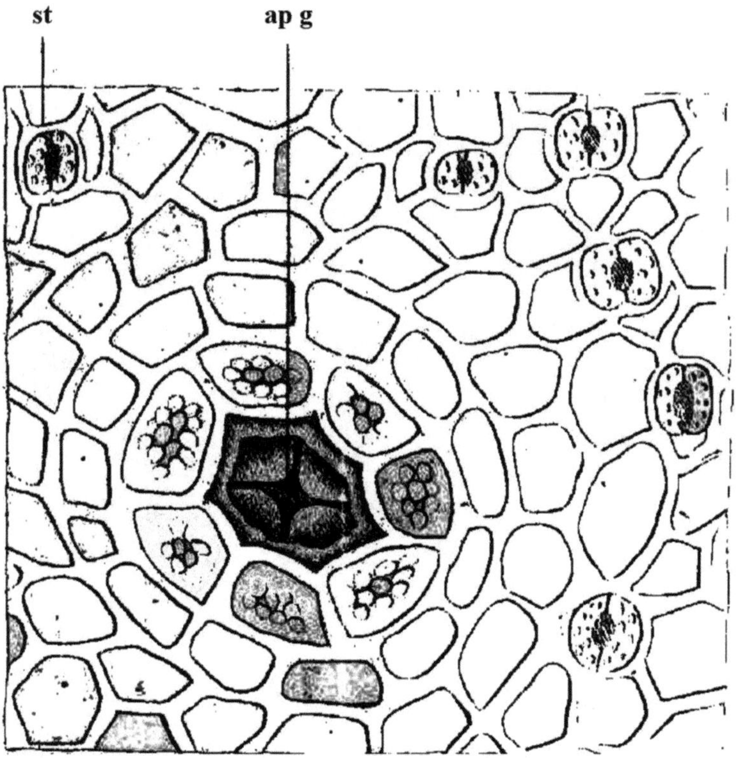

Fig. 5.6 Licopoli "organ" in the lamina of *Statice monopetala*, front view (*ap g* aperture of the gland, *st* stomata; Licopoli 1866)

De Bary (1877) described this secreting "*organ*" in a different manner; he thought that it included eight cells originating in the divisions of a single primary mother cell, which is round or square in surface section. This cell is divided into four by two cell wall divisions, perpendicularly on the surface and on each other. Each of them is in its turn divided again so that one of the new cells is triangular and internal, and the other is rectangular and peripheral.

Volkens (1884) and Woronin (1885) adopted De Bary's (1877) descriptions and interpretations. It seems that they were not aware of Licopoli's findings since no mention is made of his interpretations. This is quite unexpected even for the papers from the nineteenth century, which are usually well documented and supported by the literature, in the manner that we know nowadays. Neither Volkens' nor Woronin's papers—written in German—have no mentions about Licopoli's findings, while the Italian paper has no references. Volkens (1884) maintained the basic eight-cell structure of these glands and pointed out their irregular layout and their role in water elimination, seeing them as "safety valves" that start working when the absorption/transpiration ratio is altered. In his opinion, any excessive calcium

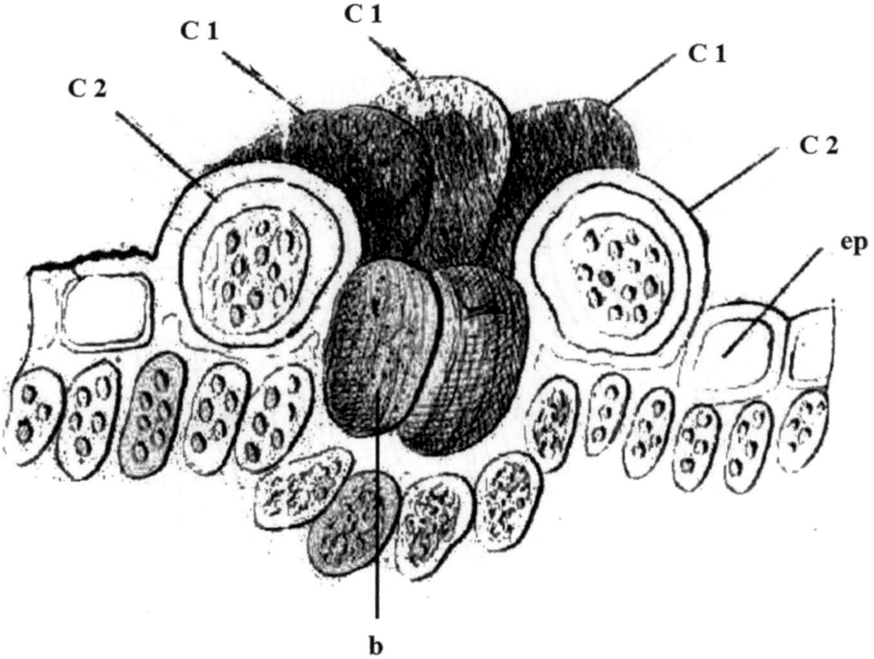

Fig. 5.7 Licopoli "organ" in the lamina of *Statice monopetala*, cross section. C 1, C 2—different types of cells projecting over epidermis level; *b borsetta*, forming the bottom of the gland, *ep* epidermis (Licopoli 1866)

salt is eliminated as carbonic acid. In *Statice limonium,* the cells adjacent to the gland become prominent and turn into conical protrusions.

Figures 5.21, 5.22, 5.23, 5.24, 5.25, and 5.26 show the graphical representations of these glands in different *Plumbaginaceae* species. However, Volkens uses the terms: "*sekretionsapparat, Kalkschuppe*, and *drüse*" corresponding to secretory structures.

Woronin (1885) investigated the leaf structure of *Statice monopetala* and evidenced chalk glands ("*Kalkdrüse*") (Figs. 5.27, 5.28, and 5.29); he also gave a drawing of these glands in *S. sareptana* (Fig. 5.30). In addition to the anatomical description of these glands, he made an interesting ecological observation: the secretion of calcium carbonate by species of *Plumbaginaceae* is conditioned by the soil composition, precisely by its content in calcium carbonate. Woronin correctly claimed that many species of this botanical family do not show an excretory process.

Maury (1886) tried to explain the structure of the *Plumbaginaceae* glands, by pointing out the possible reasons for which other authors considered that these structures rely on eight and not on four cells. When viewed from the top, on a small area of the epidermis, the "*organ*" looks like a circle divided into four sectors by two diameters perpendicular to each other. Each of these sectors *seems* (Maury's

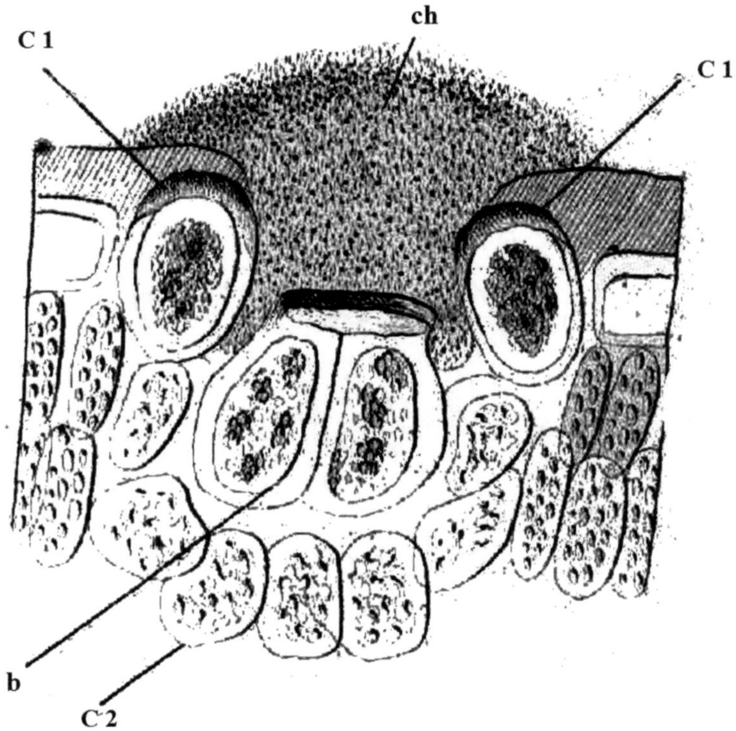

Fig. 5.8 Licopoli "organ" in the lamina of *Statice monopetala*, cross section. C 1, C 2—different types of cells; *b borsetta*, forming the bottom of the gland; *ch* chalk deposit (Licopoli 1866)

emphasis in the text) divided itself into two by a tangential line, which is more blurred than those of the other sectors. This is actually the inner wall of each secreting cell, which borders the central intercellular space; thus, it is this wall that corresponds to this line (which may be best seen on a longitudinal section of the "*organ*"). The secreting cells are curved, joined together at the bottom, and then loose on all their length. Although the substance produced is mixed in this intercellular space, it expands at mid-cell height, the upper ends of which remain close to one another, so that the amount of secreted substance if not very large. The internal pressure of these four cells causes the product to exit due to the pressure put by the inner space walls on the fluid. This fluid removal mechanism is correlated by Maury only with a structure built on four cells. In this author's opinion, if there were eight cells, the substance would simply be exuded by the outer side of the "organ." In other words, de Bary (1877), Volkens (1884), and Woronin (1885) argued that the calcium-containing fluid was eliminated by a mere osmotic phenomenon.

Maury also conducted experiments on some *Plumbaginaceae* species, which were designed especially to analyze the formation and nature of efflorescences, made up of very fine salt filaments, occurring on the surface of the *Plumbago*

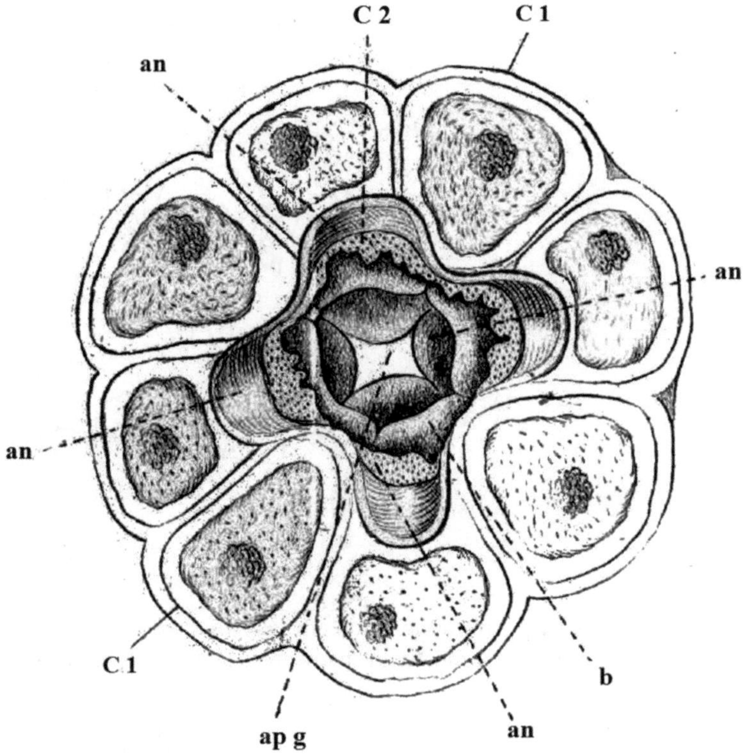

Fig. 5.9 Licopoli "organ" in the lamina of *Statice monopetala*, detail in front view. C 1, C 2—different types of cells; *an* angles formed by the intersection of different types of cells; *ap g* aperture of the gland; *b borsetta*, forming the bottom of the gland (Licopoli 1866)

capensis and *P. zeylanica* organs. These experiments also allowed the drawing of several conclusions:

1. The mineral substance secreted by the Licopoli "*organs*" is shaped like filaments due to the pressure put on the central cavity of the organ by the four secreting cells;
2. In case of humid conditions or in the presence of water (rain water, irrigation), the mineral substance becomes hydrated and the filaments turn into small disks on the epidermis;
3. The role of this mineral substance is similar to that played by hairs in other plants; the author argues that it regulates transpiration.

Maury substantiates this last aspect in the following manner: the *Plumbaginaceae* living in arid or maritime environments must make up for the absence of hairs by accumulating a mineral substance on their surface. Species living in arid environments, *Limoniastrum* species, and a specific number of *Statice* species are covered by a calcareous coating, which prevents them from a high transpiration. The data supporting his assumptions would be that the *Armeria* and

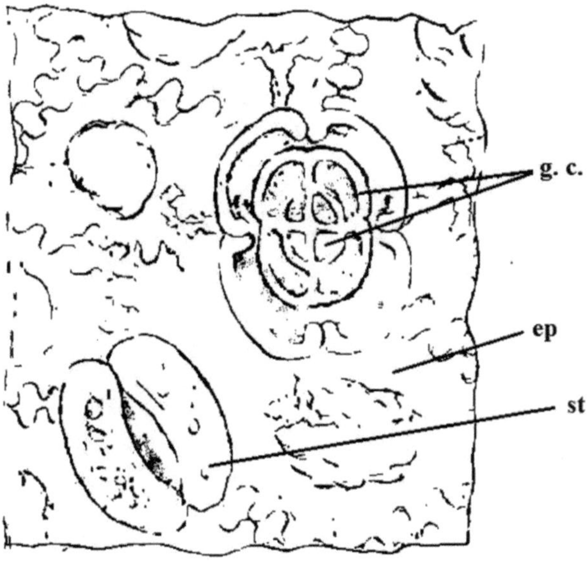

Fig. 5.10 Licopoli "organ" in the lamina of *Statice duriaei*, front view (*g. c.* gland cells, *ep* epidermis, *st* stomata) (Licopoli 1879)

Acantholimon species living in the uplands are less affected by these influences. The *Plumbago* species vegetate mostly in shady areas and therefore have a reduced number of Licopoli "organs."

Whereas Maury (1886) was very certain of its proof supporting the four-cell structure of these Licopoli "organs," Vuillemin (1887) claimed that the eight-secreting-cell structure was very easy to prove. Although thin, the walls of these cells are easily dissolved in reagents; accessory cells are persistent and their boundaries are hard and cutinized, and they are joined together at the bottom of the gland. These edges are carinated and followed by two side expansions applied directly on the connection line separating the accessory cells. The latter thus form a continuous barrier between the glandular cells, on the one hand, and the parenchyma and epidermis, on the other hand; all the substances that cross from ones to the others have to pass through the accessory cells. The cutinized ridges have a rather constant layout in the various genera of the *Plumbaginaceae* family; each of them is made up of a lateral and a deep side. The lateral side makes up a triangle pointing toward the inside of the gland; the four deep sections, which form a cross, are almost parallel to the surface of the epidermis.

Unlike Maury (1886), who claimed that the *Limoniastrum monopetalum* "organs" are full of limestone-containing substances, Vuillemin (1887), when analyzing the same species, did not point that out. Instead, he used another research method: he burned a piece of leaf in potassium; this action, even when it lasts for a long time, does not modify the limestone-containing product. The epidermis is easily dissociated and each isolated gland remains stuck to the excreted mass. The dissolution process led to the disappearance of the thin walls separating the glandular cells; the accessory cells often persist with the cutinized ridges, which support and separate them. When one examines this type of "*skeleton*"

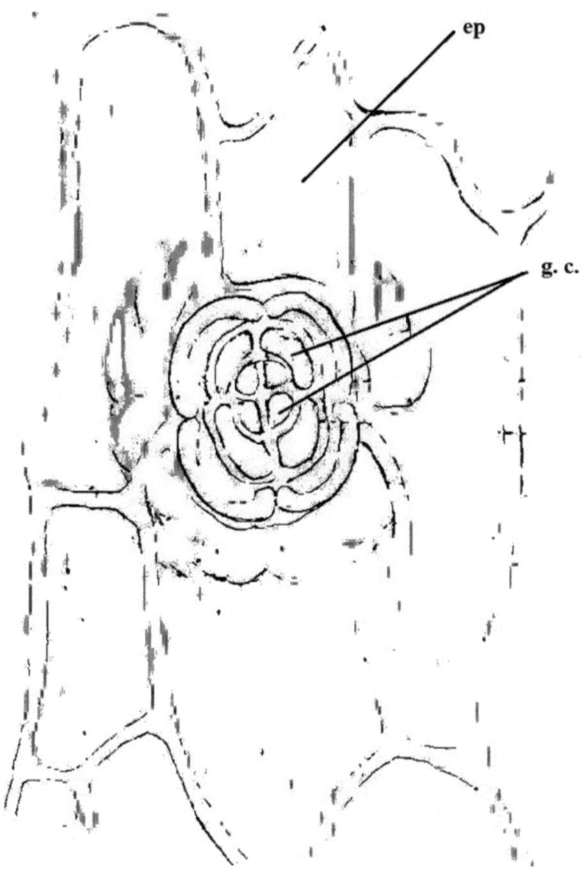

Fig. 5.11 Licopoli "organ" in the lamina of *Statice splendens*, front view (*g. c.* gland cells, *ep* epidermis) (Licopoli 1879)

(in *Limoniastrum monopetalum*—Fig. 5.31 and *Statice latifolia*—Fig. 5.32), one may notice the completely loose and empty gland, despite the limestone covering the external side. The concretion stuck to the inner chamber (inner space) diverticula, which precedes the gland, is made up of two parts joined together by a constriction: the outer part, found on the surface of the epidermis, and the inner four-lobed part, which resembles the shape of the actual gland.

In *Statice imbricata* (Fig. 5.33), six cells, separated by very thin angled walls, can be noticed. There are actually four glandular cells flanked by two accessory cells. The thin cellulosic walls stretching between the accessory cells and the secreting components are almost always partly masked by cutinized borders. Glandular cells usually stick out from the surface of the leaf, since the accessory cells sink between the gland and the adjacent portions of the epidermis.

The parenchyma cells have an oblong shape and a palisade-like layout (with much-reduced meatuses) in the gland (Fig. 5.33b). In the section joined to the epidermis, the accessory cells are often much thicker than in the deep section. The epidermal cells have punctuations both on their lateral sides and on their deep side.

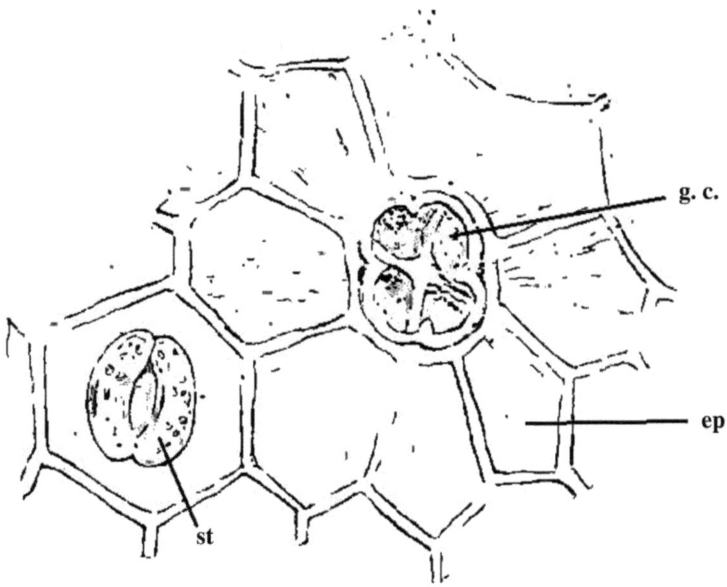

Fig. 5.12 Licopoli "organ" in the lamina of *Statice limonium*, front view (*g. c.* gland cells, *ep* epidermis) (Licopoli 1879)

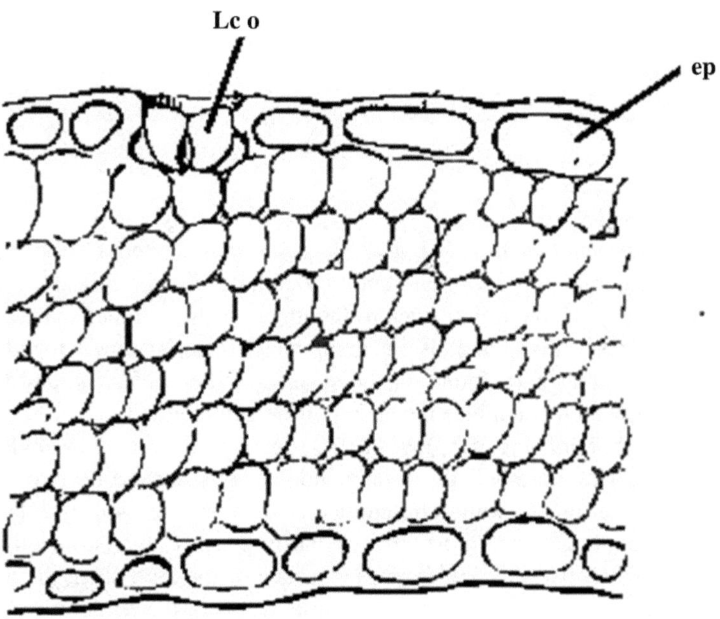

Fig. 5.13 Cross section through the lamina of *Plumbago europaea* (*ep* epidermis, *Lc o* Licopoli "organ" (Maury 1886)

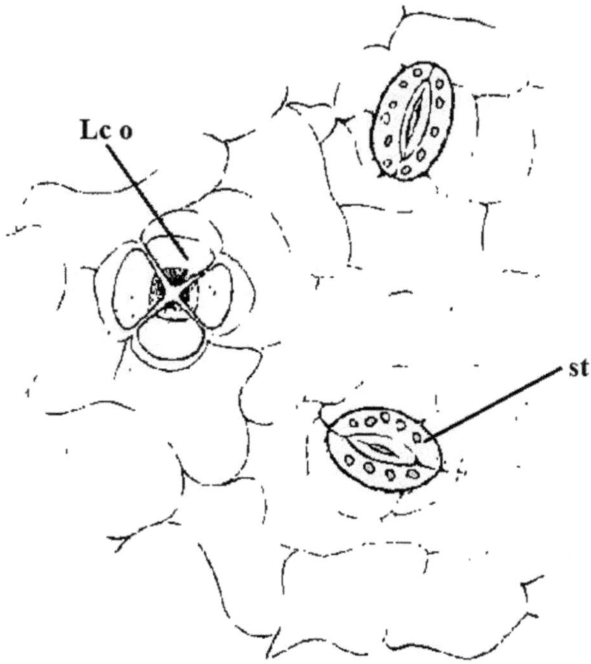

Fig. 5.14 Licopoli "organs" (*Lc o*) in the epidermis of *Plumbago europaea* (*st* stomata; Maury 1886)

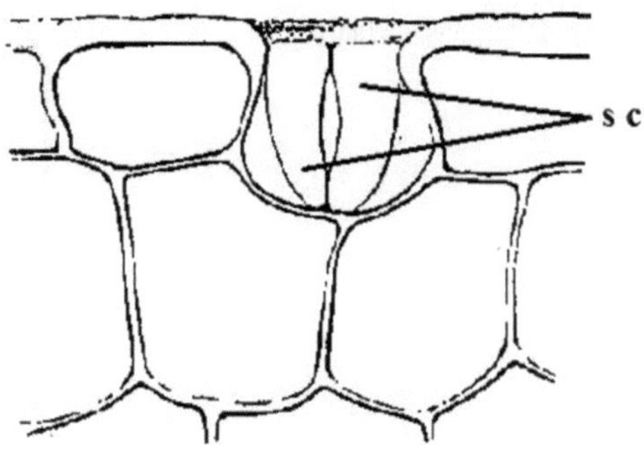

Fig. 5.15 Cross section through the lamina of *Plumbago larpentae* (*s c* secretory cells of Licopoli "organs"; Maury 1886)

These punctuations are evenly scattered on the lateral sides and grouped on the deep one in round surfaces (these surfaces correspond to parenchyma cell insertions), whereas the opaque ones correspond to intercellular meatuses.

The cuticle is interrupted in the hypostomatic chambers (Fig. 5.32) and fenestrated outside these chambers.

Generally speaking, the basic structure of the glands detected and studied by Vuillemin in the *Plumbaginaceae* species remains constant. Only four of the eight glandular cells are excretive. The two rows of cells are sometimes similar in terms of their dark and fine-grained content, which clearly sets them apart from accessory cells and from epidermal or cortical elements. Exchanges occur easily among them due to their thin walls. External secreting cells easily communicate with accessory

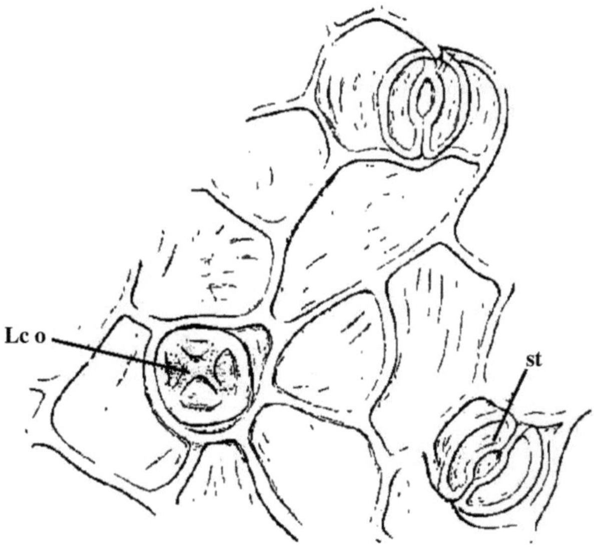

Fig. 5.16 Licopoli "organs" (*Lc o*) in the epidermis of *Armeria plantaginea* (*st* stomata; Maury 1886)

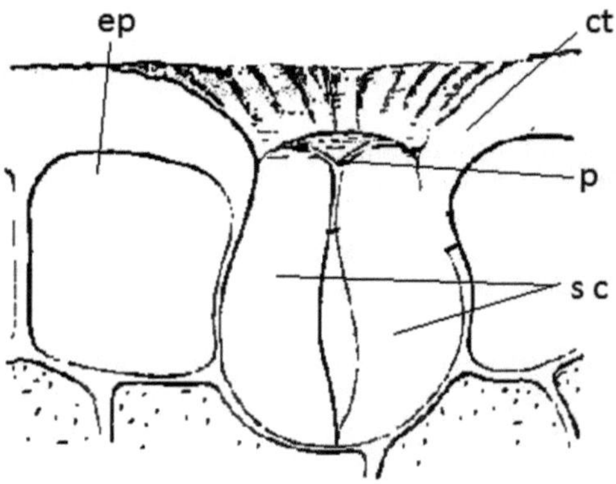

Fig. 5.17 Cross section through the lamina of *Statice limonium* (*ep* epidermis, *ct* cuticle, *p* pore, *s c* secretory cells of Licopoli "organs"; Maury 1886)

cells by osmosis, along their walls, which are also thin, but they are separated from the latter by other leaf tissues. Cutinized ridges prevent any communication between parenchyma and glandular cells in the interstice separating accessory cells; they prevent the formation of any meatus by providing the proper sealing of the latter. In the species the accessory cells of which are very well developed and partly sealed on the sides, like *Limoniastrum guyonianum* (Fig. 5.34), a cuticle sheet grows between them and bifurcates on their outer side, so as to prevent wall detachment. Accessory cells are connected with epidermis and parenchyma cells; they are bridge connecting leaf tissues and gland; from this point of view, they behave like the basal cells of the glandular hairs.

The author above considers the two anatomic structures, i.e., gland and hair, as homologous. The accessory cells would correspond to the foot, whereas the

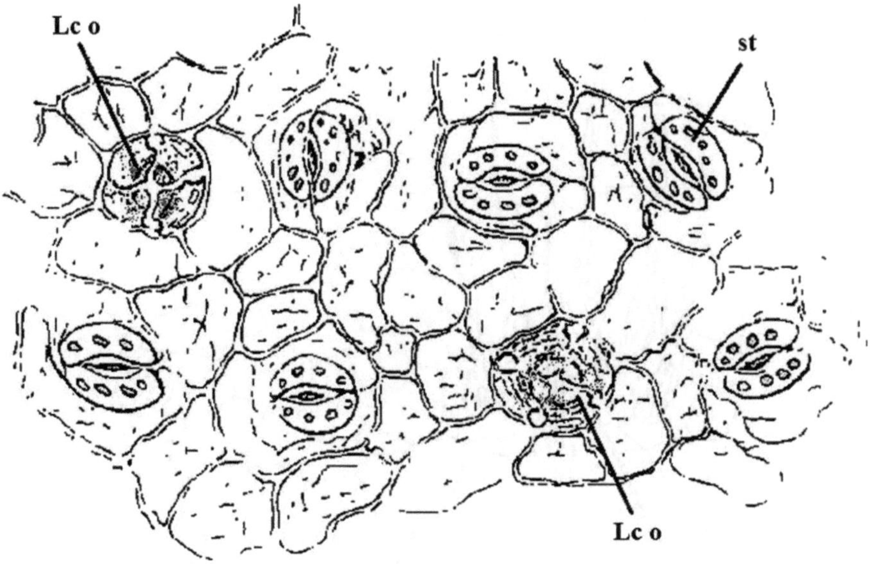

Fig. 5.18 Licopoli "organs" (*Lc o*) in the epidermis of *Statice limonium* (*st* stomata; Maury 1886)

secreting cells to the head of a glandular hair, yet one that underwent an extreme shortening.

The surface section of glandular cells differs from the other walls due to its complete cutinization. The cutinized plate was best noticed on the front view of an epidermis. In *Statice tatarica* (Fig. 5.35), the depth of the chamber preceding the gland (which is almost as thick as the epidermis) and the plate are in the depths of this layer. After having treated the epidermis with a chlorine–iodine solution, the author viewed it as a violet lamella covered with yellow disks (representing glands). Each disk still leaves the impression of two dividing walls in a cross-like layout and other four walls in a rhombus-like layout. The surface is also divided into four triangles close to the middle and four neighboring trapezoids close to the borders.

Solereder (1908) classified the *Plumbaginaceae* glands into two categories, according to their structure:

1. Chalk glands (Licopoli or Mettenius glands), located on lamina of the leaf or on the branch in all members of families, including *Aegialitis* (Fig. 5.36); these structures are not of the nature of hairs, they only consist of small groups of a few epidermal cells and in most cases excrete carbonate of lime on their external surface;
2. Mucilage glands, confined to the upper surface of the leaf sheaths. The mucilage glands consist of a larger number of cells than chalk glands and excrete a mucilaginous substance; in rare cases, they are only composed of groups of epidermal cells shaped like a palisade, thus resembling the chalk glands. In most cases, they are true trichomes, according to Solereder (1908).

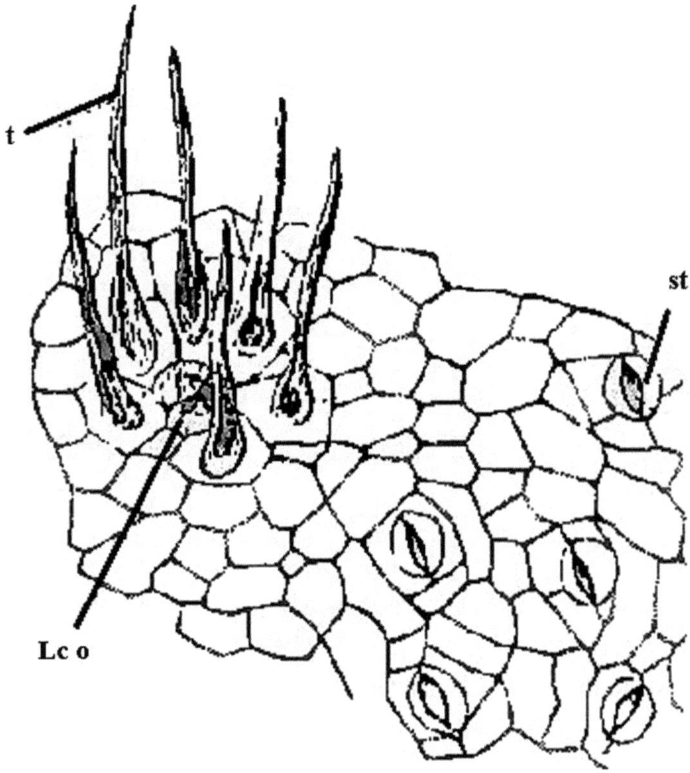

Fig. 5.19 Licopoli "organs" (*Lc o*) in the epidermis of *Statice elata* (*st* stomata, *t* trichome; Maury 1886)

Solereder (1908) gave a detailed explanation of the structure of these two types of glands.

The chalk glands (Fig. 5.37a–c) consist essentially of almost hemispherical groups of eight cells shaped like a palisade (not four, as was incorrectly stated by Mettenius and recently repeated by Maury); according to de Bary, these eight cells originate from an epidermal cell which is rounded quadrate in surface view. This cell becomes divided into four by two walls at right angles to one another and perpendicular to the surface; each of the cells thus formed is then divided once more by a vertical wall into two cells, one of which is very narrow and forms the inner corner while the other is peripheral. The walls of these glandular cells are extremely thin, with the exception of those which separate the internal surface of the gland from the neighboring tissue; these latter are suberized. The contents of the glandular cells consist of dense, finely granular protoplasm. In those cases in which the structure has been thoroughly investigated, the eight-celled group of glandular cells is cut off from the internal tissue by a double cap, each layer of which is composed of four subsidiary cells, so arranged that they appear as semilunar appendages of the glandular cells when examined in surface view and at a

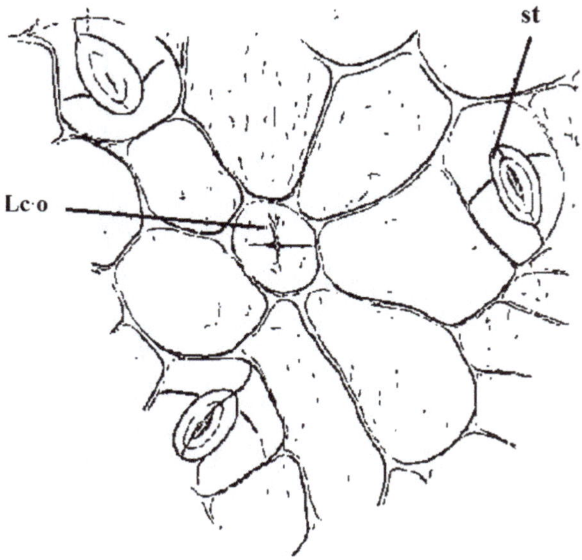

Fig. 5.20 Licopoli "organs" (*Lc o*) in the epidermis of *Statice lychnidifolia* (*st* stomata; Maury 1886)

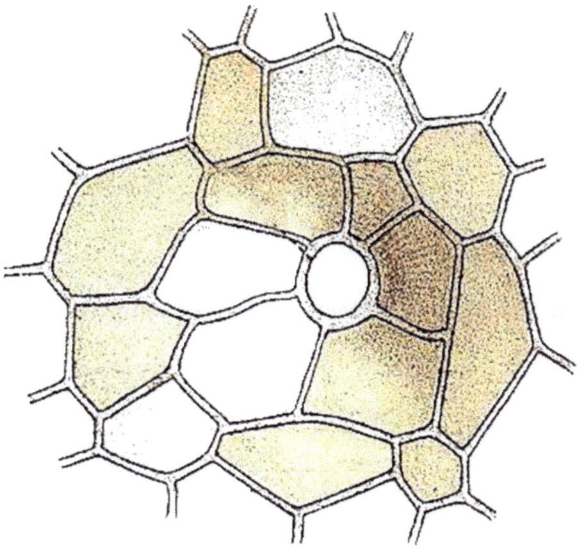

Fig. 5.21 Salt-secreting "apparatus" of *Statice limonium* (Volkens 1884)

sufficiently low focus. The cells of the upper cap (which directly encloses the group of glandular cells) have suberized walls and occasionally (Fig. 5.37a, b) reach to the level of the glandular cells, so that they appear in a surface view of the gland as a four-celled ring, surrounding the circular group of glandular cells; neither of these two characters applies to the "subsidiary" cells of the second cap.

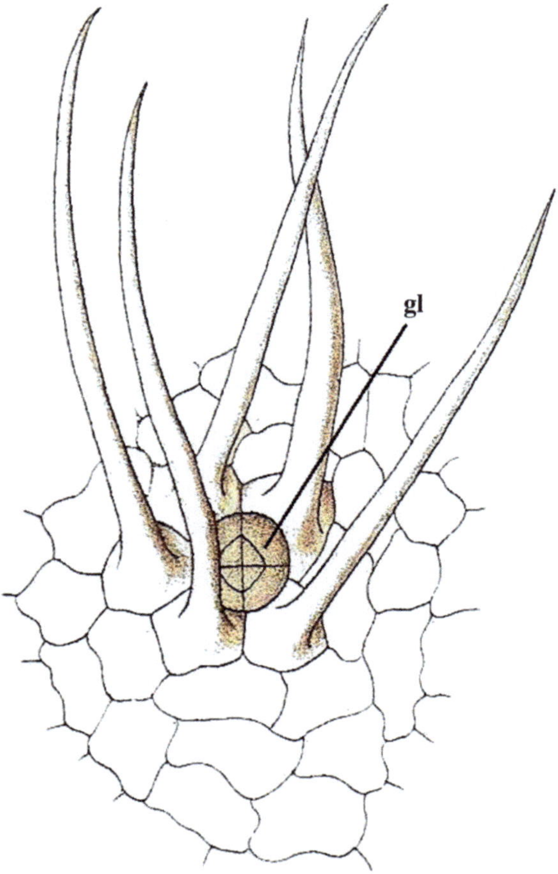

Fig. 5.22 Gland (*gl*) of *Statice latifolia* (Volkens 1884)

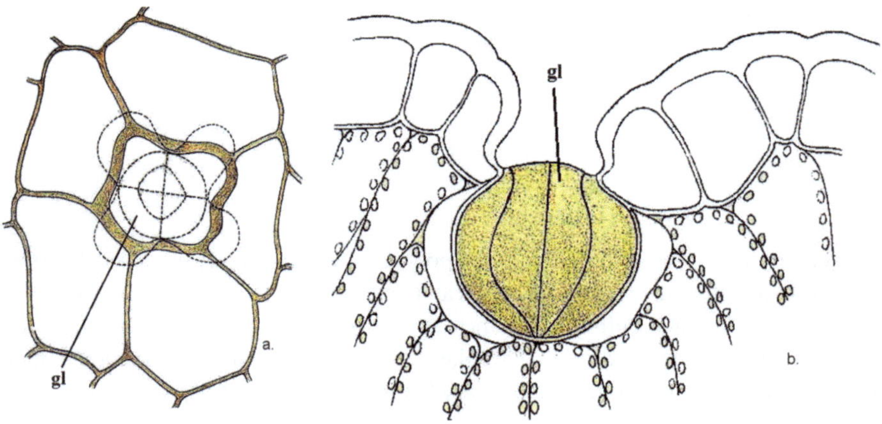

Fig. 5.23 Chalk glands (*gl*) of *Limoniastrum monopetalum* (**a**) surface view (**b**) cross section (Volkens 1884)

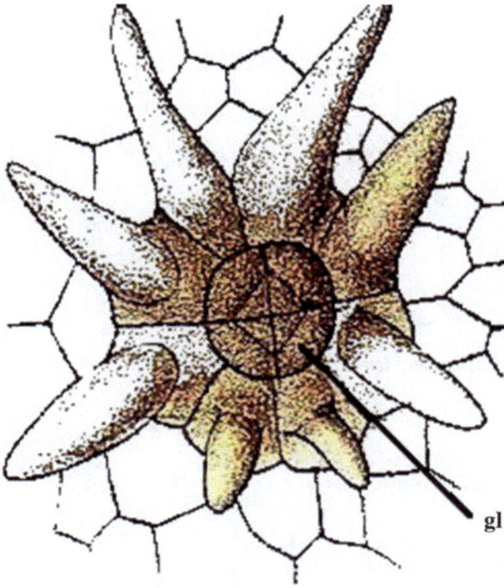

Fig. 5.24 Salt-secreting "apparatus" of *Statice pruinosa* (Volkens 1884)

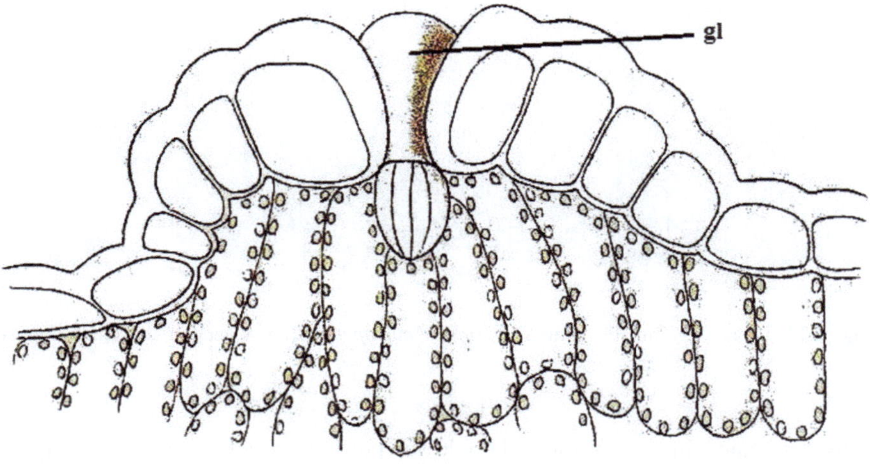

Fig. 5.25 Gland (*gl*) of *Statice rhodia* (Volkens 1884)

The function of the chalk glands is the excretion of water. The water is not conducted to the gland by means of tracheids but by the ordinary cells of the tissue lying nearest to it; consequently, these cells are sometimes grouped radially about the gland.

The mucilage glands were divided into two series, viz., those which are developed as hairs and those which are not so. The mucilage glands of the latter type are found in *Aegialitis;* in their structure, they approach nearest to the chalk glands. Each mucilage gland consists of (1) of a group of thin-walled epidermal

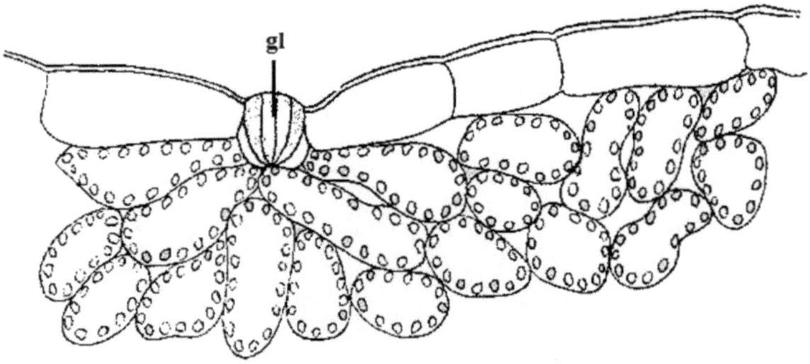

Fig. 5.26 Gland (*gl*) of *Statice occidentalis* (Volkens 1884)

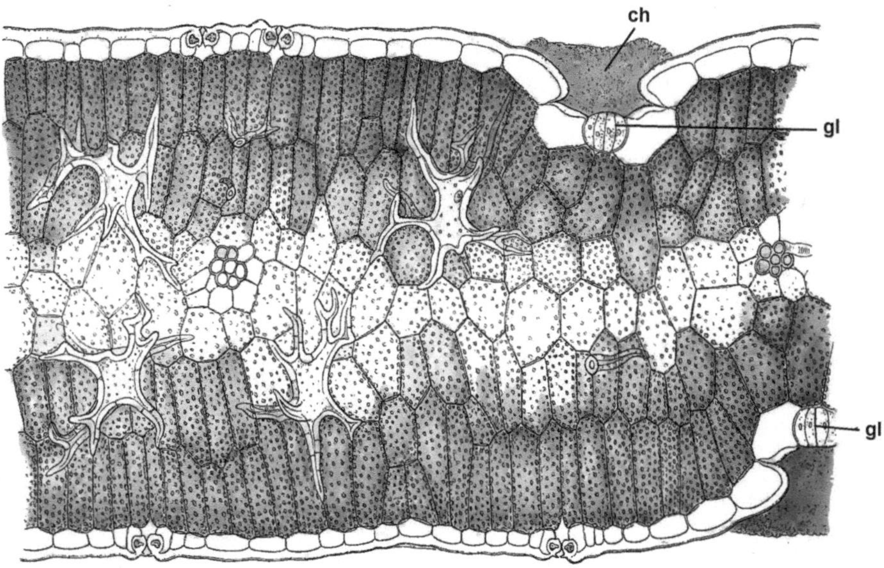

Fig. 5.27 Glands (*gl*) in the lamina of *Statice monopetala* (cross section; *ch* chalk deposit; Woronin 1885)

cells, elongated like a palisade, their lower portions penetrating deeply into the tissue of the leaf sheath, and (2) a double layer of subsidiary cells, of which those adjacent to the palisade cells have suberized cell walls. Seen from the surface, the gland has a circular outline (Fig. 5.37e–f).

According to other opinions (Ruhland 1915), the *Limonium* gland would include a complex of 16 cells, four of which are secreting cells, displayed as a circle, each of them taking up a quarter of that circle. Each cell has an outward counterpart, a small adjacent cell. Both secreting and adjacent cells are surrounded by two layers of cup-shaped cells, each layer containing four big cells laid out similarly to the secreting cells (Ziegler and Lüttge 1967). The top of the gland and the surrounding

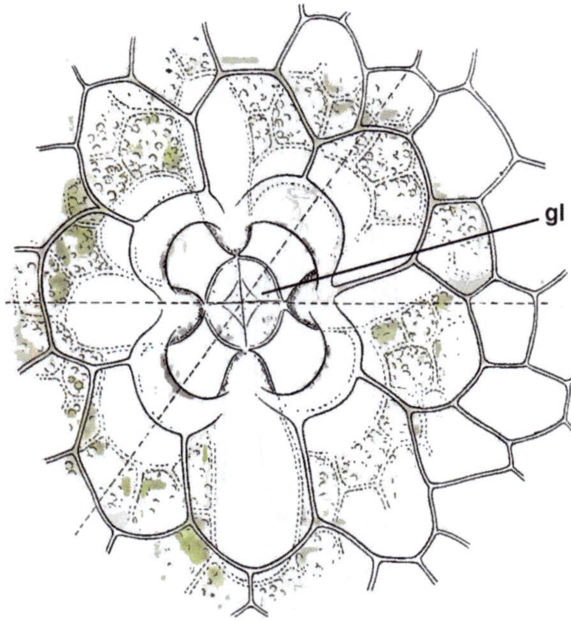

Fig. 5.28 Glands (*gl*) in the lamina of *Statice monopetala* (surface view; Woronin 1885)

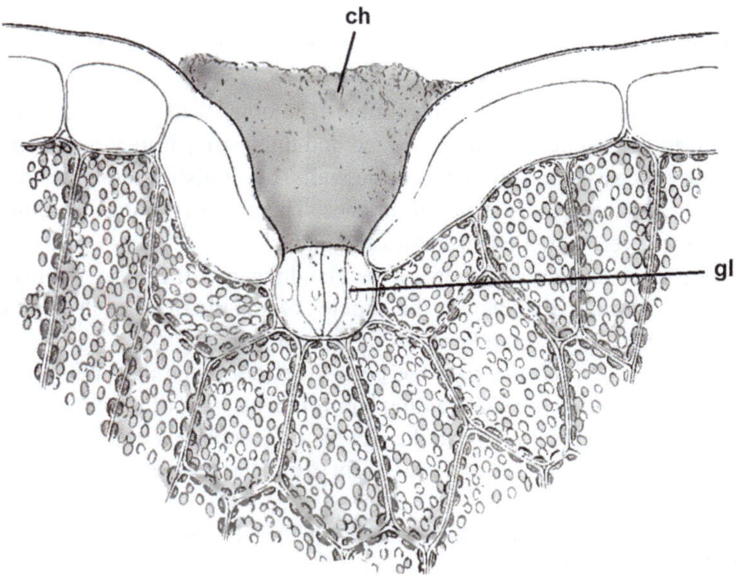

Fig. 5.29 Glands (*gl*) in the lamina of *Statice monopetala* (cross section, magnified image; Woronin 1885)

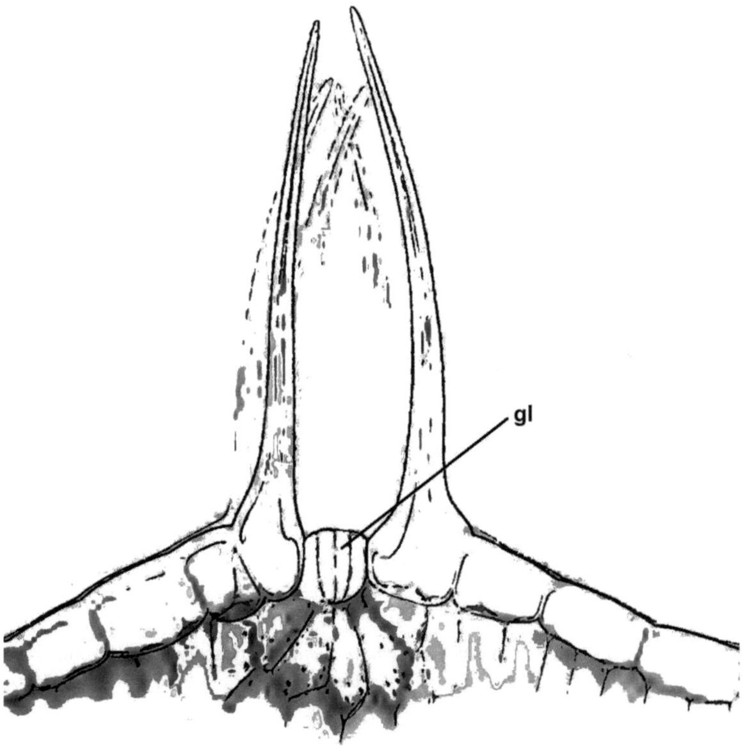

Fig. 5.30 Gland (*gl*) in the lamina of *Statice sareptana* (cross section; Woronin 1885)

epidermal cells are coated by a thick cuticle. The outer wall of the outermost layer of cupuliform cells is also highly cutinized and thus partly insulates the gland. The cutinization process is not limited to these walls, but it also extends, yet to a lesser extent, to the adjacent walls which it touches. The result is a rigid structure below the epidermis level, which encapsulates the gland.

Pores are scattered here and there on the glandular cell; there is generally a single pore (of about 1 mμ in diameter) in the cuticle coating the tip of each secreting cell. The fluid secreted by the gland crosses these small pores.

Nevertheless, the presence of these pores does not mean that the secreting cell cytoplasm is exposed to the surrounding environment; it is still protected by the cell wall. In fact, it has been proven that the chemical composition of the cellulose in the cell wall surrounding the pores is different from the rest of the cellulosic mass (Helder 1956).

Large pores are also found in the walls of the glandular cells adjacent to the assimilating tissue of leaves. The contact with the four large extraglandular cells, also called collecting cells, is possible through these pores. Each of these glandular cells is usually in contact with a few regular mesophyll cells. Ion transport from the mesophyll to the gland is probably the main function of these cells. Glandular cells differ from regular mesophyll cells from the viewpoint of their shape and layout.

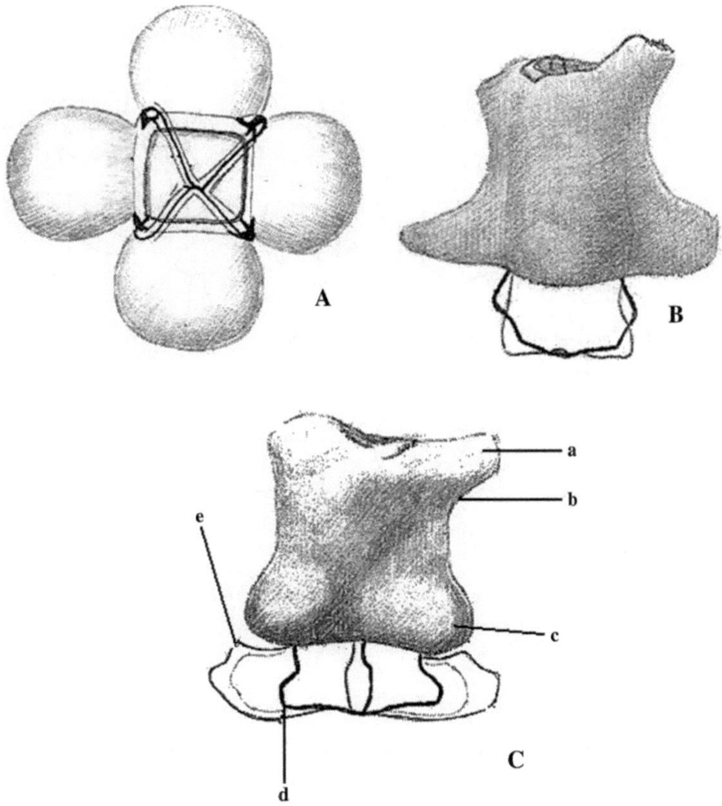

Fig. 5.31 Structure of gland in *Limoniastrum monopetalum*. (**A**) Gland observed in front view, without chalk mass; (**B**) skeleton of gland, without accessory cells; (**C**) **a**, external limit of cutinized frame that forms the edge of internal chamber; **b**, the orifice of chamber in which basis gland opens; **c**, basis of chambers diverticula; **e**, extremity of free side of accessory cells (Vuillemin 1887)

They have densely granulated cytoplasm, a big nucleus, and thin walls. A high number of small vacuoles and various organelles exist instead of a central vacuole; this is especially the case with the four secreting cells. Nuclei are often found near the pores, especially in the locations susceptible of being involved in ion transport. From this point of view, they resemble epidermal cells and may be easily distinguished from the fundamental tissue mass.

In most cases, the glands of the *Plumbaginaceae* species are scattered almost on all the shoots of the plants and especially on the leaves. In *Statice pruinosa*, the number of glands on the stem exceeds the number of glands on the leaves, where they are outnumbered by stomata. Moreover, the glands of this species are located at the top of certain special structures. In *Statice gmelini*, the stomata are about ten times more numerous than the glands (Ruhland 1915).

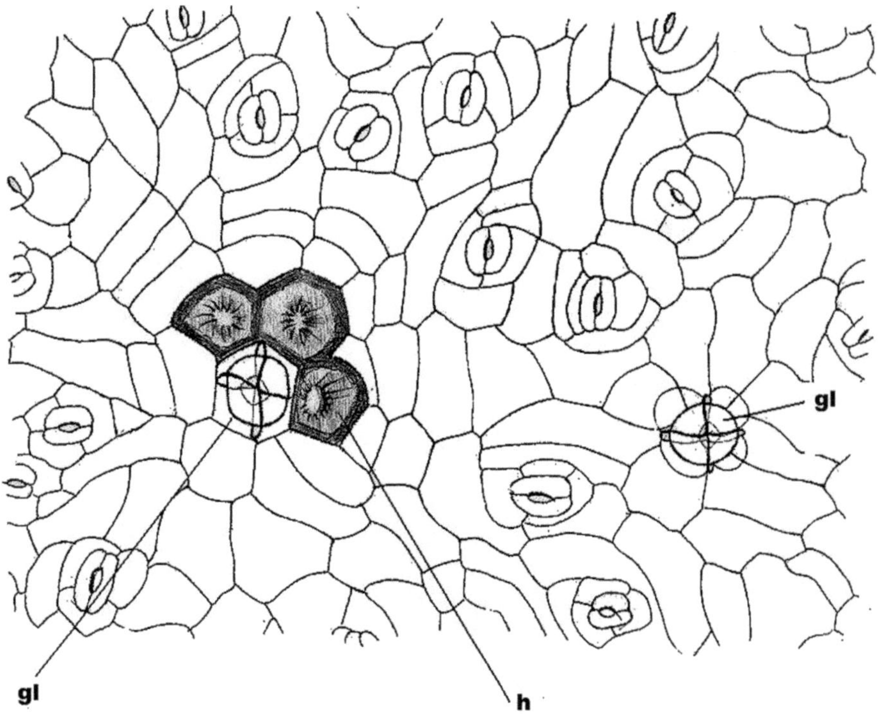

Fig. 5.32 *Statice latifolia*. Epidermis, in surface view (*gl* glands, *h* hairs) (Vuillemin 1887)

Table 5.2 shows the number and distribution of glands in various species belonging to the *Plumbaginaceae* family.

In all the species analyzed by different authors, gland formation is initiated in the early leaf development stages and their differentiation ends much earlier than the differentiation of other foliar tissues. This may suggest their special importance in "organ" development (Helder 1956).

Salt gland cells differ from the surrounding epidermal and parenchyma ones from several points of view. Glandular cells lack a central vacuole, and the number of mitochondria and other organelles is much higher. These traits may suggest that glandular cells do not initially function as "accumulating" organs but rather as transit cells. Salts are carried outside by specific energy-consuming mechanisms, in which energy is produced by the activity of mitochondria.

In some cases, however, ion concentration in glandular cells may be higher than in adjacent cells. There are actually many resemblances between the active transport processes occurring in the glands and those occurring in other organs or tissues.

Ion transport control structures, which are analogous to the Casparian strips in the root endodermis, were also detected in salt glands. Some glandular cells have cutinized and suberized walls, especially those dividing regular parenchyma cells. In some places, the cytoplasm is tightly bound in these strips, just like in Casparian cells.

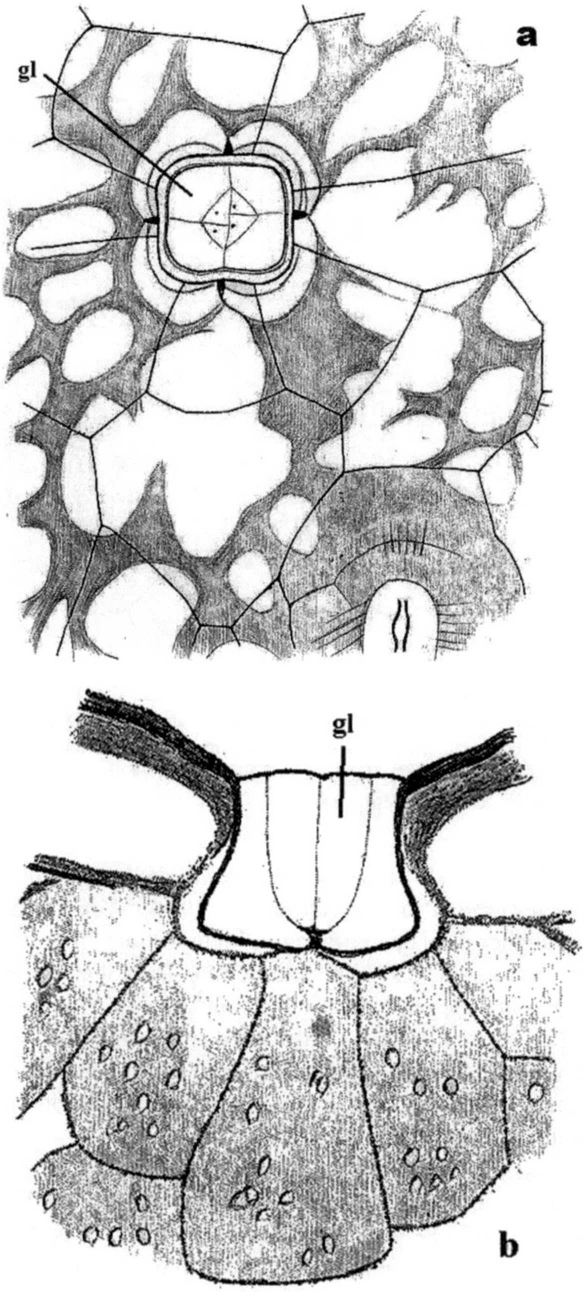

Fig. 5.33 *Statice imbricata*. (**a**) Cuticular network of deep side of epidermis, continued in the proximity of a stoma, (**b**) gland, in cross section, with four secretory cells and two accessory cells; *gl* gland (Vuillemin 1887)

Two types of glands, namely, chalk (salt) and mucilage glands, were also detected in *Limoniastrum monopetalum*, a species growing in the saline and arid regions of Egypt (Batanouny and Abo Sitta 1977).

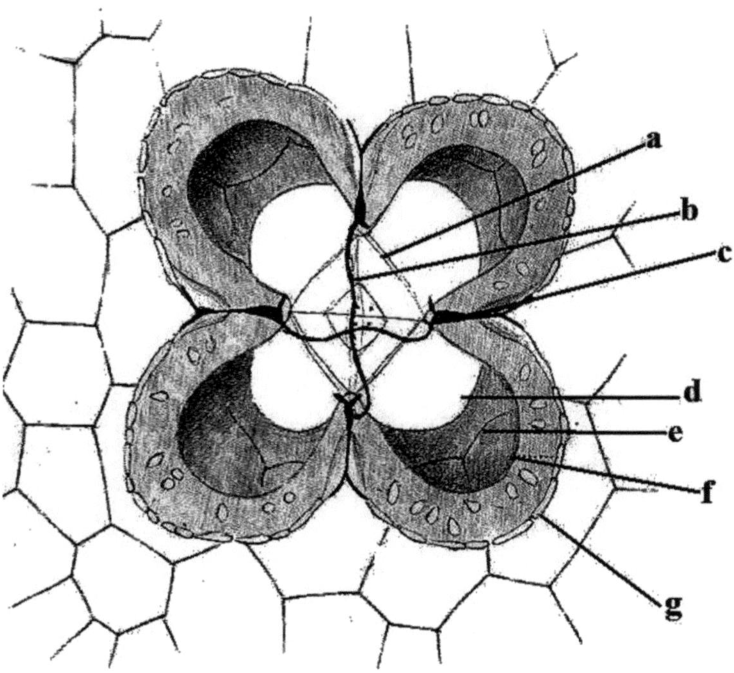

Fig. 5.34 *Limoniastrum guyonianum.* (**a**) Frame that delimitates the free surface of secretory cells, (**b**) cutinized edges that support the gland, (**c**) projection of edges between accessory cells, (**d**) the orifice at which basis the glands opens, (**e**) external limit of cutinized frame that forms the limit of chamber, (**f**) the basis of diverticula of chamber, (**g**) the most external segment of accessory cells (Vuillemin 1887)

Chalk glands occur on both sides of the leaf and on young stems. Their numbers reach 2135/cm^2 on the bottom and 1955/cm^2 on the top of the leaf. These numbers are higher in plants growing in highly saline environments, namely, 2406/cm^2 on the bottom and 1979/cm^2 on the top. The foliar surface varies among individuals living in different habitats: 108 mm^2 in highly saline habitats and 196 mm^2 in less saline ones. In plants transplanted in greenhouses, in low salinity, and adequate humidity conditions, the leaf area is 205 mm^2, and there are 1759 glands/cm^2 on the bottom epidermis and 1373 glands/cm^2 on the top epidermis. Fewer glands than stomata were also detected.

Salt (chalk) glands are located below the epidermal cell level, which seems to facilitate excreted mass retention. Batanouny (1973) concluded that they were made up of 12 cells surrounded by 4 accessory cells ("*Nebenzellen*," in original). The 12 secreting cells are displayed in groups of 4, divided by thick walls in right angles onto one another and perpendicular to the surface.

The four accessory cells surround the secreting cells. A pore surrounded by a transverse aperture at the surface of the leaf is located at the tip of the gland. The 12 glandular cells have extremely thick walls except, to a lesser extent, the walls adjacent to the foot of the gland; glandular cells are deprived of a central vacuole

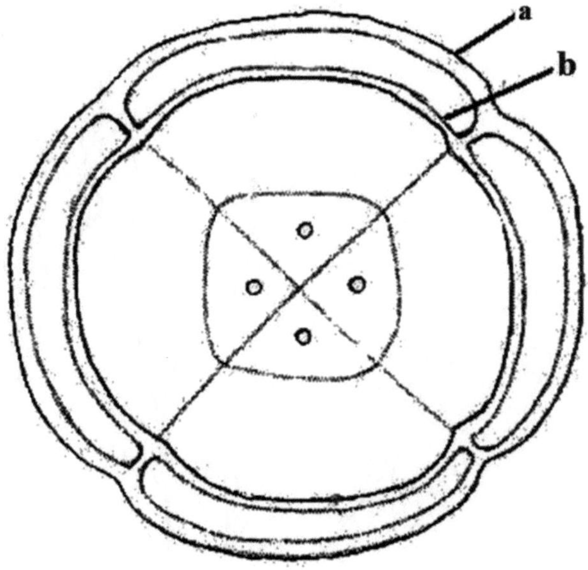

Fig. 5.35 *Statice tatarica*. (**a**) Orifice of excavation in the depth of which the gland opens, (**b**) frame that delimitates the free surface of secretory cells (Vuillemin 1887)

but have granular cytoplasm and a big nucleus. The outer wall of the accessory cells is highly cutinized. The top of the gland and the surrounding epidermal cells are also covered by a fine cuticle.

In the same species, the mucilage glands are scattered at the foot of the foliar sheath, on its top side, and touch the stem. Chalk glands occur on the bottom (outer) side. Mucilaginous glands abound at the foot of the sheath and their number decreases as we move away from it, until they are replaced by salt (chalk) glands. Mucilaginous glands are located above the epidermis level on a base (stand) made up of several solid cells. The number of secreting cells, which may be prism, column, or cone shaped, varies. Glandular cells are surrounded by big non-secreting accessory cells. The gland is outlined by a cutinized layer. In front view, the mucilaginous gland has a circular or oval outline.

Secretion produced by mucilage glands appears on the surface of the leaf in the form of white "*tubers*," which will disappear if they are treated with dilute hydrochloric acid. By means of these glands, the excess of salts from plant organs is eliminated. The presence of these secretions on the surface of organs involved in transpiration could be a mechanism by which plant reduces water consumption.

In Romania, salt glands were analyzed by Moţiu et al. (1987) in *Limonium gmelinii* (Fig. 5.38) and Grigore and Toma (2010). Moţiu et al. (1987) showed that the epidermis of the shoots incorporates Licopoli "*organs*" (playing a role in calcium carbonate secretion), represented by octo-cellular complexes, and very scarce mucilage glands.

Salama et al. (1999) investigated salt glands in several halophytes from Egypt and gave very accurate descriptions for salt glands of *Limoniastrum monopetalum* (Fig. 5.39), *Limonium pruinosum* (Fig. 5.40), and *Limonium axillare*(Fig. 5.41).

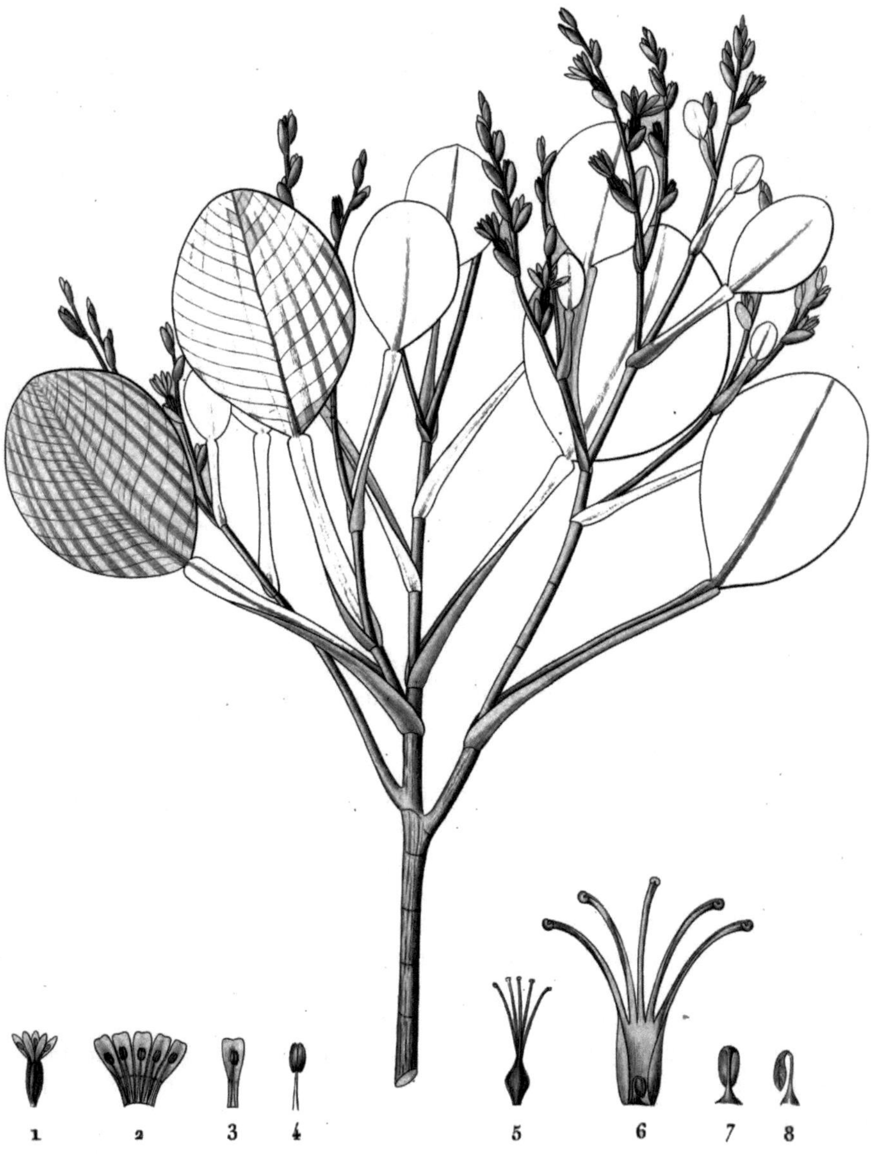

Fig. 5.36 *Aegialitis annulata*, mangrove plant (1), complete flower (2), corolla formed of five petals, fused at the basis (3), petal (4), stamen (5), ovary, style, and stigma (6), expanded ovary, revealing the ovules (7, 8) (de Freycinet and Gaudichaud 1826)

Grigore et al. (2014) found salt glands in several *Limonium* species, collected from Spain, *L. furfuraceum, L. girardianum, L. narborense*, and in *L. gmelinii*, collected from Romania (Figs. 5.42, 5.43, 5.44, and 5.45) (Grigore and Toma 2010).

Fig. 5.37 Structure of the glands of *Aegialitis annulata*. (**a–c**) Chalk glands, (**a**) cross section, (**b**) surface view from outside, (**c**) surface view, from within, (**d–f**) mucilage glands, (**d**) cross section, (**e**) surface view from outside, (**f**) surface view from inside (Solereder 1908)

Table 5.2 Number of salt glands/cm^2 in different species of the *Plumbaginaceae* family

		Leaf			
Family	Species	Upper epidermis	Lower epidermis	Stem	References
Plumbaginaceae	*Statice gmelini*	722	644		a
	S. bellidifolia	960	830		b
	S. binervosa	750	1240		b
	S. sinuata	700	1200		c
	S. graeca	1100	900		c
	S. pruinosa	1900	1300	4000	c
	Limonium latifolium	3300	2900		c
	Plumbago capensis	2900	2100	300	c
	P. europaea	1500			c
	Limonium vulgare	3066 ± 272	2952 ± 246		e
	Limoniastrum monopetalum	1979	2406		d
	Armeria maritima	565 ± 37	548 ± 35		e

(a) Ruhland (1915), (b) De Fraine (1916), (c) Waisel (1972), (d) Batanouny and Abo Sitta (1977), (e) Rozema and Gude (1981)

5.2 Tamaricaceae

Other salt glands that have held attention, in addition to those of *Plumbaginaceae* species, are those that occur in ***Tamarix*** (*Tamaricaceae*) species, and in particular *Tamarix aphylla* (Brunner 1909; Paulsen 1912; Warming and Graebner 1914; Decker 1961; Campbell and Strong 1964; Thomson and Liu 1967; Fahn 1967; Waisel et al. 1966; Shimony and Fahn 1968; Thomson et al. 1969; Campbell et al. 1974).

Usually, species from *Tamaricaceae* have salt glands with six secretory cells (Figs. 5.46 and 5.47) (Paulsen 1912; Volkens 1887).

In *Tamarix aphylla*, salt gland consists of six secreting cells (with dense cytoplasm and of two extraglandular, collecting cells, intensely vacuolated. Secreting cells are provided with a cuticular "*sheath*," except for some portions of cellular walls, which are involved in connecting with collecting cells. Through these portions, plasmodesmata pass and connect cytoplasm from two groups of cells. These portions of cellular walls, which are penetrated by plasmodesmata, form what is called transfusion area (Fahn 1988).

In *Tamarix pentandra*, the gland is formed from six secreting cells, with granular cytoplasm and big nucleus, and two internal collecting cells, highly vacuolated; in this case, gland originates from a single proto-dermis cell (Campbell and Strong 1964).

Vesque (1883) studied the histology of leaves of species from *Caryophyllales*; he described the foliar anatomy of several species from *Tamaricaceae*. Interestingly, he did not explicitly designate the glands with this term, as we know it

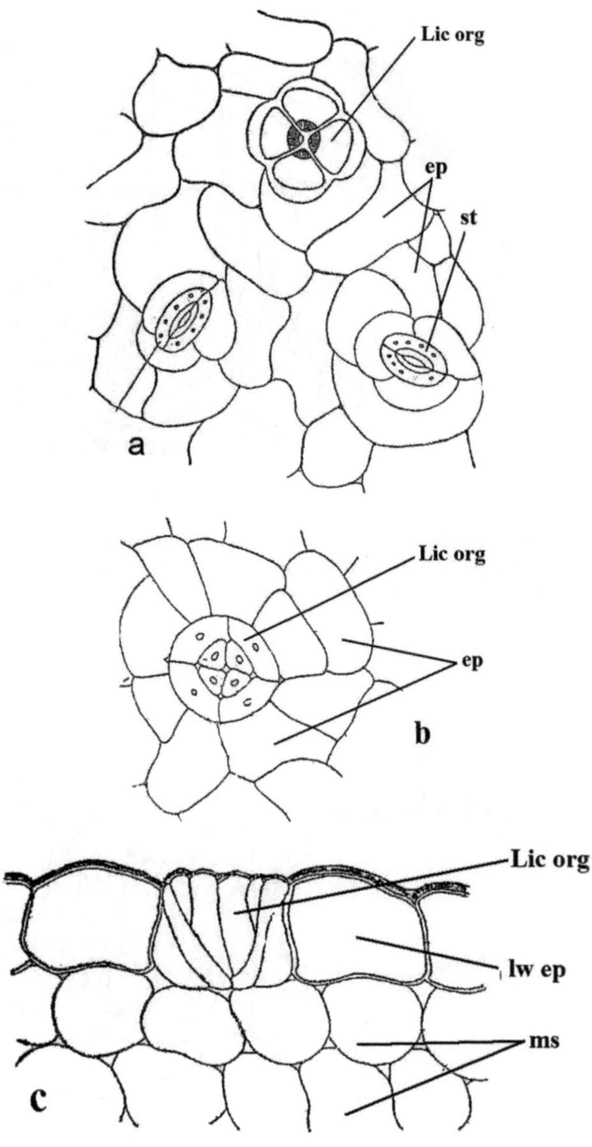

Fig. 5.38 Salt glands in lamina of *Limonium gmelinii*. (**a**) Lower epidermis, surface view (**b**) Licopoli "organ" (**c**) cross section through the lamina; *ep* epidermis cells, *Lic org* Licopoli "organ," *ms* mesophyll, *lw ep* lower epidermis, *st* stomata (Moţiu et al. 1987)

nowadays in plant anatomy; instead, he uses the terms: *bicellular glands (consisting of two collateral cells)* or *sessile glandular hairs, located in the depth of cavities*. But from his drawings of investigated species, as well from the text explanations, it is easily understood that there are actually proper salt glands (see Paulsen's explanations, 1912). For instance, he said that *R. hyrcanica* has "*glandular hairs, deeply sunken in large cavities from epidermis*"; comparing this with the given drawing of the same species, we can conclude that he actually refers to salt glands.

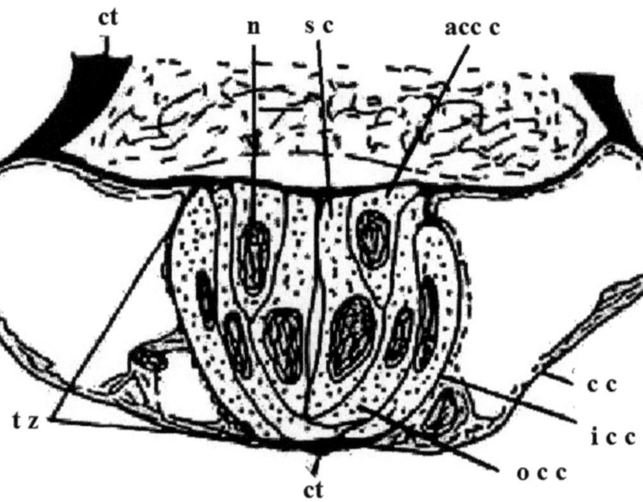

Fig. 5.39 Salt gland of *Limoniastrum monopetalum* (*ep* epidermis, *ct* cuticle, *c c* collecting cells, *i c c* inner cup cells, *o c c* outer cup cells, *s c* secretory cells, *n* nucleus, *acc c* accessory cells, *t z* transfusion zone) (Salama et al. 1999)

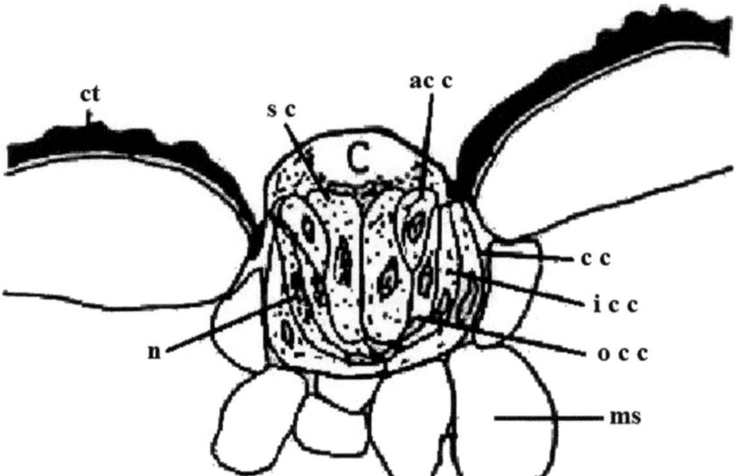

Fig. 5.40 Salt gland of *Limonium pruinosum* (*ct* cuticle, *c c* collecting cells, *i c c* inner cup cells, *o c c* outer cup cells, *s c* secretory cells, *n* nucleus, *acc c* accessory cells, *C* collecting compartment, *ms* mesophyll) (Salama et al. 1999)

This is available for *Tamarix gallica* (Fig. 5.48), *T. articulata* (Fig. 5.49), *Reaumuria persica* (Fig. 5.50), *R. hyrcanica* (Fig. 5.51), and *R. oxiana* (Fig. 5.52).

The structure of salt gland in *Reaumuria oxiana* (Paulsen 1912) is very similar to that of the glands figured by Volkens (1887) from *R. hirtella*. Volkens (1887) was of

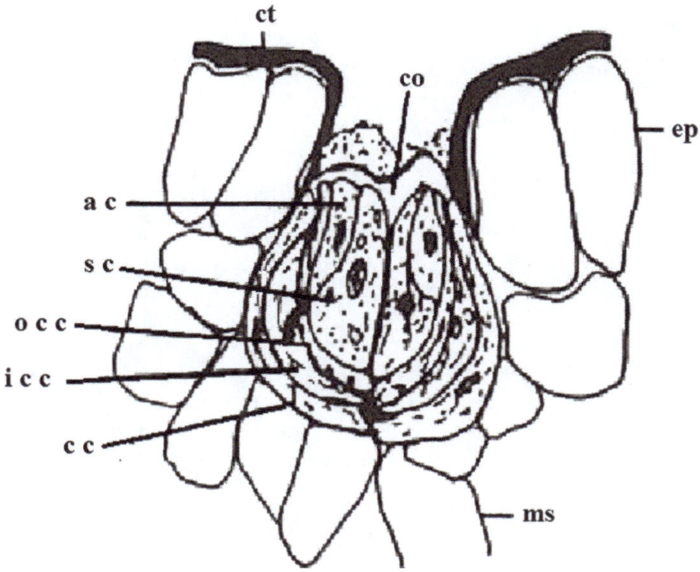

Fig. 5.41 Salt gland of *Limonium axillare* (*ct* cuticle, *c c* collecting cells, *i c c* inner cup cells, *o c c* outer cup cells, *s c* secretory cells, *acc c* accessory cells, *co* collecting compartment, *ms* mesophyll) (Salama et al. 1999)

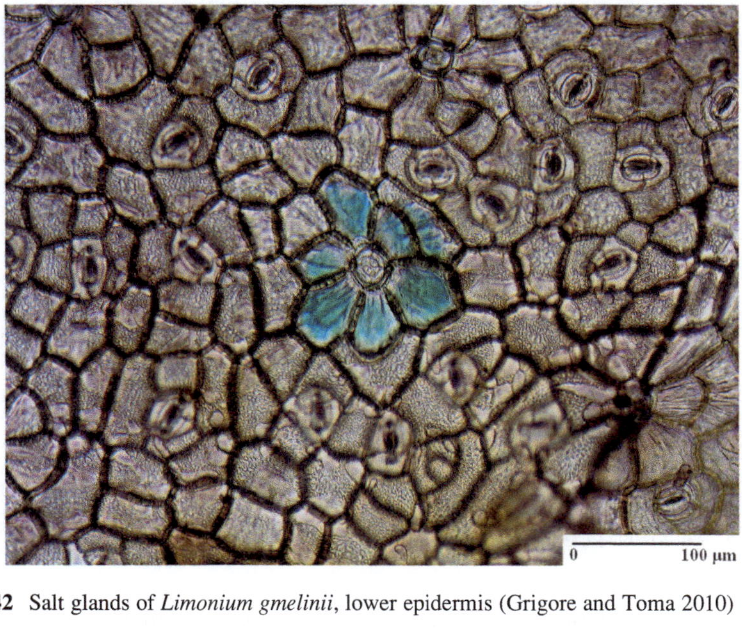

Fig. 5.42 Salt glands of *Limonium gmelinii*, lower epidermis (Grigore and Toma 2010)

Fig. 5.43 Salt glands of *Limonium gmelinii*, lower epidermis (Grigore and Toma 2010)

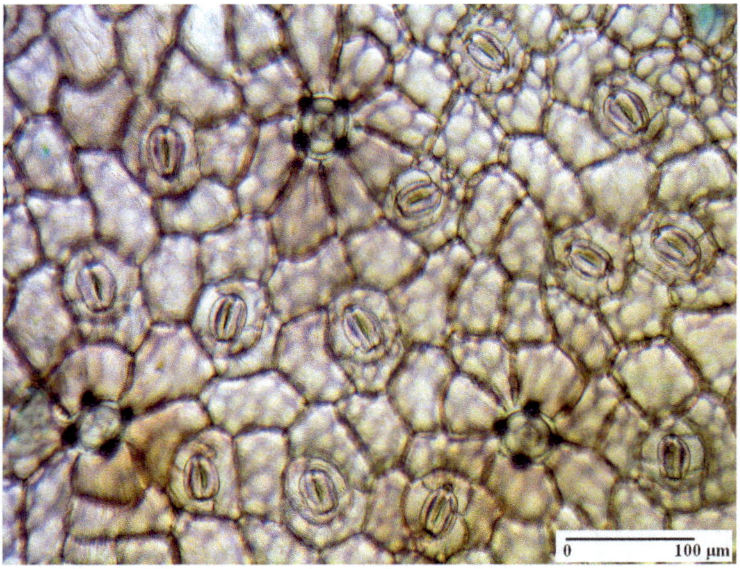

Fig. 5.44 Salt glands of *Limonium gmelinii*, upper epidermis (Grigore and Toma 2010)

the opinion that during the night the excreted salts absorbed water from the atmosphere (dew), which might then be absorbed by the gland and thus be used by the plant.

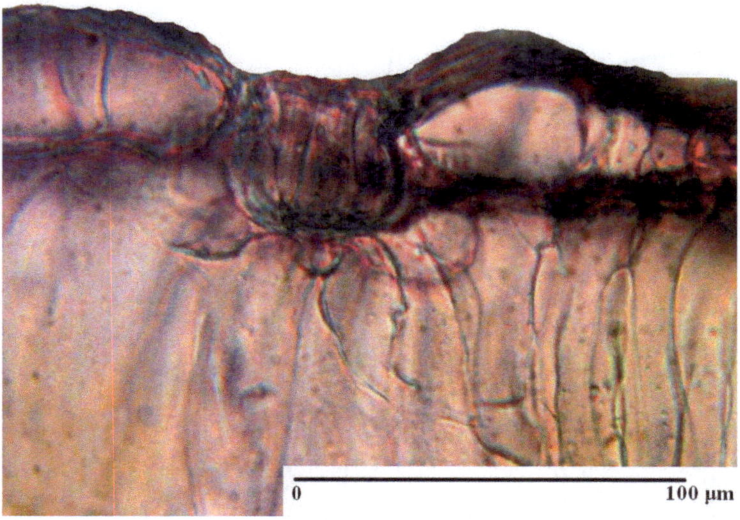

Fig. 5.45 Salt glands of *Limonium gmelinii*, cross section through the lamina (Grigore and Toma 2010)

Marloth (1887) rejected this opinion and states that it is impossible for the glands of the leaf to absorb water from the surface without at the same time absorbing the salts.

On the contrary, the salt solution on the surface must absorb water from the gland, and according to impressive study of Fitting (1911), this is what takes place. And still more important, Fitting concluded that plants in the desert store salt internally up to a certain specific maximum varying for different species, and that in this physiologically determined limitation of salt storage, they have a sufficient means for securing the high osmotic pressure which Fitting has pointed out in the desert plants and which enables them to obtain water from the soil. In addition to sodium chloride, there is also excretion of carbonate of lime often in great quantities, so it is not at all certain that the excretion of sodium chloride is of any special importance.

Myricaria germanica (Fig. 5.53) is another species of the *Tamaricaceae*, where salt glands (Fig. 5.54) were evidenced by Vuillemin (1887); he has described as consisting of a pair of secretory cells and one pair of accessory cells, separated by the first by oblique walls, which isolates them from the leaf parenchyma and epidermis.

In Fig. 5.55, salt drops secreted by salt glands of *Tamarix canariensis* are noticeable (indicated by black arrows—original picture, salt marsh from Alicante, Spain, 2010).

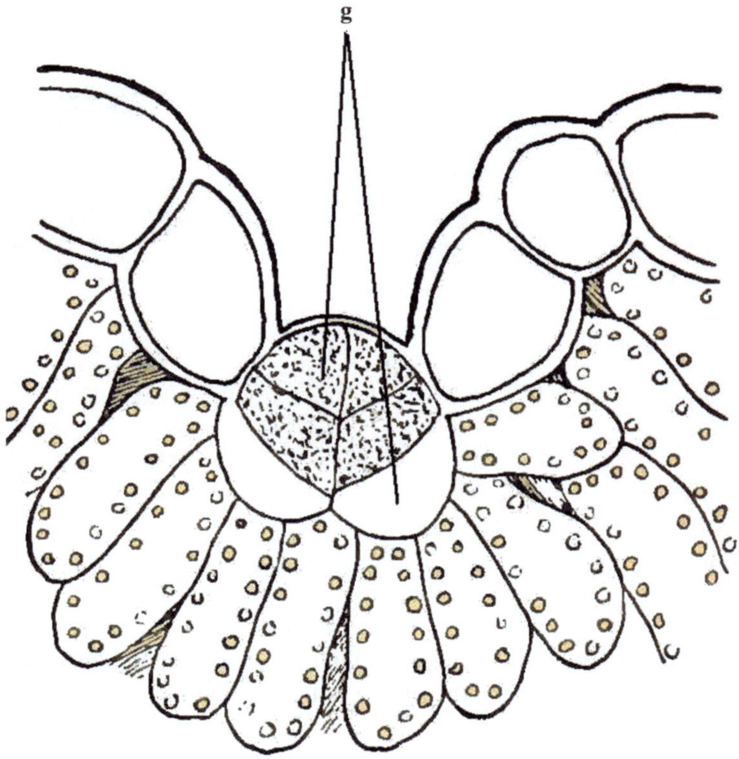

Fig. 5.46 Salt gland (*g*) in the leaf of *Tamarix mannifera* (Volkens 1887)

5.3 Frankeniaceae

Perhaps less studied than the *Limonium* or *Tamarix* glands were those of *Frankenia* species. As in the case of salt glands from *Plumbaginaceae*, salt secretion in *Frankeniaceae* halophytic species has been suggested a long time before the description structure of salt glands; for instance, Deslongchamps (1820) described the leaves of *Frankenia pulverulenta* as being covered *by dust*, which indicates, more likely, the presence of salt deposits on their surface.

They have been included by Thomson (1975) in the same category as those of *Tamarix* and *Limonium,* namely glands that are completely surrounded by a cuticular layer. Vuillemin (1887) concluded that the initial epidermal cell giving rise to glandular cells undergoes a single division perpendicular to the surface of the epidermis, rather than two, as in the case of *Plumbaginaceae* species. Each cell is divided by an oblique wall, resulting in one secretory and one accessory cell. The two secretory cells are separated by a thin wall; accessory cells have thicker walls and more pits than the cells of the epidermis. This thickening is maintained even at the oblique wall that separates them from the glandular cells. Accessory cells continue directly to the epidermal cells, with whose structure resembles; they are

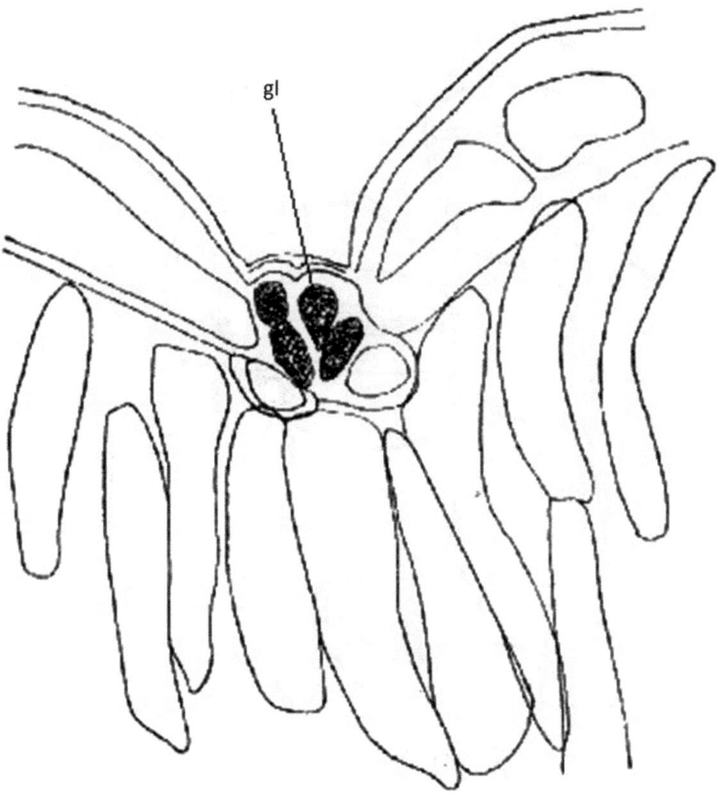

Fig. 5.47 Salt gland (*gl*) in the leaf of *Reaumuria oxiana* (Paulsen 1912)

only slightly straighter and placed at a slightly deeper level (Fig. 5.56). In cross section, they have a shape of a triangle with convex edges. At both ends, the cells are widely applied to each other, but in the middle are in contact only in their basal part (Fig. 5.57).

Seen from the front, these glands are very much like some stomata (Figs. 5.58 and 5.59). However, they are distinguished from these by accessory cells that are well noticeable and by larger sizes of glandular cells compared with guard cells.

In *Frankenia floribunda*, there is an anomaly, in the sense that two glands have fused with one of their ends, in such a way that their openings are placed side by side. The fusing line has become thickened and gathered the organization of normal accessory cells.

Glands are widespread in regions where palisade tissue develops, while stomata are in the right of spongy tissue.

Solereder (1908) also evidenced salt glands in *Frankenia pulverulenta* (Fig. 5.60) and gave them an accurate description.

Salt glands of *Frankenia grandiflora* have been investigated by Campbell and Thomson (1976), in plants grown both in normal conditions and under the influence

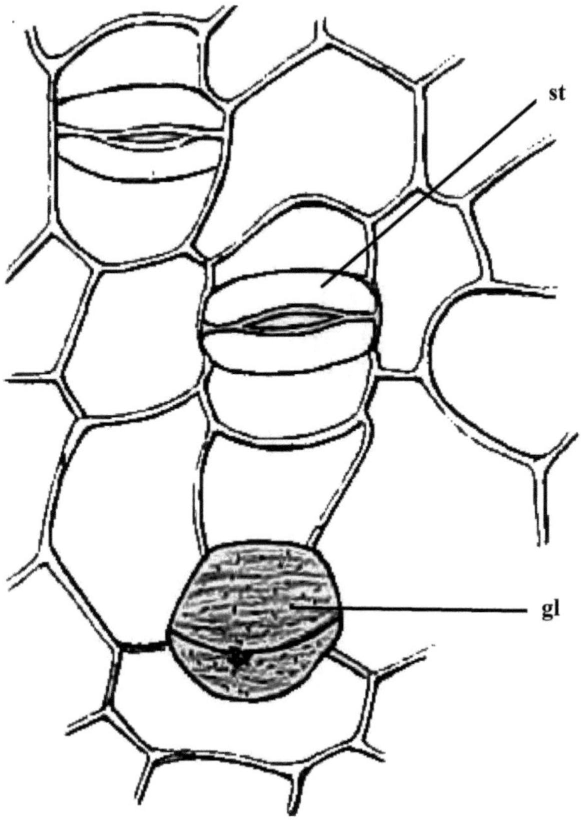

Fig. 5.48 Glands (*gl*) from the epidermis of lamina of *Tamarix gallica* (*st* stomata) (Vesque 1883)

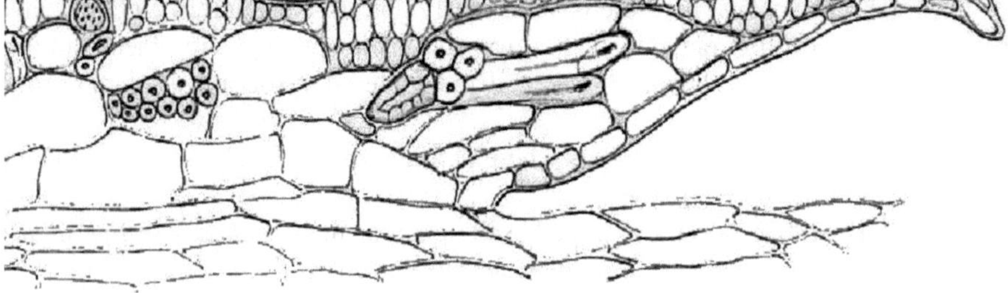

Fig. 5.49 Glands (*gl*) from the epidermis of lamina of *Tamarix articulata* (Vesque 1883)

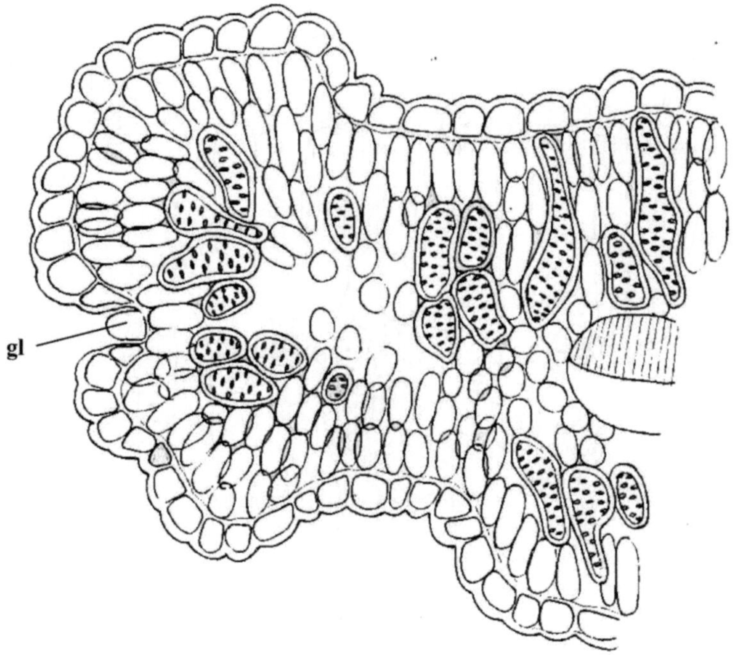

Fig. 5.50 Cross section through the lamina of *Reaumuria persica* (*gl* gland) (Vesque 1883)

of NaCl treatment. These glands consist of a bilateral complex of six secretory cells. The entire complex of cells is almost completely covered by cuticle. Only a part of the periphery of the gland lacks a cuticle and represents a portion of the wall between each inner secretory cell and neighboring cells of mesophyll, which form, as we mentioned, the transfusion zone. In this species, this area has many plasmodesmata, though seem to be less abundant than in *Tamarix* (Thomson and Liu 1967). The most characteristic aspect of the area of transfusion in this species is the bulbous enlargements.

Chermezon (1910) evidenced salt glands in *Frankenia laevis* (Fig. 5.61).

Paulsen (1912) described the salt glands of *Frankenia hirsuta* (Fig. 5.62), specifying that they are located on both sides of lamina.

Salama et al. (1999) evidenced salt glands in *Frankenia revoluta* and gave a very accurate description of them (Fig. 5.63).

Grigore et al. (2014) found salt glands in Mediterranean species *Frankenia laevis* (Fig. 5.64).

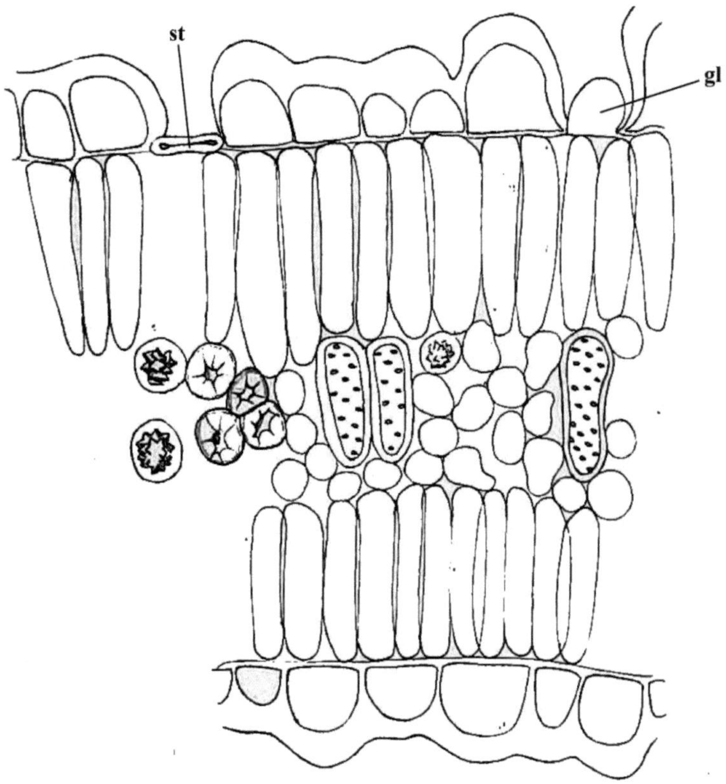

Fig. 5.51 Cross section through the lamina of *Reaumuria hyrcanica* (*gl* glands, *st* stomata) (Vesque 1883)

5.4 Salt Glands of Mangroves

Mangrove species have salt glands that represent a major mechanism allowing these species to deal with elevated and changeable concentrations of salt.

A plethora of literature treated the botany of mangroves in general (Pora 1969; Blasco 1991; Tomlinson, 1995; Packham and Willis 1997; Dawes 1998; Lal 2002; Saenger 2002; Lüttge 2002, 2008; Hogarth 2007) and salt glands in particular; the latter has been reviewed by Grigore and Toma (2010).

It is well known that mangroves species exhibit a series of strategies for or tolerance of salinity stress (Feller and Sitnik 1996a, b):

Avoidance Strategies

1. Exclusion of salts by the plant roots
2. Excretion of salts from salt glands in the leaves
3. Dilution of salts by increased water content in tissues (succulence)
4. Elimination of salt-saturated organs

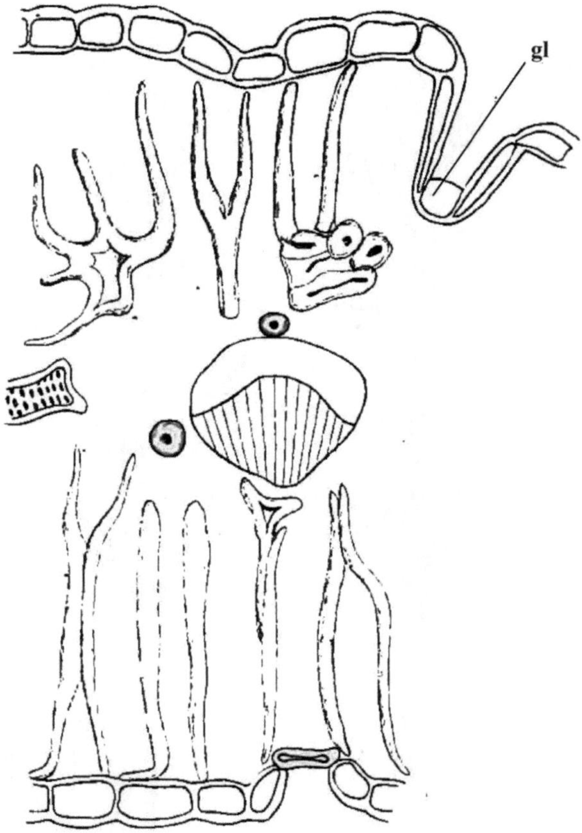

Fig. 5.52 Cross section through the lamina of *Reaumuria oxiana* (*gl* glands) (Vesque 1883)

Tolerance strategies

1. Compartmentalization of salts in the vacuole—removes toxic ions from metabolically active portions of the cell
2. Synthesis of organic (compatible) solutes—to balance inorganic ions in the vacuoles

Structural/Anatomical Modifications to a Saline Environment

1. Stomata on lower leaf surface—decreases water loss from plant
2. Thickened cuticle on leaf surface—decreases water loss
3. Salt glands in leaf epidermis

Secretory structures in mangrove species have been generally discussed by Waisel (1972), Fahn (1988), Tomlinson (1995), Dawes (1998), Kathiresan and Bingham (2001), Balasubramanian and Khan (2002), Lüttge (2002), Hogarth (2007), Koyro and Lieth (2008); many authors refer to salt glands of *Avicenniaceae* representatives: Wilie (1883), van Tieghem (1898), van der Bakhuizen (1921),

Fig. 5.53 *Myricaria germanica*. (**a, h**) Branches with leaves (**b**) branch with fruits (**c**) flower (**d**) floral diagram (**e**) opened capsule (**f**) capsule free of hairs (**g**) haired seed (**i**) magnified leaf (Andersson 1849)

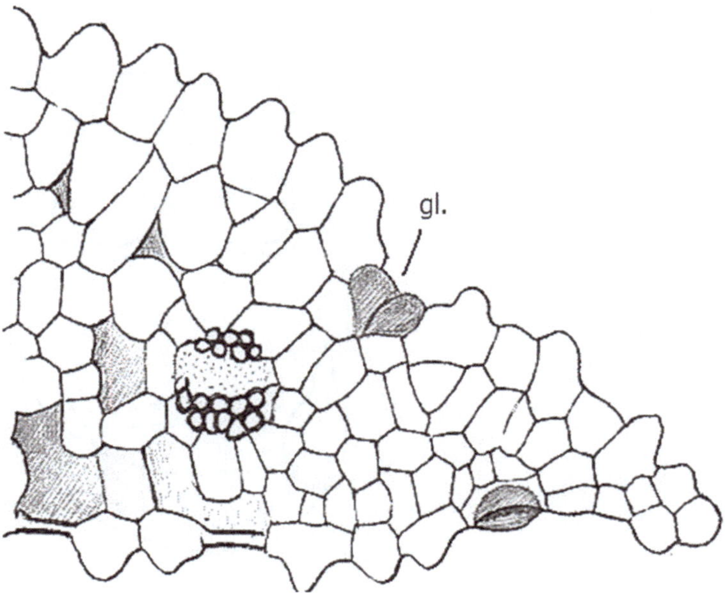

Fig. 5.54 Cross section through the lamina of *Myricaria germanica* (*gl* gland) (Vuillemin 1887)

Fig. 5.55 Salty drops secreted by salt glands of *Tamarix canariensis* (original photo, salt marsh from Alicante, Spain)

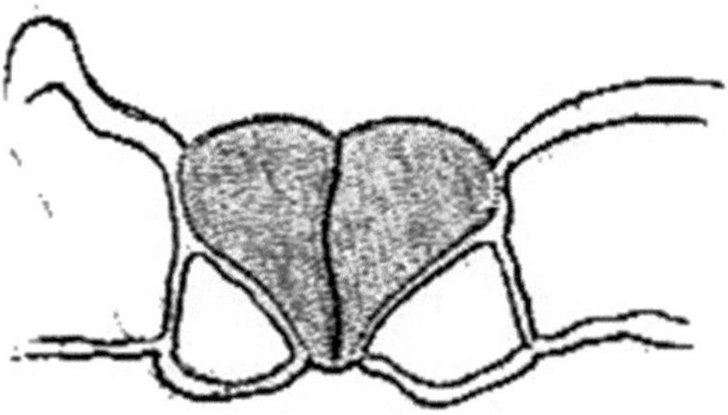

Fig. 5.56 Salt gland of *Frankenia ericifolia*

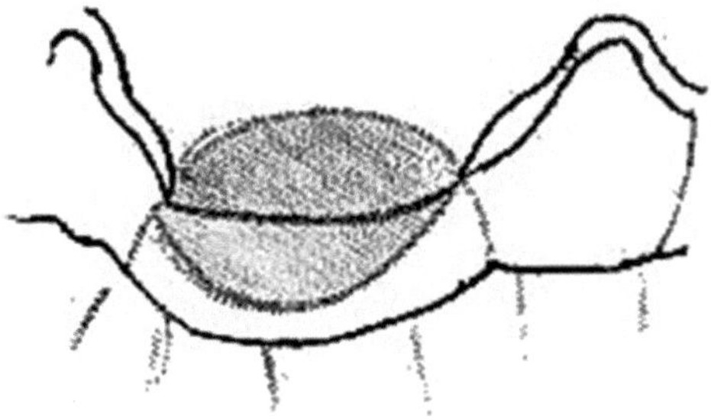

Fig. 5.57 Salt gland of *Frankenia laevis* (Vuillemin 1887)

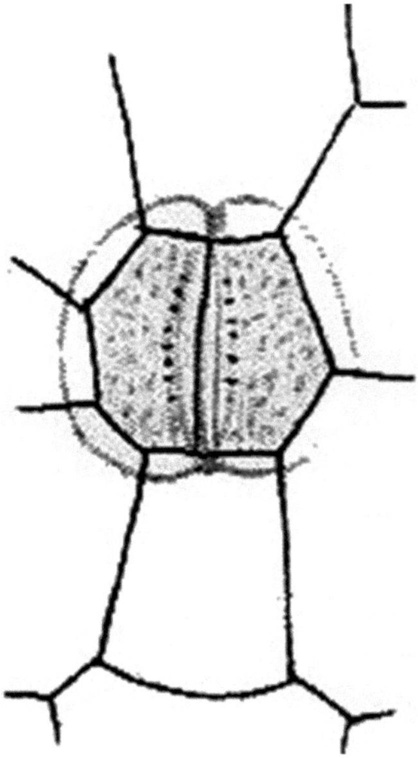

Fig. 5.58 Salt gland of *Frankenia capitata*

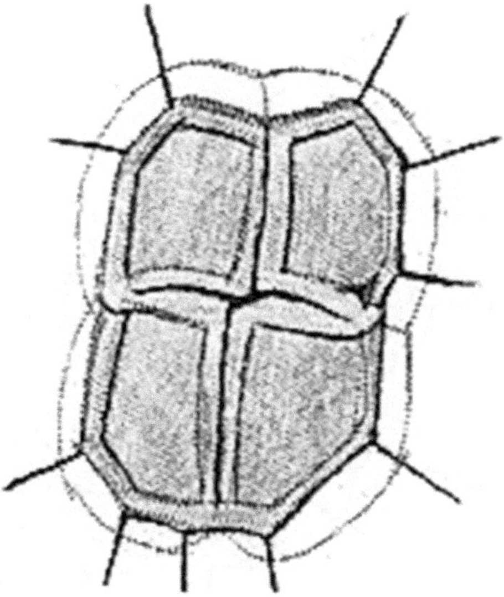

Fig. 5.59 Salt gland of *Frankenia floribunda* (surface view) (Vuillemin 1887)

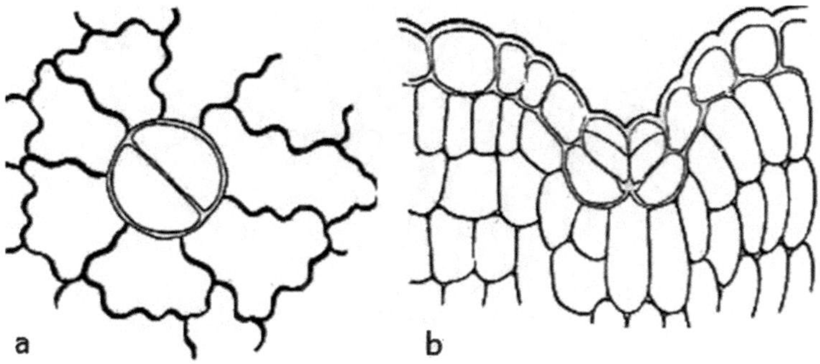

Fig. 5.60 Salt glands of *Frankenia pulverulenta*. (**a**) Surface view and (**b**) cross section (Solereder 1908)

Mullan (1931), Baylis (1940–1941), Shimony et al. (1973), Fahn and Shimony (1977), Drennan and Berjak (1982), Drennan and Pammenter (1982), Carpenter (1983), Drennan et al. (1987), Ish-Shalom-Gordon and Dubinsky (1990), Smith et al. (1989), Fitzgerald and Allaway (1991), Fitzgerald et al. (1992), Deschida et al. (1992), Das (2002), Nandy (Datta) et al. (2005), and Griffiths et al. (2008).

In *Avicennia marina* (Fig. 5.65), the glands consist of an indefinite number of cells (usually five to nine), arranged in a group of four or more cells located at the top of a pedicel and two to four collecting cells (Waisel 1972).

Fahn (1988) found that the glands of *Avicennia marina* are formed of two to four collecting cells, a disk (pedicel cell), and, usually, eight, and sometimes 12 secretory cells, radially arranged.

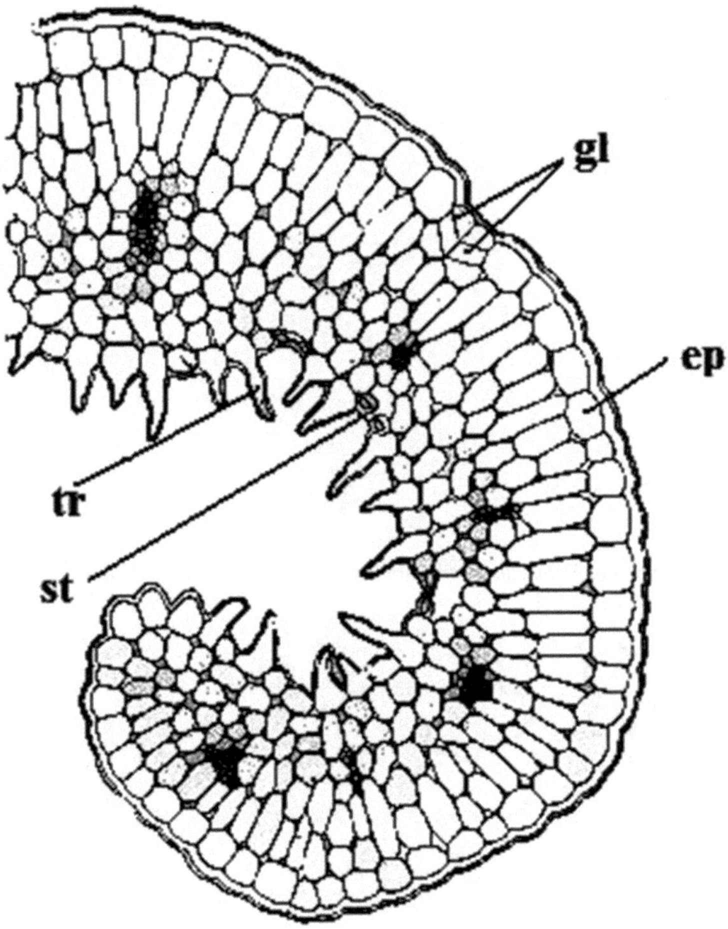

Fig. 5.61 Cross section through the lamina of *Frankenia laevis* (*ep* epidermis, *gl* gland, *tr* trichome, *st* stomata) (Chermezon 1910)

It is known that mangrove species can be divided into two groups based on their mechanism of resistance to salt: those that eliminate salt at the root level and those that accumulate it (Levitt 1972). *Avicennia marina* has a relatively high concentration of salts in the sap: about 24 mM (Scholander et al. 1966), and the salt excess is removed from the plant through the removal of salts at the surface of the leaf. Scholander (1968) showed that salt removal from root level is a process that does not depend directly on the respiration metabolism; he called this phenomenon ultrafiltration. He identified a similar system in the leaves, whose functioning depends only on the integrity of semipermeable membranes. It is generally accepted that the excretion in *Avicennia* is attributed to glandular structures of the leaf surface. However, it seems that only young leaves possess such glands; in mature leaf, they degenerate (Drennan and Berjak 1982).

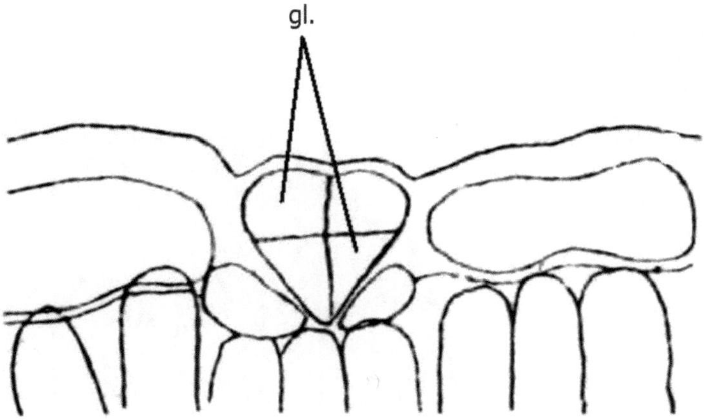

Fig. 5.62 Salt gland (*gl*) of *Frankenia hirsuta* Paulsen (1912)

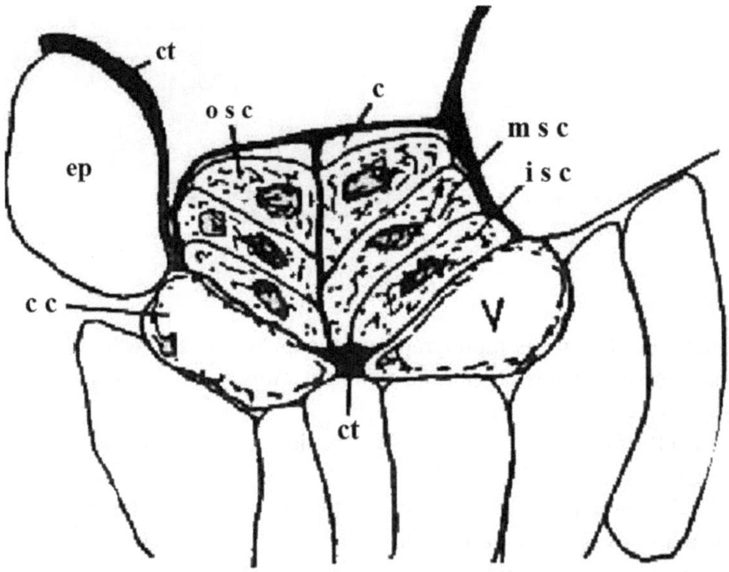

Fig. 5.63 Salt gland of *Frankenia revoluta* (*ct* cuticle, *c c* collecting cells, *ep* epidermis, *I s c* inner secretory cells, *o s c* outer secretory cells, *m s c* middle secretory cells, *c* collecting compartment, *v* vacuole) (Salama et al. 1999)

Baylis (1940–1941) evidenced salt glands (nominated by him as glandular hairs) in *Avicennia officinalis* (Figs. 5.66, 5.67, and 5.68).

Das and Ghose (1996) studied the anatomy of several Indian mangrove species. They evidenced glands (glandular hairs) in *Avicennia alba* (Fig. 5.69) and *Acanthus ilicifolius* (Fig. 5.70).

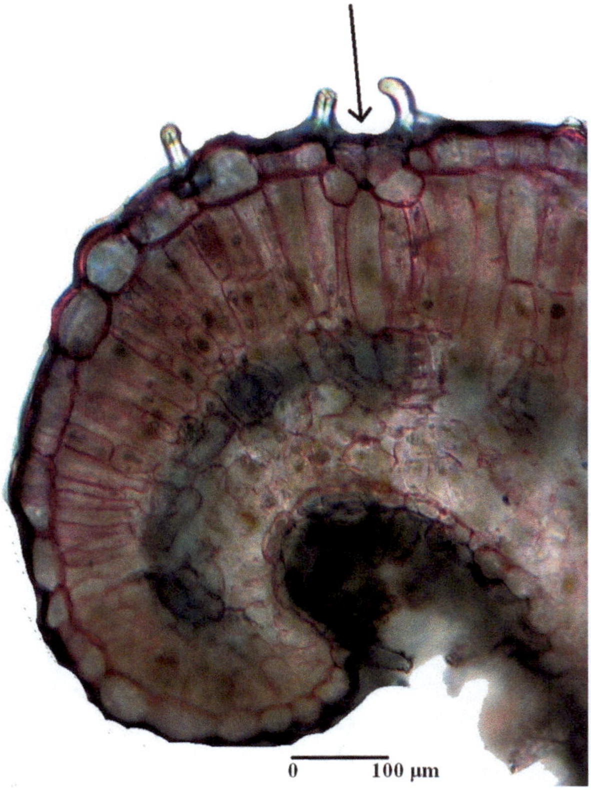

Fig. 5.64 Cross section through the lamina of *Frankenia laevis* (*black arrow* indicates the localization of salt gland) (Grigore et al. 2014)

Tomlinson (1994) in his monograph on mangrove botany described glands in several mangrove species: *Aegialitis annulata, Aegiceras corniculata, Acanthus ilicifolius,* and *Avicennia marina.*

Schmidt (1905) gave a detailed description of salt glands of *Aegiceras corniculata* (Figs. 5.71, 5.72, and 5.73).

Das (2002) studied the ontogenesis of stomata and glandular hairs of several species of mangroves in India. Glandular hairs only occur on the adaxial side of *Acanthus ilicifolius* (Fig. 5.74) (*Acanthaceae* family) and *Aegialitis rotundifolia* (*Plumbaginaceae* family) leaves.

In *Acanthus* (Fig. 5.74, top), mature glandular hairs are pear shaped, whereas *Aegialitis* hairs (Fig. 5.74, bottom) are located in cup-shaped cavities (pits). In both species, mature hairs basically include four to eight terminal cells in radial layout, two stalk cells, and one basal cell. The primary hair cell differs from the other epidermal cells by its bigger size, well-defined nucleus, and a high number of vacuoles. In cross section, the primordium juts out from the layer of epidermal cells in *Acanthus*, whereas in *Aegialitis* it remains at the same level as the pit formed by the epidermal cells. In both species, the first primordium division occurs transversely and forms the terminal and basal cells. The terminal cell undergoes a second

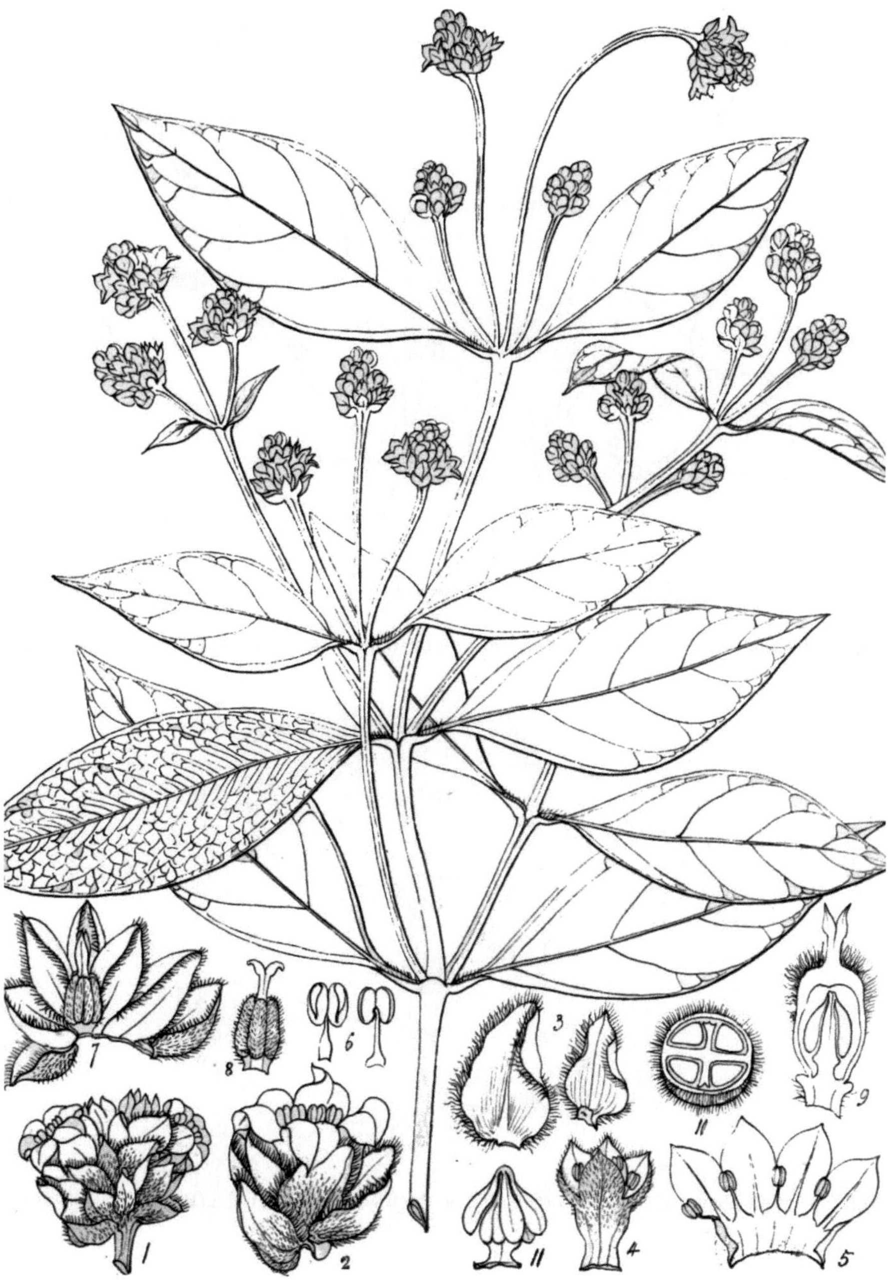

Fig. 5.65 *Avicennia marina* (Wight 1850)

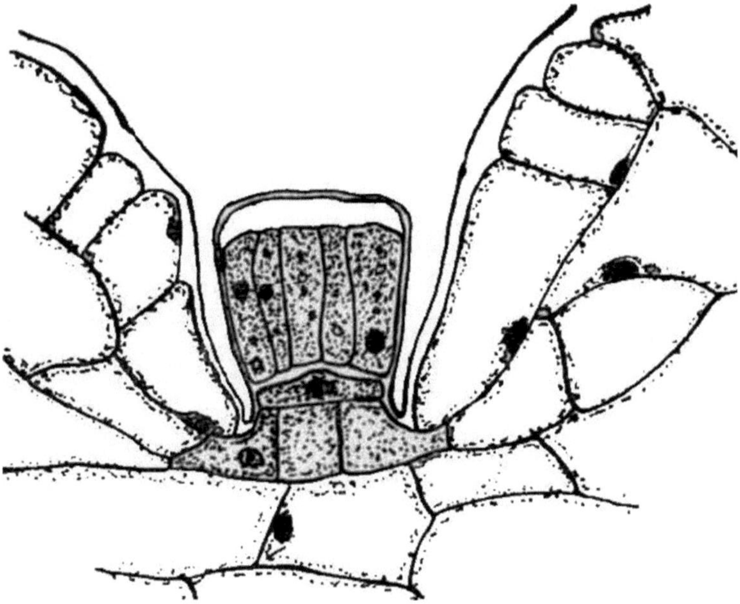

Fig. 5.66 Salt gland of *Avicennia officinalis*, on the upper side of lamina (Baylis 1940–1941)

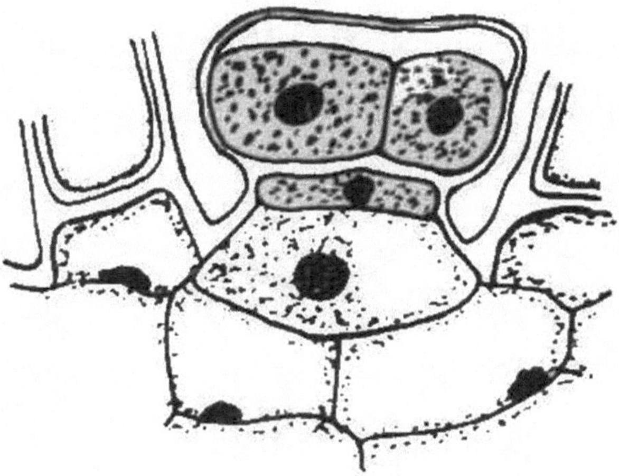

Fig. 5.67 Salt gland of *Avicennia officinalis*, on the lower side of lamina (Baylis 1940–1941)

transverse division, thus leading to the formation of a terminal cell and of a stalk cell, whereas the basal cell remains undivided.

In *Acanthus,* the stalk cell is divided transversely once and generates two stalk cells, whereas, in *Aegialitis*, the third division occurs both in the stalk cell and in the terminal cell, longitudinally and in a straight angle on each side, thus forming two stalk cells and two terminal cells. The stalk cell then remains undivided and the

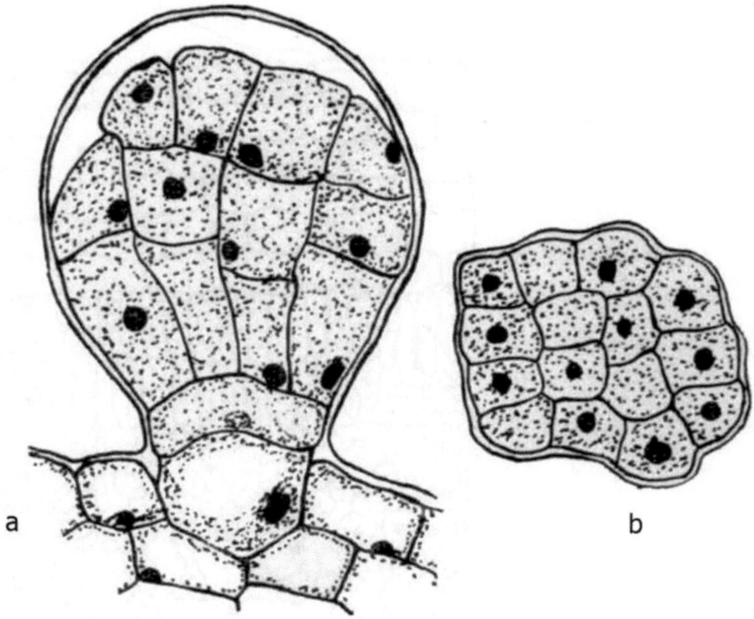

Fig. 5.68 Salt gland of *Avicennia officinalis*, located at the basis petiole (Baylis 1940–1941)

terminal cell undergoes a sequence of longitudinal divisions, thus forming eight terminal cells with a final radial layout. The glandular hair stays "embedded" into a continuous cutinized layer. In front view, the terminal cells of both species have a radial layout.

The presence of salt-secreting structures in very different taxa from the viewpoint of their belonging to a family or another may be the proof of a convergence: although they are heterogeneous from many points of view, different species possess the same adaptation mechanisms to a high salt content. It is most tempting to determine the exact implications of these structures in the life of a plant, the environmental advantage (if any) of that species, the factors, and the manner in which these factors influence salt secretion. These are issues on which a holistic understanding of these structures depends.

5.5 Primulaceae

Salt glands were also evidenced in ***Primulaceae*** species, where from those of *Glaux maritima* (Fig. 5.75) were the most intensely studied, followed by those of *Samolus repens* (Fig. 5.76). Glands could, therefore, be an ecological adaptation to marine habitats.

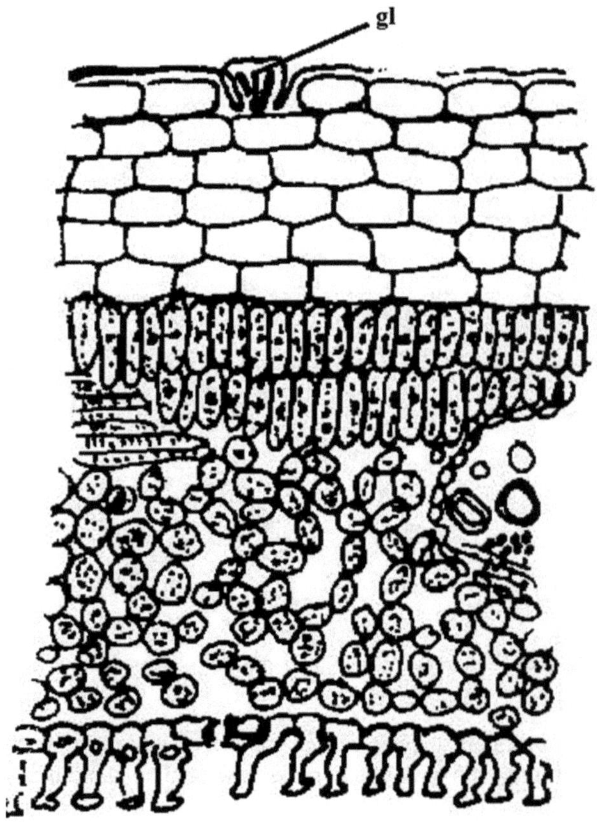

Fig. 5.69 Glands (*gl*) of *Avicennia alba* (Das and Ghose 1996)

Glands of *Glaux maritima* are sunk in depressions of the epidermis. On the surface of the leaf section, glands appear to be surrounded by six epidermal cells arranged regularly (Rozema et al. 1977). No links between vascular elements and salt-secreting glands were evidenced. From inside to outside the gland, three types of cells can be described: a large basal cell, with central vacuole and a thin layer of parietal cytoplasm containing a large nucleus and chloroplasts. This cell, similar to some extent to those of mesophyll, corresponds to collecting cell of the gland. Near to it, there is a stalk cell, with the large nucleus; lateral walls of this cell are encrusted with suberin and cutin. The external part of the gland consists of four to eight gland secretory cells, whose base is in continuation of the upper part of the stalk cell. Secreting cells are characterized by the relatively large nucleus and dense cytoplasm.

As already stated, the ecological function of the activity of these glands is to remove excess of absorbed ions, especially sodium and chloride. But this statement only covers a part of the complex reality that is established between the plant and the environment. In fact, there are many questions that need to be considered: what is the amount of salts removed with respect to that absorbed? To which extent salt secretion manages to maintain the internal salt concentration at a constant level?

Fig. 5.70 Glands (*gl*) of *Acanthus ilicifolius* (Das and Ghose 1996)

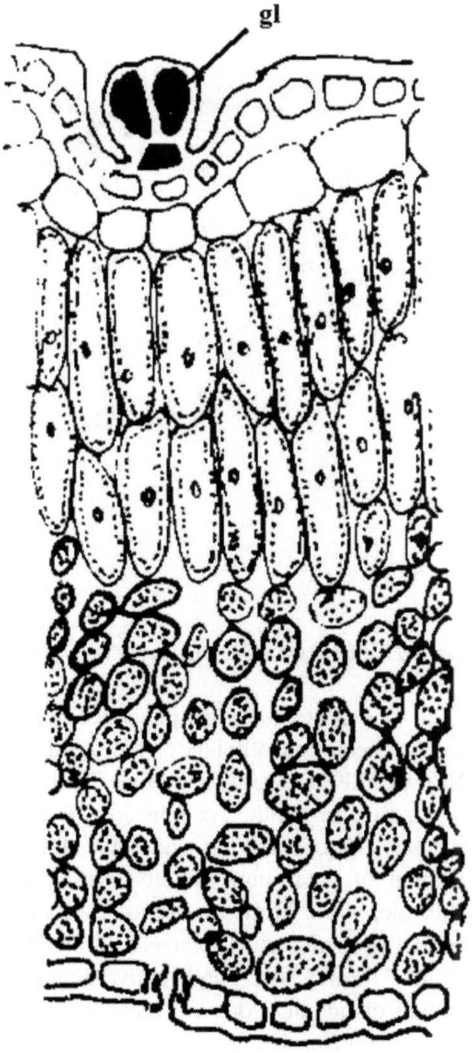

Which is the specificity of the mechanism of salt secretion and particularly NaCl? It seems that ecological relevance of secretion mechanism can be established only when taking into account both the importance of secretory processes and water balance of plant (Waisel 1972); comparisons between other mechanisms of adaptation to salinity, such as ions removing through the roots, and succulence can be also considered.

It seems, therefore, that the ecological importance and relevance of ion secretion are determined not only taking into account the amount of secreted ions but also the increase of the internal concentration of ions. Pollak and Waisel (1979) used the parameter "relative excretion" (ratio of the amount of secreted ions and ionic internal content) as a criterion for the efficiency of secretory mechanisms.

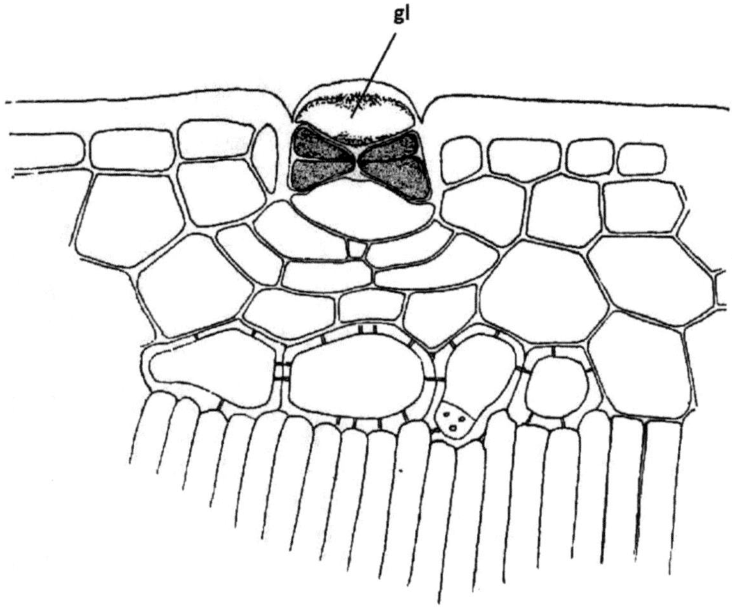

Fig. 5.71 Salt gland (*gl*) in *Aegiceras corniculata*—cross section (Schmidt 1905)

In this context, an ecophysiological study showed that for *Glaux maritima*, sodium secretion efficiency (the ability to maintain a constant level of internal Na) was 20% (Rozema and Gude 1981).

Finally, in another ecophysiological study, it has been considered the same species, *Glaux maritima*, and it was found that in media with 300 mM NaCl, the amount of potassium excreted was 2 mmol, 30 mmol sodium, and 31 mmol chloride, while in media free of NaCl, the secreted amounts were 3–4 mmol Na and Cl (Rozema and Riphagen 1977). Therefore, a distinction between “”inactive glands” (0 mM NaCl) and “active glands” (300 mM NaCl) has been made.

Salt glands in *Glaux maritima* were evidenced by Warming (1897) (Fig. 5.77) and by von Minden (1899) (Figs. 5.78 and 5.79).

Kamienski (1880) evidenced salt glands in *Samolus littoralis* (Fig. 5.80), while Cross (1909) evidenced them in *Samolus repens* (Figs. 5.81 and 5.82).

Grigore et al. (2014) found salt glands in *Glaux maritima* (Fig. 5.83), collected from a salt marsh from Poland.

5.6 Poaceae

Salt glands occurring in ***Poaceae*** species were intensely studied in relation to their resistance to salt stress; they have been largely reviewed by Grigore and Toma (2010).

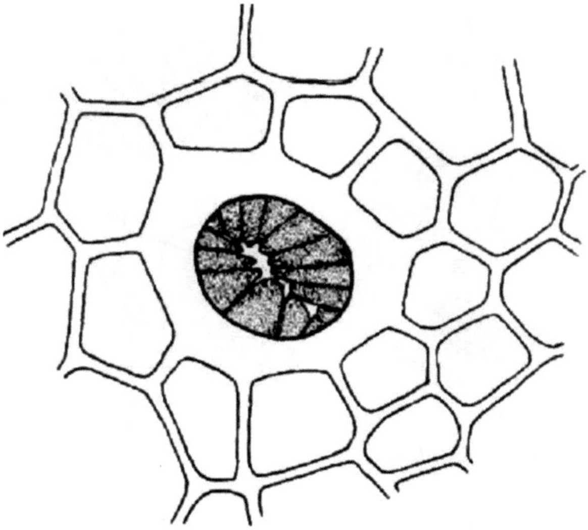

Fig. 5.72 Salt gland in *Aegiceras corniculata*—surface view (Schmidt 1905)

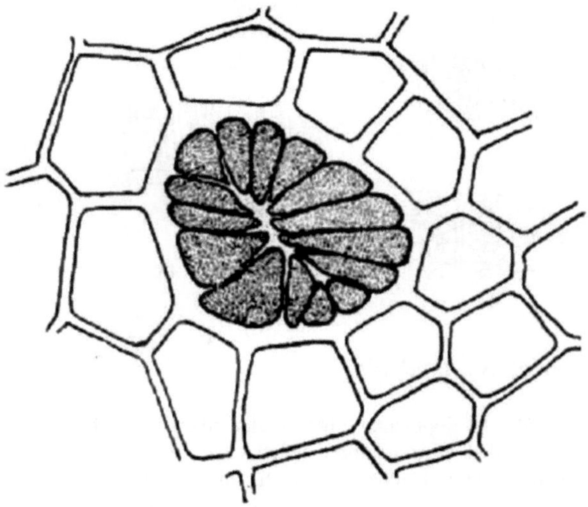

Fig. 5.73 Salt gland in *Aegiceras corniculata*—surface view (Schmidt 1905)

Salt glands of grasses were evidenced by Skelding and Winterbotham (1939), Helder (1956), Levering and Thomson (1971), Taleisnik and Anton (1988), Amarasinghe (1989), Flowers et al. (1990), Marcum and Murdoch (1990), and Marcum (1999).

Within the *Poaceae*, bicellular epidermal glands are the most characteristic and occur in more than 30 species of tribes *Chlorideae, Eragrosteae, Aeluropodeae,* and *Pappophoreae*.

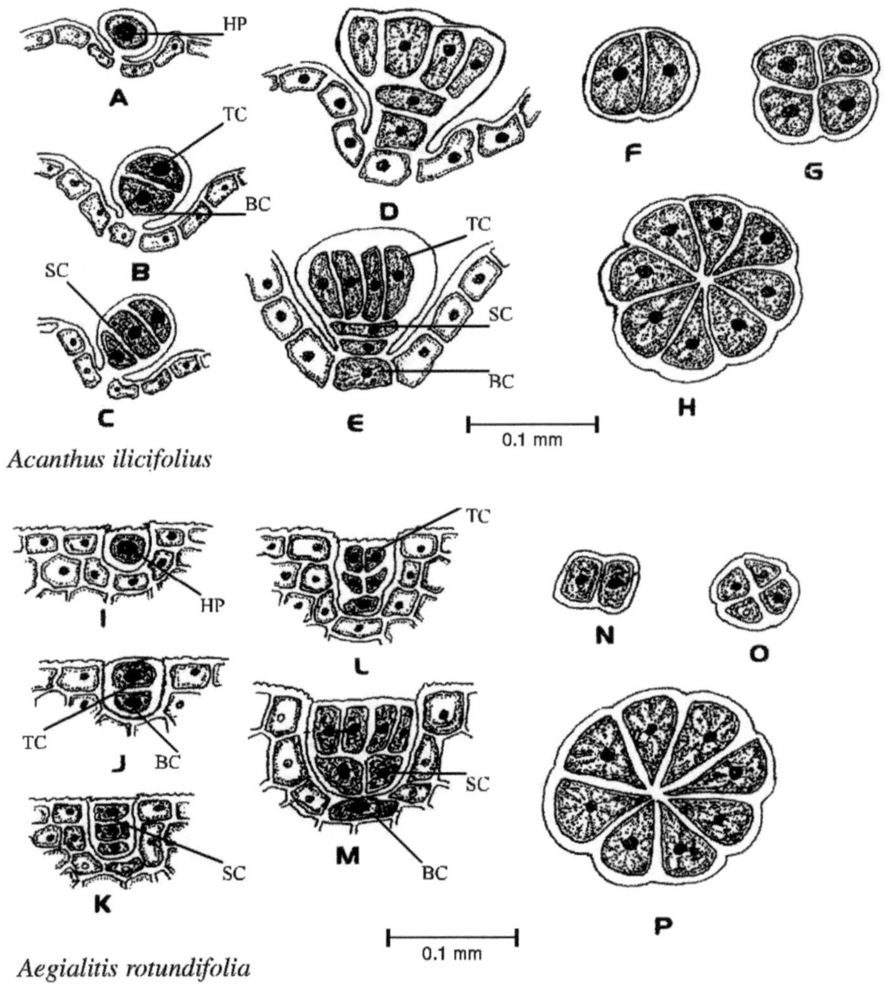

Fig. 5.74 Developmental stages of glandular hairs. **a–h***Acanthus ilicifolius*; (**a**) hair primordium, (**b**) two-celled stage, (**c**) three-celled stage, (**d**) one basal cell, one stalk cell, and four terminal cells, (**e**) mature glandular hair, (**f–h**) surface view of gradual developmental stages of terminal cells. **i–p***Aegialitis rotundifolia*; (**i**) hair primordium, (**J**) dividing stages with one terminal and basal cell, (**k**) three-celled stage, (**l**) undivided basal cell and longitudinal division of stalk and terminal cells, (**m**) undivided two stalk cells and one basal cell with the longitudinal division of terminal cells, **n–p** surface view of gradual developmental stages of terminal cells (*HM* hair primordium, *BC* basal cell, *SC* stalk cells, *TC* terminal cells) (Das 2002)

One of the simplest structures is the bicellular one from *Aeluropus littoralis*, consisting of one large basal cell and a terminal (cap) cell, with cutinized walls. As in *Spartina, the* gland is not sunken into the epidermis but is relatively above its level. Its structure is rather similar to that of a hair, than that of a proper gland.

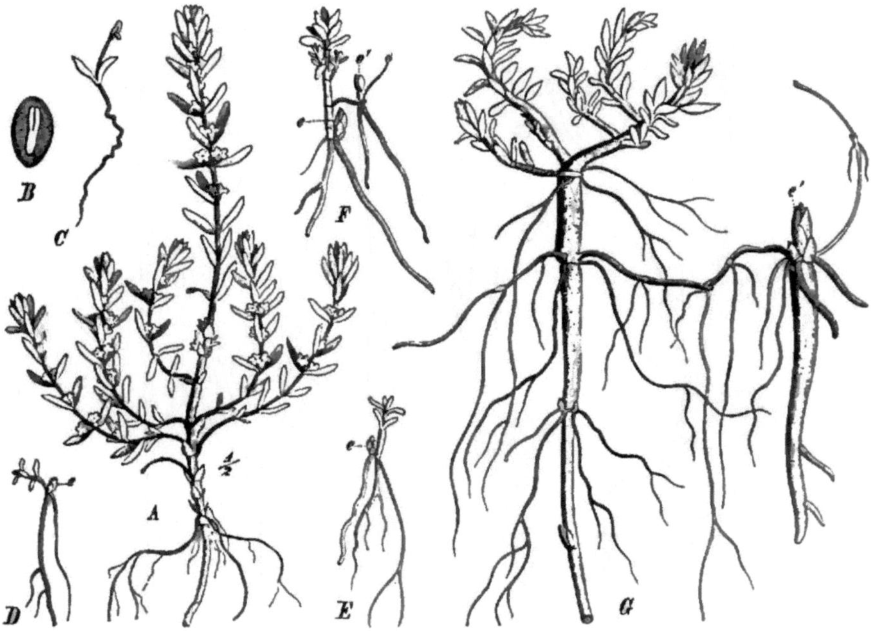

Fig. 5.75 *Glaux maritima*. (**a**) Mature plant, (**b**) seed, (**c–f**) different stages of seedling plant, (**g**) more advanced developed plant) (Pax and Knuth 1905)

On *Chloris gayana*, gland consists of three cells, one large collecting, one stalk cell, and an upper, presumably with the secretory role (Waisel 1972).

At *Spartina townsendii*, the structure of gland is also of bicellular type (Fig. 5.84) (Sutherland and Eastwood 1916; Skelding and Winterbotham 1939).

Liphschitz and Waisel (1974) conducted an extensive but concise study regarding the structure of salt glands in grasses. They pointed out that the glands are present on both sides of the epidermis of the investigated species, arranged in longitudinal rows parallel to the nervures. Each gland consists of two cells: a basal and terminal one. Basal cell corresponds to collecting one, while that terminal (upper) corresponds to secretory one. The cells contain dense cytoplasm and a prominent nucleus but lack a central vacuole. Both cells have cells with suberified and cutinized walls. Cutinization is more pronounced in the external wall of the secretory cell and in the walls of the lower cell, bordering adjacent epidermal cells. The walls of the basal cell are lignified in their upper part, as in the region of the "bottle neck" of gland. The basic structure of the gland is maintained in all investigated species. However, there are also variations in the structure and functioning of the glands; they refer to the shape of the basal cell and to that of the secretory cell.

It was found that some glands are sunk into the epidermis (*Spartina*), or their cells are located above the epidermal cell level (*Bouteloua*). There are also transitional forms between semi-sunken glands (*Coelachyrum, Dinebra, Tetrapogon*); in

Fig. 5.76 *Samolus repens*. (**a**, **b**) General habit, (**c**, **d**) flower, (**e**) longitudinally sectioned corolla) (Pax and Knuth 1905)

some species, gland has the appearance of a hair with narrow and elongated secretory cell, located above the narrow base (*Bouteloua*, *Tetranche dregei*).

Sunken glands also occur in *Sporobolus*, while in *Crypsis* they have a hair-like aspect.

It has been shown that under NaCl exposure, sodium content increased in the gland, while the potassium was stopped; secretion depends on the duration of exposure to chloride treatment. Species with sunken glands and large, oval basal

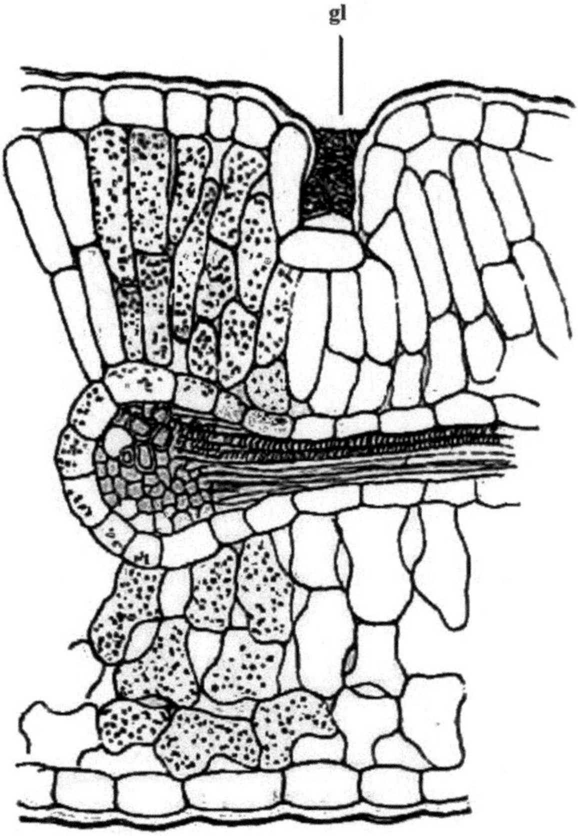

Fig. 5.77 Salt gland (*gl*) in the lamina of *Glaux maritima* (Warming 1897)

cell seem to secrete more salt than species with hair-shaped glands that have narrow, elongated basal secretory cell.

All these observations correlated with the distribution of habitats and species. It should be noted that salt secretion occurs most often when plants are exposed to saline habitat conditions. When to the plants medium, NaCl is not added, secretion does not occur. In these circumstances, glands can be confused with brushes, especially when formed by a basal cell and an elongated secretory cell.

This observation is of great importance since it could explain why other authors do not describe salt glands in grasses they investigated.

But more interesting are the possible inferences to be drawn here. The existence of salt glands on the leaves of *Chlorideae* species suggests that these tribes evolved from common or very close ancestors. Some of them must have occupied saline habitats. This is related to the fact that under experimental conditions, plant growth improves when salt is administrated. Such data suggest that all species of this subfamily would have possessed salt glands. Likely, some glands would have evolved in the direction of microhairs in those species that would have migrated to nonsaline environments. The existence of semi-sunken glands in plants that

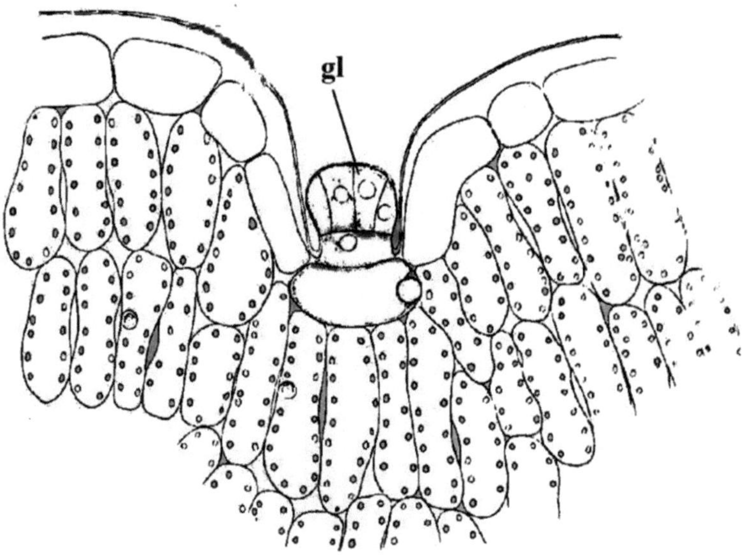

Fig. 5.78 Salt gland (*gl*) in the lamina of *Glaux maritima* (von Minden 1899)

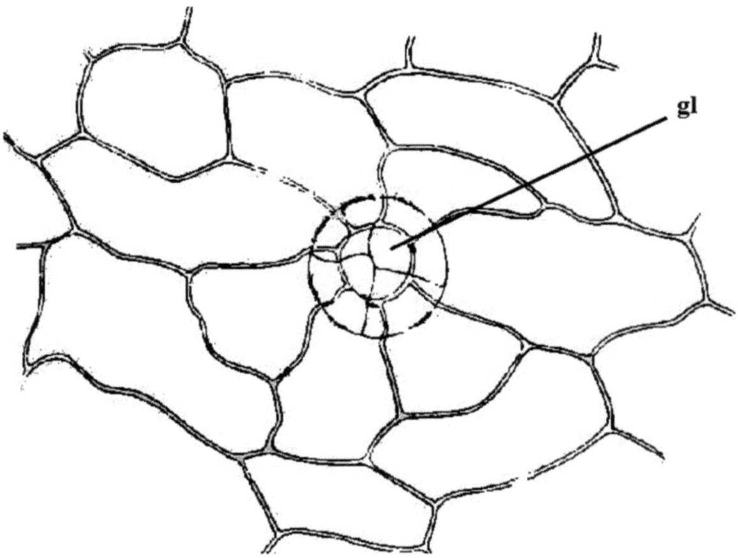

Fig. 5.79 Salt gland (*gl*) in the lamina of *Glaux maritima*, surface view (von Minden 1899)

currently occupy nonsaline habitats might suggest that the change of the halophytic nature to glycophytic one had occurred recently.

It has been shown that in one species of wild rice, *Porteresia coarctata*, salt secretion is an important factor in the balance of salts in leaves (Flowers et al.

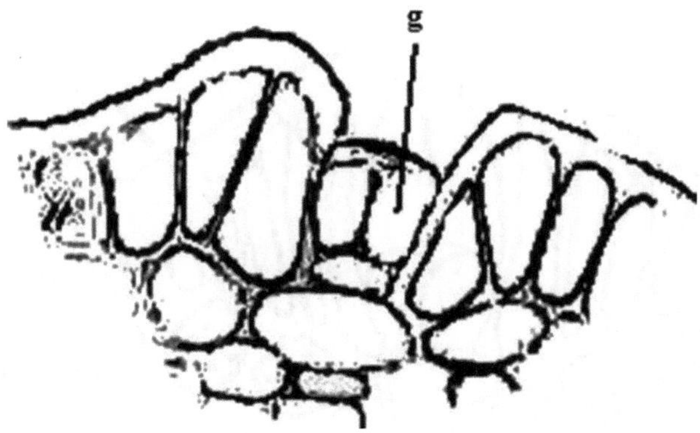

Fig. 5.80 Salt gland (*g*) in the lamina of *Samolus littoralis* (Kamienski 1880)

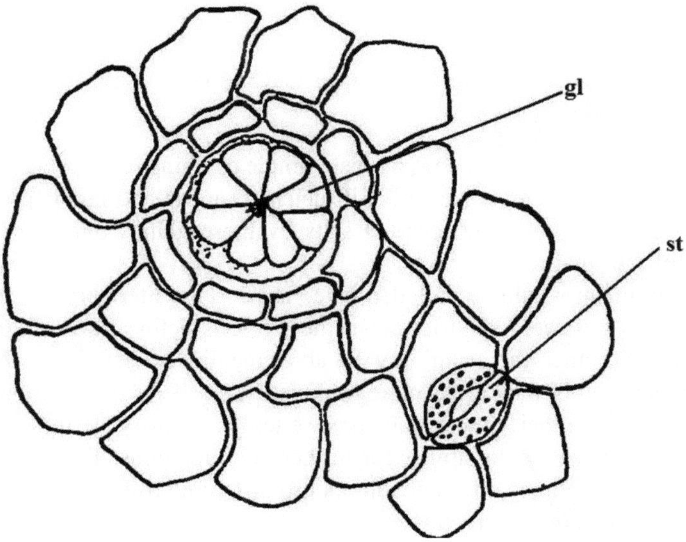

Fig. 5.81 Salt gland (*gl*) in the lamina of *Samolus repens* (*st*. stomata) (Cross 1909)

1990). This mechanism is facilitated by secretory hairs, sometimes bifurcated, located at the adaxial side of the leaf. They arise from the epidermis and are unicellular with an electron-dense vacuole covered with a cuticle.

Experimentally, it has been suggested that under saline conditions, the number of hairs increases. With an increase in the external concentration of the salts, the vacuolar concentration of sodium and chloride increases; this is higher than that of the mesophyll cells. X-ray analysis confirmed that in plants exposed to NaCl, the ratio of Na:K was higher than in mesophyll cells, which was 7:3, in contrast to 0:9, under the conditions of salinity of 100 mM NaCl.

Thus, perhaps this is the simplest architecture of salt secretory structures; in this sense, there is no distinction between a basal cell and a secretory one. However, it is

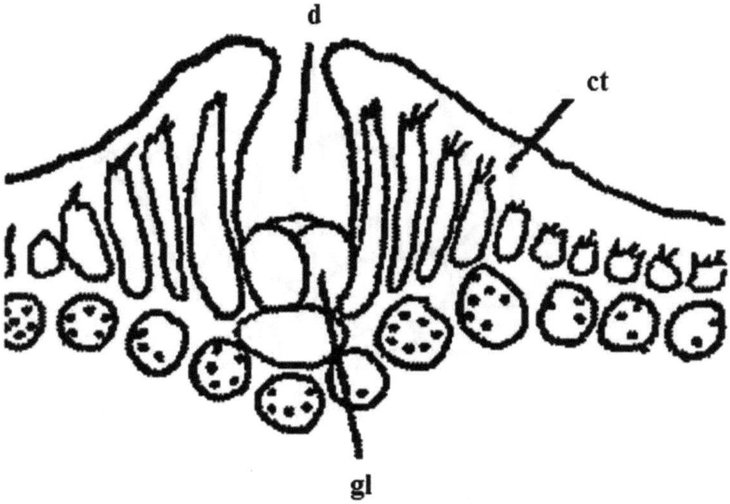

Fig. 5.82 Salt gland (*gl*) in the stem of *Samolus repens* (Cross 1909)

not known if these hairs release ions by their simple crushing (collapsing), as in the case of hairs found in *Atriplex* species, or if such hairs are able to simply provide ion secretion. The fact is that the secreted salts by these glands represent a major proportion of ions that reach the leaf.

A short comment would be included here. The foreign literature (meaning not Romanian) does not seem so strict about the rigorous delineation of terms, in a didactic way. Whether the Fahn' (1988) abovementioned definition about salt glands is being used or they focus more on functional aspects of secretion—many authors do not use in their descriptions precise terms; in this way, they open new perspectives and gave the possibility for further interpretations. This is the reason explaining why sometimes different authors use different terms but talking in fact about the same anatomical structure.

This is the reason we have used sometimes throughout this book apparently vague expressions; we tried, therefore, to consciously avoid an excessively strict language. The text analysis that has been often conducted when consulting large and heterogeneous old literature requires a respect for the (relative and sometimes imprecise) language that botanists have used. In this conceptual framework, even the term "salt glands" may be confusing; it might suggest that glands per se contain and finally eliminate salts to the exterior. But salt glands do not produce salt, sensu stricto, but they are rather specialized devices that concentrate and transport salt to the outside of the plant. "Salts" (another generic term) are, in fact, a product of the metabolism of the whole plant.

Following this idea, Skelding and Winterbotham (1939), studying the secretory formations of *Spartina townsendii*, use the term "hydathodes" instead that of "glands." Of course, there are no hydathodes and salt glands in the same species,

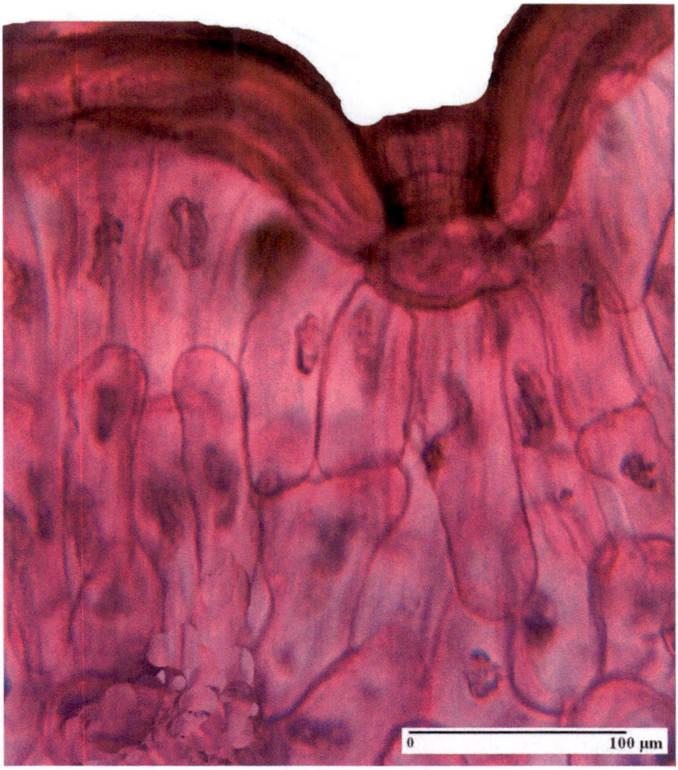

Fig. 5.83 Salt gland in the lamina of *Glaux maritima* (Grigore et al. 2014)

but only salt glands, which are yet called "hydathodes," although it is stated quite clearly and unequivocally that they "secrete a salt solution consisting mainly of sodium chloride NaCl [..]" (op. cit, pp. 78).

The salt glands are therefore epidermal structures, located on both sides of the lamina, not in direct contact with the plant conducting system. They are always in contact with the chlorenchymatic tissue on the adaxial side, while on abaxial side, they are separated from it only by a layer of large colorless cells.

Mature gland comprises four epidermal cells arranged in a way that leaves a cylindrical opening in the epidermis (opening of the gland) and a specialized structure consisting of two cells, which represent the proper gland.

Basal cell is large, attached to the four neighboring epidermal cells, but is sunken in leaf tissues, so forms a small cylindrical depression over it. It is bounded by walls of epidermal cells from the same longitudinal row and by a part of the lateral walls of epidermal cells from neighboring rows.

Upper cell from the top gland cell is the cap cell and can be seen as an extension to continue the wall of the basal cell; apart from its attachment to the basal cell, it isquite free in the opening of the gland which it fills.

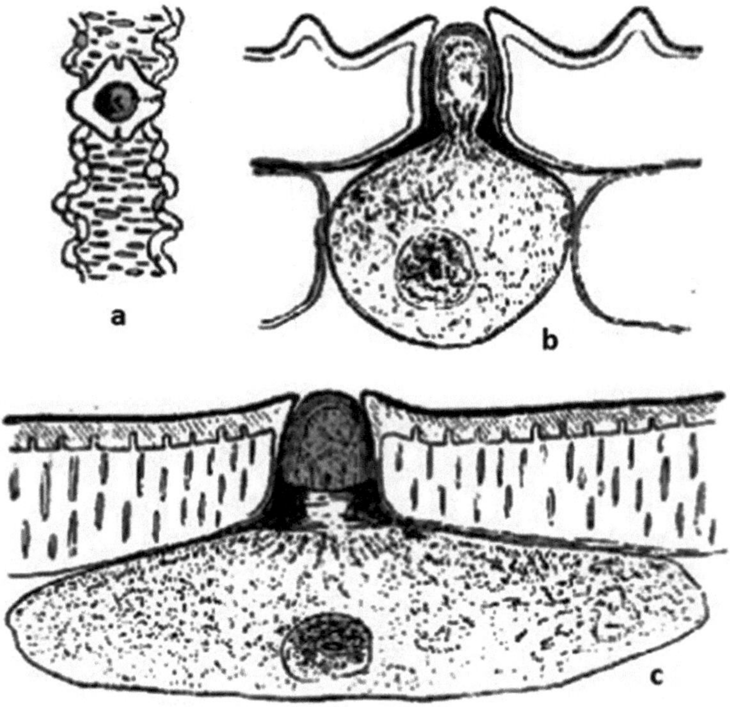

Fig. 5.84 Salt gland of *Spartina townsendii* (leaf, **a**—surface view; **b**—cross section; radial longitudinal section) (Sutherland and Eastwood, 1916)

The authors also provide information on the development of the gland. Thus, the two types of gland cells derived from a single original epidermal cell, which is different from other cells from the epidermis even when the young leaf is in the bud stage. Initial gland grows faster than the other epidermal cells and owes its shape due to the broadening of a basal part in mesophyll and to the external projection of external part toward the exterior of the epidermis. The nucleus is usually large.

The nucleus of an initial cell undergoes a normal mitotic division; a daughter nucleus remains in the basal region of the young gland, while another goes in the external region of the gland. During cell division, following immediately, a cell wall formed in the plane of the leaf surface protrudes in the initial cell, in order to form the secretory cell. The basal cell grows rapidly and soon reaches its final shape and size, but the cap cell grows very little after cell division. Shortly after cell division, the well of hydathode appears as a result of an overgrowth of the epidermal cells and the consequent sinking of the glandular cells into the leaf tissue until the cap cell no longer projects from the surface.

Marcum (2001) thought that, at least at first glance, glands of grasses are similar to epidermal bicellular microhairs. Although microhair-like glands were observed in all subfamilies of grasses, except for *Pooideae*, functional glands were found only in *Chloridoideae*.

Glands of grasses differ from those of dicots, being formed, as already mentioned, from a basal cell attached or sunken to/in the epidermis and a cap cell.

The glands have cutinized cell walls and are often surrounded by papillae. Although there is a basic bicellular structure to almost all species of chloridoideae, the appearance of glands may vary.

Sometimes glands are longitudinally arranged in parallel rows, at the top of intercostal regions of the leaf, adjacent to rows of stomata. In some cases, the glands are sunken into the epidermis, with the basal cell completely embedded, e.g., desert salt grasses; in others, the basal cell is semi-embedded. Rarely, the basal cell may extend out from the epidermis. The size of gland varies between 25 and 70 μm.

Plasmodesmata occur in the thick wall of the basal cell and secretory cell, as well between basal cell and mesophyll cells. There were no plasmodesmata in the wall between basal cell and adjacent epidermal cells.

The nucleus of a basal cell is quite large. Plastids contain plastoglobules, dense stroma, some peripheral vesicles, and membranes. Dictyosomes, ribosomes, and endoplasmic reticulum were also observed. The main feature is the large nucleus of the secretory cell and dense cytoplasm containing mitochondria and plastids with plastoglobules.

5.7 Salt Hairs (Vesicular Hairs, Salt Bladders, Bladders) from Chenopodiaceae

Secretory structures from *Chenopodiaceae* have been intensely studied and discussed, as many other features of very halophytic species of this intriguing botanical family.

These salt hairs (also called in the literature: vesicular hairs, vesiculated hairs, salt bladders, bladder hairs, or simply, bladders) are present in many species of *Chenopodiaceae*, especially in *Atriplex* species. These structures have been reviewed and largely discussed by Grigore and Toma (2010). Many authors studied them, from different points of view, especially in *Atriplex* genus (de Bary 1884; Volkens 1887; Arcangeli 1890; Warming 1909; Chermezon 1910; Wood 1925; Black 1954; Pyykkő 1966; Osmond et al., 1969; West 1970; Goodin and Mozafar 1970; Pallaghy 1970; Mozafar and Goodin 1970; Smaoui, 1971; Campbell et al. 1974; Osmond 1974; Troughton and Card 1974; Thomson and Platt-Aloia 1979; Jeschke and Stelter 1983; Bennert and Schmidt 1983; Aslam et al. 1986; Fahn 1988; Karimi and Ungar 1989; Breckle et al. 1990; Ungar 1991; Freitas and Breckle 1992, 1993a, b; Gorham 1995; Mohr et al. 1995; Glenn et al. 1997; Wickens 1998; Jacoby 1999; Dickison 2000; Lambert and Turner 2000; Orcutt and Nilsen 2000; D'Ambrogio et al. 2000; Batanouny 2001; Breckle 2002; Walter and Breckle 2002; Wahid 2003; Singh 2004; Pandey and Sinha 2005; Khan and Qaiser 2006; Khan and Gul 2006; Evert et al. 2006; de Araujo et al. 2006; Redondo-Gomez et al. 2007; Dajic 2006; Ingrouille and Eddie 2006; Frayssinet et al. 2007; Grigore and

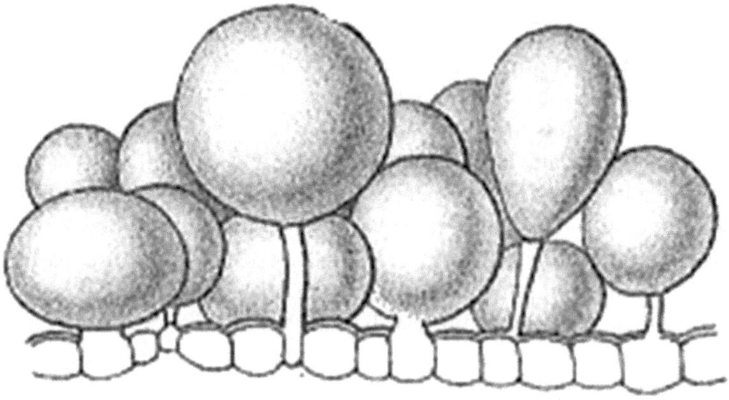

Fig. 5.85 Bladders of *Atriplex leucoclada* (Volkens 1887)

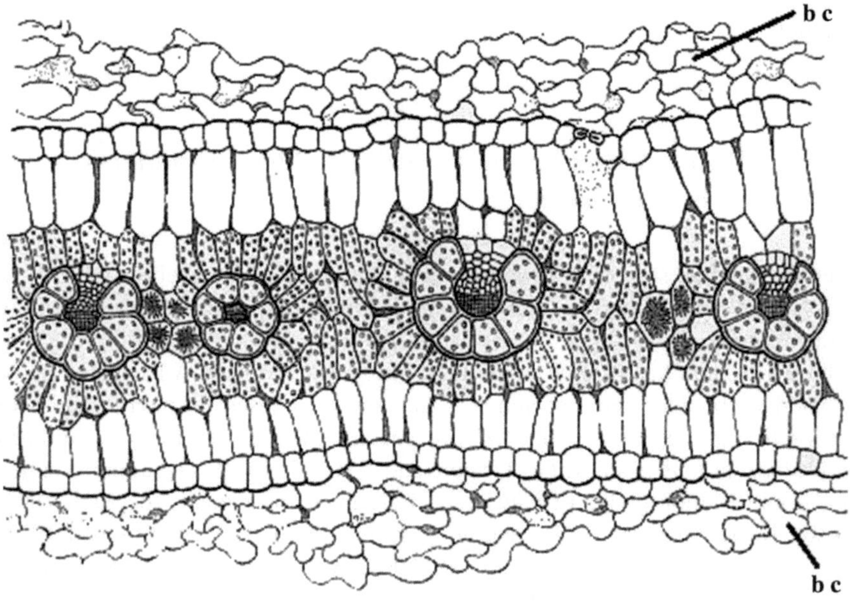

Fig. 5.86 Bladder cells (*b c*) of *Atriplex halimus* (Volkens 1887)

Toma 2007, 2008; Grigore 2008a, b; Lambers et al. 2008; Grigore et al. 2014; Jáuregui et al. 2014; Mahi et al. 2015; Kabbash 2016).

These hairs consist of two cells: a small stalk cell and one large cell, the vesicular (bladder) one (Fig. 5.85). The structure of stalk cell is similar to that of various other types of gland cells. It consists of a dense cytoplasm rich in mitochondria, endoplasmic reticulum, and numerous small vesicles; it also contains chloroplasts. Stalk

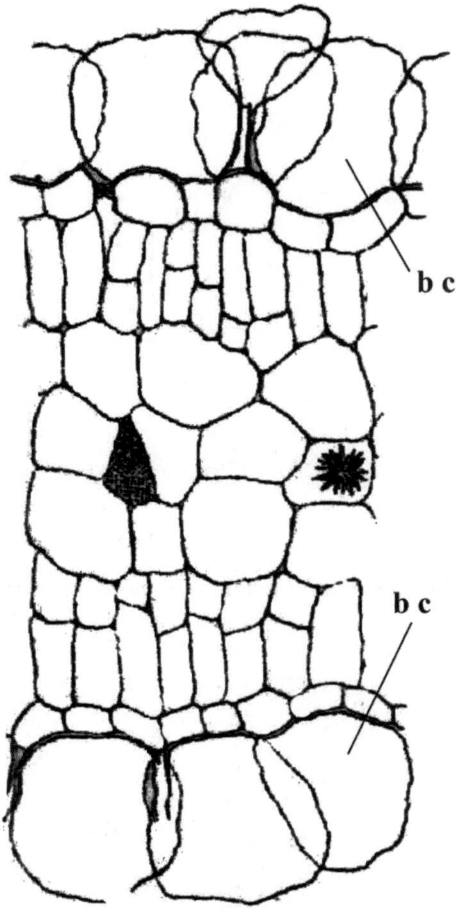

Fig. 5.87 Bladder cells (*b c*) of *Obione pedunculata* (Warming 1890)

cells are interconnected with vesicular cells or those of mesophyll by numerous plasmodesmata.

It has been shown that in these bladder cells, sodium and chloride concentration is higher than in mesophyll cells and higher than that of the external environment. Also, vesicles exhibit a higher electron negativity than the rest of mesophyll cells and external solution. Therefore, salt flow direction is from internal media, through mesophyll, toward these secretory structures. Since this process takes place against the concentration gradient, it seems that the process requires energy consumption.

These hairs were evidenced in *Atriplex leucoclada* (Fig. 5.85) and in *Atriplex halimus* (Fig. 5.86) by Volkens (1887). Bonnier and Du Sablon (1905) believed that these hairs fill with water in wet periods and during dry periods; these water reserves are consumed; the walls of these cells, emptied and flattened, form a coating that protects the plant against dehydration.

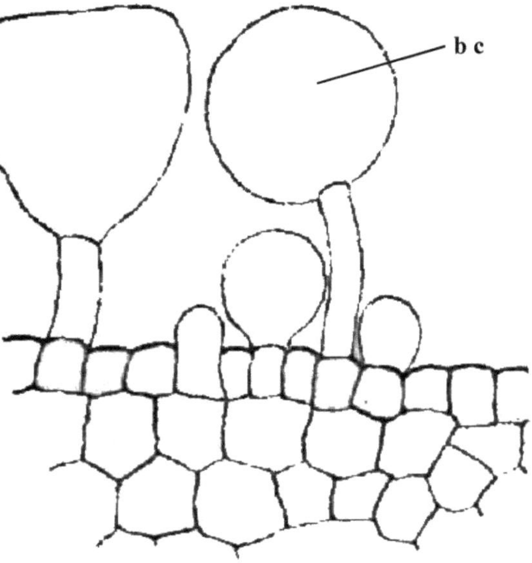

Fig. 5.88 Bladder cells (*b c*) of *Obione portulacoides* (Warming 1890)

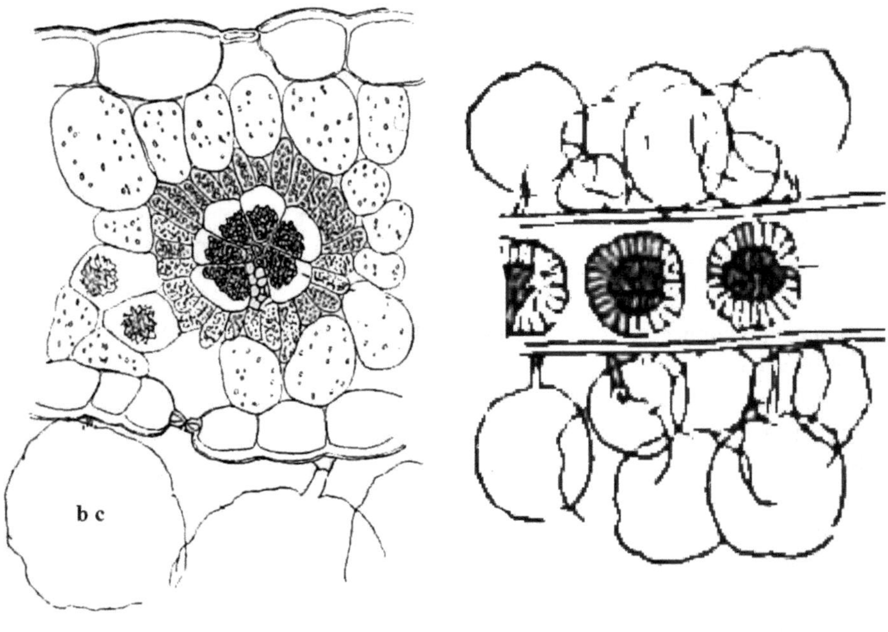

Fig. 5.89 Bladder cells (*b c*) of *Atriplex farinosa*(*left*—detail; *right*—general appearance) (Warming 1897)

Warming (1890) found these bladders in *Obione pedunculata* (Fig. 5.87), *O. portulacoides* (Fig. 5.88), and *Atriplex farinosa* (Fig. 5.89) (Warming 1897).

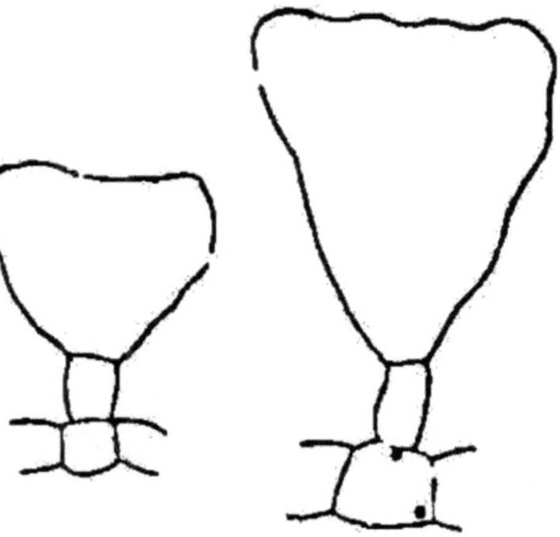

Fig. 5.90 Bladder cells in *Atriplex laciniata* (Monteil 1906)

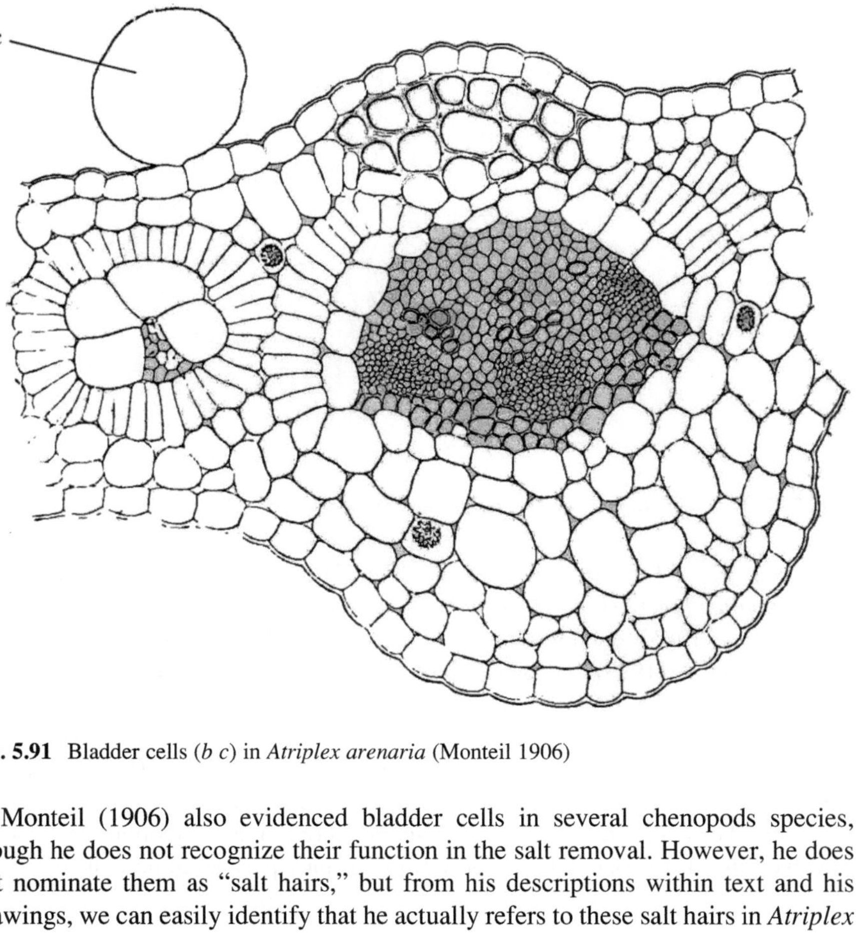

Fig. 5.91 Bladder cells (*b c*) in *Atriplex arenaria* (Monteil 1906)

Monteil (1906) also evidenced bladder cells in several chenopods species, though he does not recognize their function in the salt removal. However, he does not nominate them as "salt hairs," but from his descriptions within text and his drawings, we can easily identify that he actually refers to these salt hairs in *Atriplex*

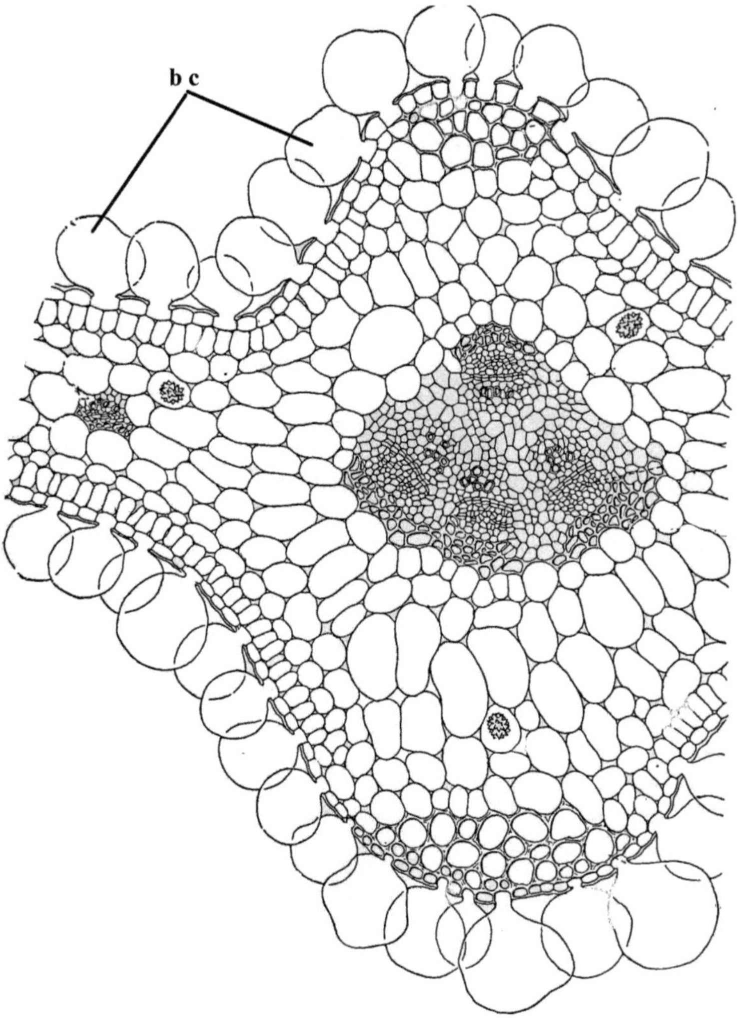

Fig. 5.92 Bladder cells (*b c*) in *Atriplex portulacoides* (Monteil 1906)

laciniata (Fig. 5.90), *Atriplex arenaria* (Fig. 5.91), *A. portulacoides* (Fig. 5.92), *Chenopodium botrys* (Fig. 5.93), and *C. hybridum* (Fig. 5.94).

Gamaley (1985) found these bladder cells in *Atriplex sibirica*.

Salama et al. (1999) evidenced these salt hairs in *Atriplex farinosa* (Fig. 5.95) and *A. halimus* (Fig. 5.96).

Grigore et al. (2014) found salt hairs in several halophytes from *Chenopodiaceae*: *Halimione verrucifera* (Figs. 5.97 and 5.98), *H. portulacoides* (Fig. 5.99), *Atriplex tatarica* (Fig. 5.100), and *A. halimus* (Fig. 5.101).

It has been shown that vesicular hairs represent an important mechanism involved in salt tolerance. In *Atriplex halimus*, they play an important role in the

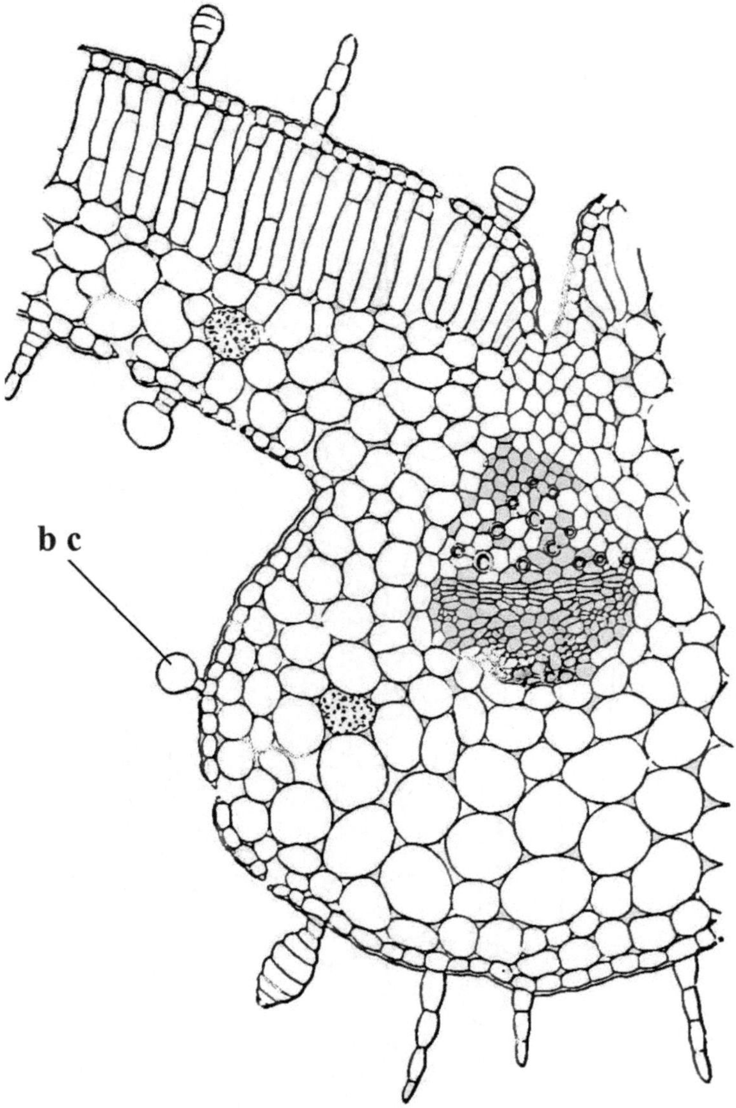

Fig. 5.93 Bladder cells (*b c*) in *Chenopodium botrys* (Monteil 1906)

removal of salt from the leaves, thus preventing the accumulation of toxic ions in the parenchyma and vascular tissues. In this way, a constant level of salt in the leaf cells is maintained (Mozafar and Goodin 1970). In addition, inside these bladders, concentrations of Na and K are very high and increase when plants are subjected to saline treatment. The chloride content also increases under these conditions, which balances the greater part of Na and K ions in the bladders. It is also supposed that

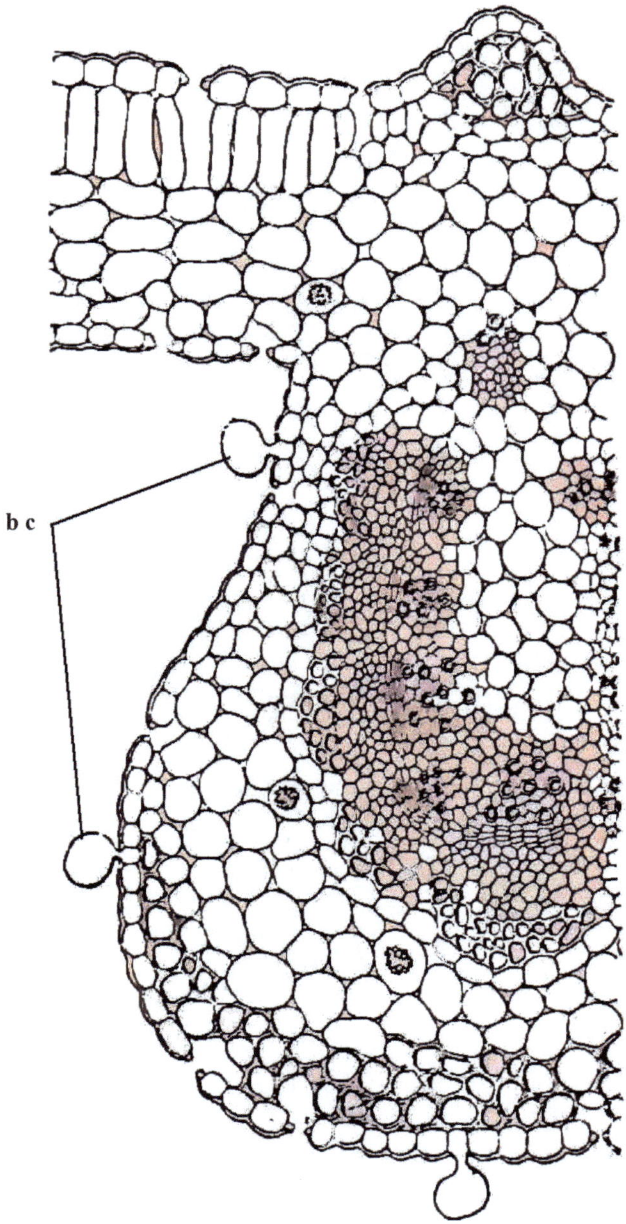

Fig. 5.94 Bladder cells (*b c*) in *Chenopodium hybridum* (Monteil 1906)

the hairs increase transpiration, thus leading to the accumulation of salts inside them.

In *Atriplex buchananii* (Troughton and Card 1974), vesicular hairs are involved in salt tolerance of plants growing in saline habitats. In addition, to this tolerance

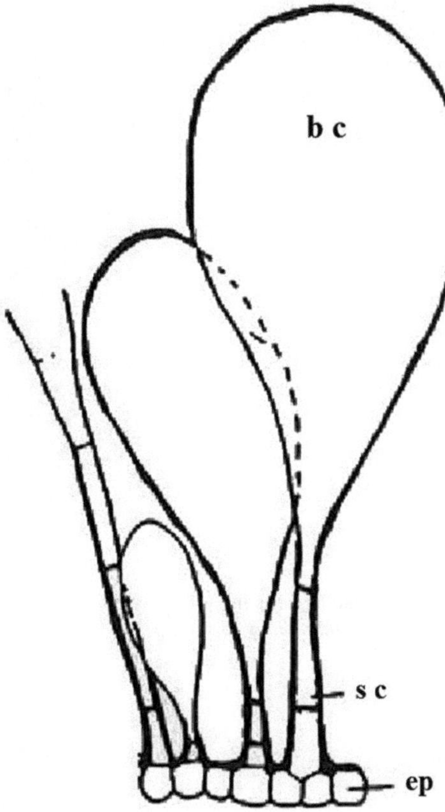

Fig. 5.95 Salt hairs in *Atriplex farinosa* (*b c* bladder cell, *ep* epidermis, *s c* stalk cell) (Salama et al. 1999)

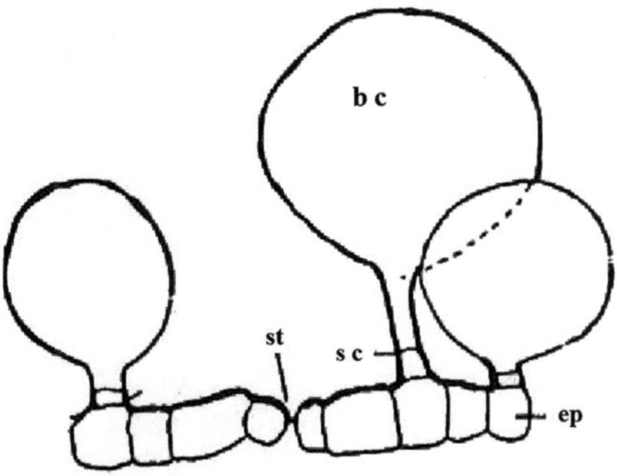

Fig. 5.96 Salt hairs in *Atriplex halimus* (*b c* bladder cell, *ep* epidermis, *s c* stalk cell, *st* stomata) (Salama et al. 1999)

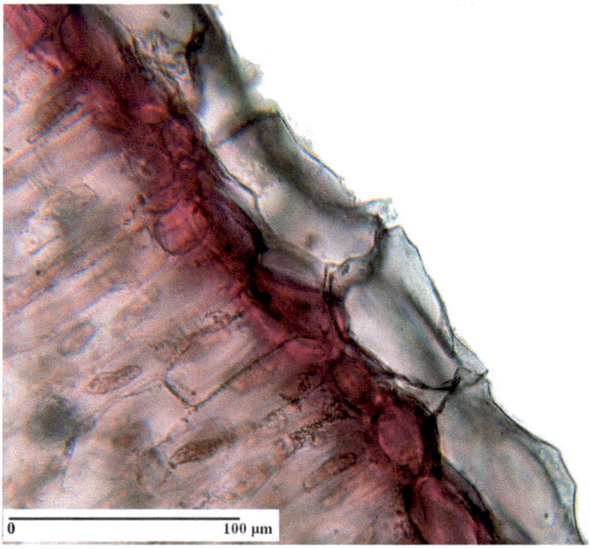

Fig. 5.97 Bladder cells in *Halimione verrucifera* (lamina, RO)

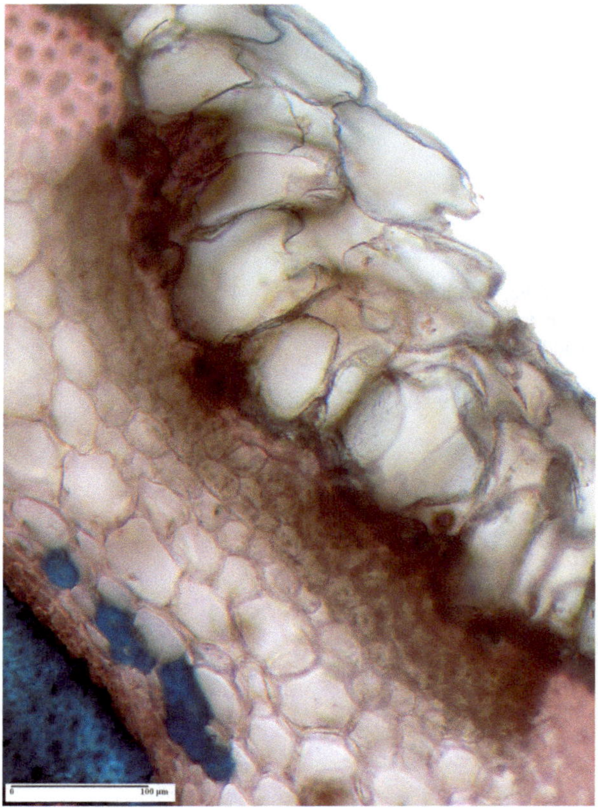

Fig. 5.98 Bladder cells in *Halimione portulacoides* (stem, ESP)

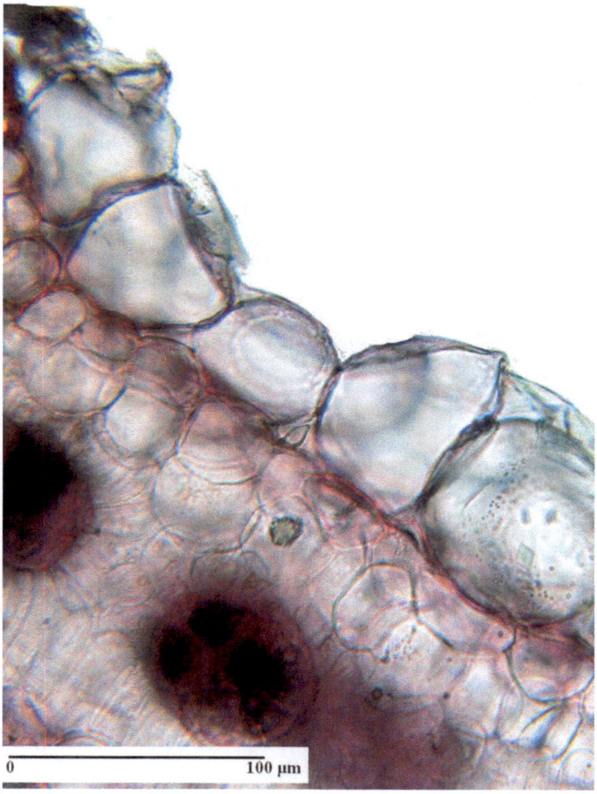

Fig. 5.99 Bladder cells in *Atriplex tatarica* (lamina, RO)

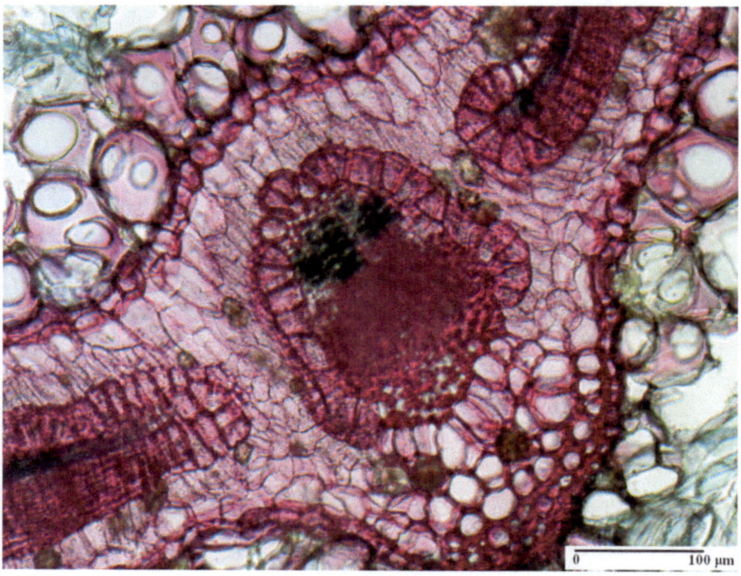

Fig. 5.100 Bladder cells in *Atriplex halimus* (lamina, ESP)

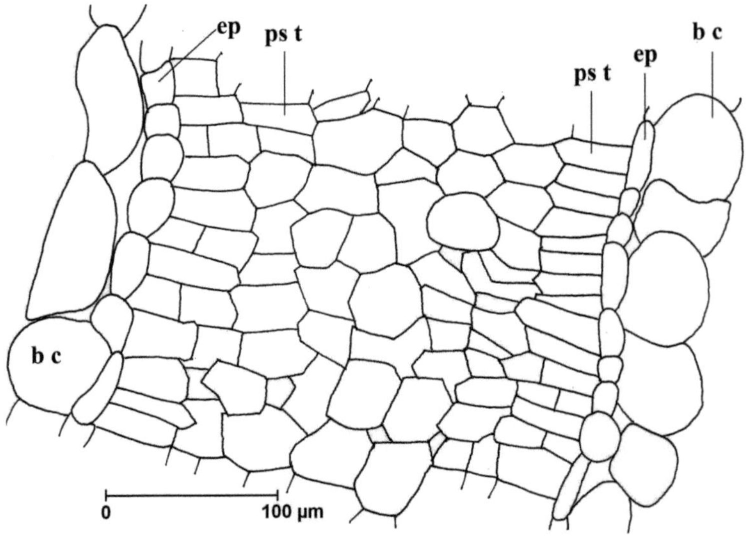

Fig. 5.101 Drawing of cross section through the lamina of *Halimione verrucifera* (ep – epidermis; b c – bladder cell; ps t – palisade tissue) (Grigore and Toma, 2010)

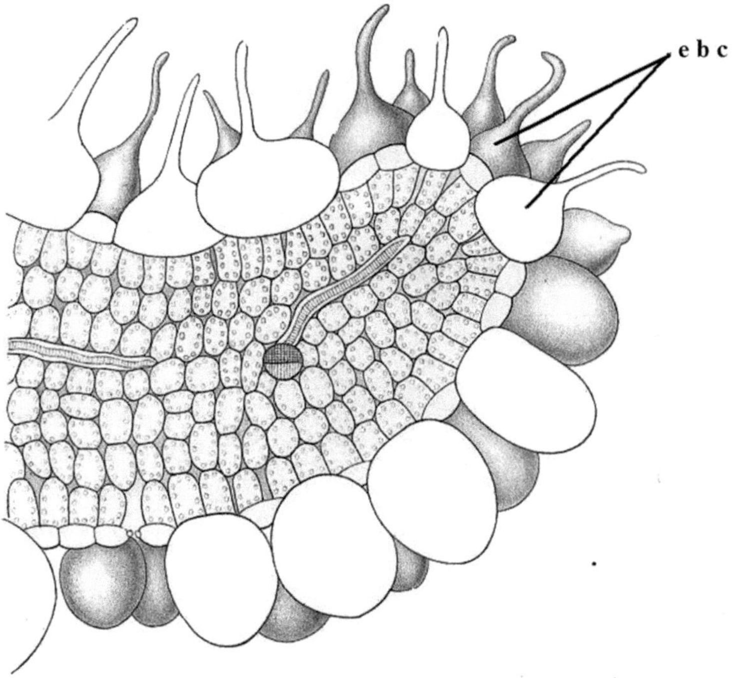

Fig. 5.102 Epidermal bladder cells (*e b c*) from the leaf of *Mesembryanthemum crystallinum* (Volkens 1887)

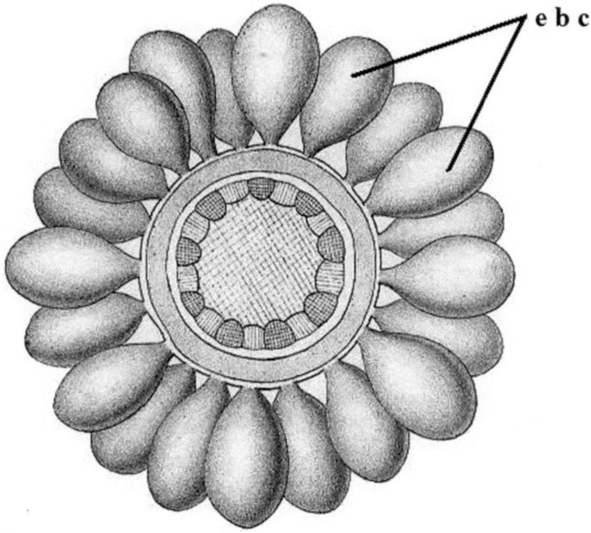

Fig. 5.103 Epidermal bladder cells (*e b c*) from the stem of *Mesembryanthemum crystallinum* (Volkens 1887)

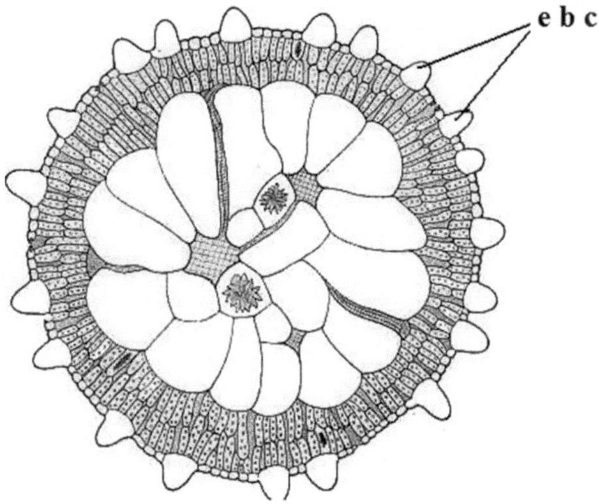

Fig. 5.104 Epidermal bladder cells (*e b c*) from the leaf of *Mesembryanthemum forskaolii* (Volkens 1887)

small leaves with an epicuticular wax layer and high stomatal resistance may contribute.

This type of secretory structure was also evidenced in other genera of the *Chenopodiaceae* family: *Chenopodium* (Brian and Cattlin 1968) and *Halimione* (Baumeister and Kloos 1974; Grigore et al. 2014). However, salt hairs of *Atriplex* were the more intensely studied.

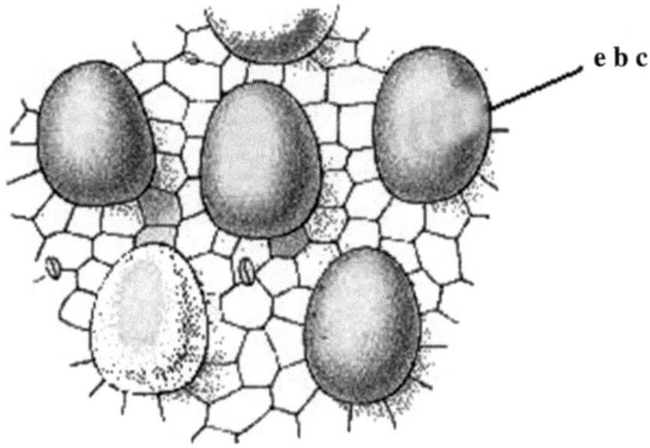

Fig. 5.105 Epidermal bladder cells (*e b c*) from the leaf of *Mesembryanthemum forskaolii* (Volkens 1887)

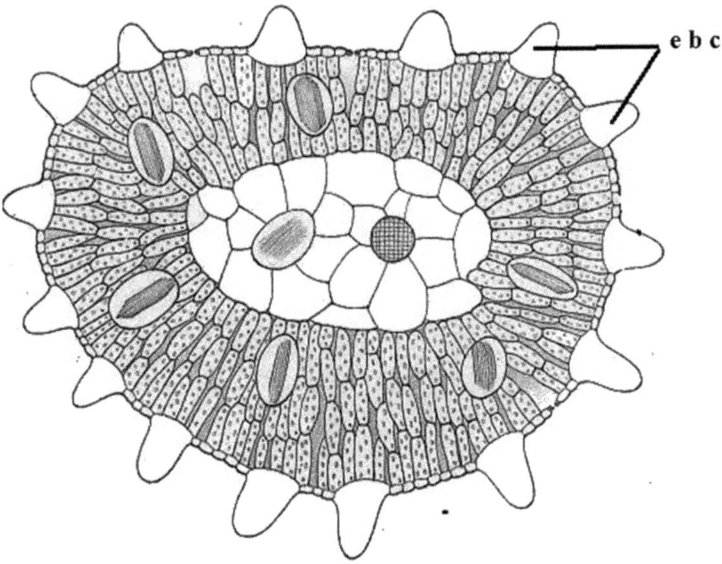

Fig. 5.106 Epidermal bladder cells (*e b c*) from leaf of *Mesembryanthemum nodiflorum* (Volkens 1887)

Salt secretory (glandular) structures have been also reported in halo- and metallophyte groups populating the arid and saline/metal-contaminated soils of Uzbekistan (Toderich et al. 2002).

Such structures were found in *Salsola pestifer*, *S. orientalis*, *S. paulsenii*, and *S. leptoclada*. However, from the electron microscopy micrographs, the authors included in their study, it is not very clear which kind of secretory structures they really represent. Moreover, the authors themselves use different terms for designating these secretory structures: "various gland structures," "salt glands,"

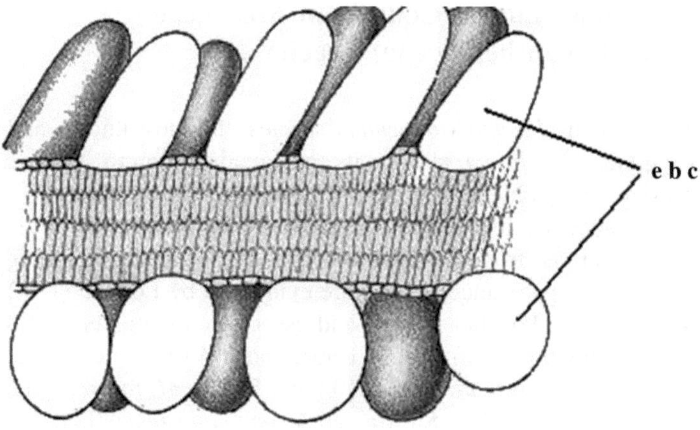

Fig. 5.107 Epidermal bladder cells (*e b c*) from the leaf of *Aizoon canariense* (Volkens 1887)

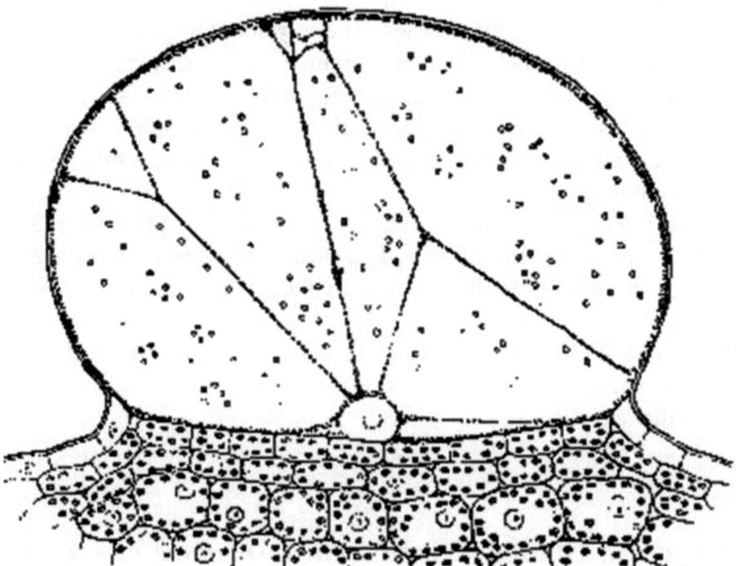

Fig. 5.108 Epidermal bladder cell, filled with water on leaf surface of *Mesembryanthemum crystallinum* (Haberlandt 1914)

"epidermal vesicles," "papillae," "salt glandular structures," "vesiculate hairs," "trichomes," and "papillas."

In fact, it has been stated that *Salsola* species from these regions are grouped into two categories: species with salt-secreting glandular structures and species that accumulate salts, which corresponds in fact to two major strategies: species that secrete and species that accumulate salts.

5.8 Epidermal Bladder Cells from Aizoaceae (Mesembryanthemaceae) Species

These cells occur in *Mesembryanthemum* species, the most known and intensely studied being *M. crystallinum*, also for its special adaptation to switch from C_3 to CAM photosynthetic pathway.

Epidermal bladders cells were recognized for a long time, due to the name of plant, *ice plant*, derived from the presence of these special accumulating cells that confer its particular appearance. They were evidenced by Loudon (1842), Lindley (1846), Wood (1861), Pouchet (1883), and especially by Volkens (1887) who is perhaps the first drawing them and by Haberlandt (1914). Thus, Volkens (1887) found them in *M. crystallinum* (Figs. 5.102 and 5.103), *M. forskaolii* (Figs. 5.104 and 5.105), *M. nodiflorum* (Fig. 5.106), and *Aizoon canariense* (Fig. 5.107).

Schimper (1903) considered the epidermal bladder cells from *M. crystallinum* as a xerophytic adaptation; they would represent isolated water storage cells (forming a peripheral water storage tissue), which are, however, a feature much rarer than proper water tissue. In the mentioned species, some epidermal cells project above epidermis, forming large water vesicles. These water storage cells remain always filled with protoplasm and cell vacuolar content, in any case, free of air. The volume of stored water varies widely. According to Schimper (1903), when transpiration is weak, these vesicles are filled with water during the night or when the sky is cloudy, while when transpiration is intense, these vesicles supply assimilating cells with water and then collapse.

Haberlandt (1914) gave to these cells an accurate description (Fig. 5.108) and discussed those of *M. crystallinum* as a typical adaptation of xerophytes, whose epidermal cells can sometimes increase their volume, thus acquiring a water storage role.

Zemke (1938–1939) found these bladder cells in *Mesembryanthemum salicornioides*, vegetating in the Namib Desert; they were evidenced both on the stem and on the leaf surface.

5.9 Other Salt Glands (*Cressa cretica*, *Ipomoea pes-caprae*, and *Lavatera arborea*)

Very little known and described quite recently, glands of *Cressa cretica* and *Ipomoea pes-caprae (Convolvulaceae)* and those of *Lavatera arborea (Malvaceae)* have been discussed by Grigore and Toma (2010).

Warming described the anatomical structure of *Ipomoea pes-caprae*, designating these glands as either "glandular hairs" or "hydathodes" (Figs. 5.109 and 5.110).

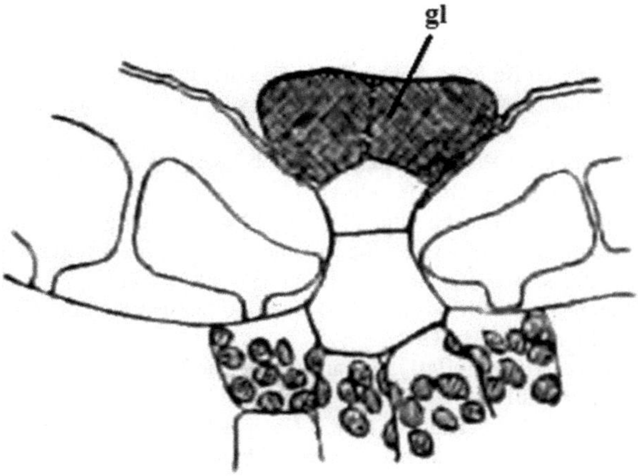

Fig. 5.109 Salt gland (*gl*) of *Ipomoea pes-caprae* (Warming 1897)

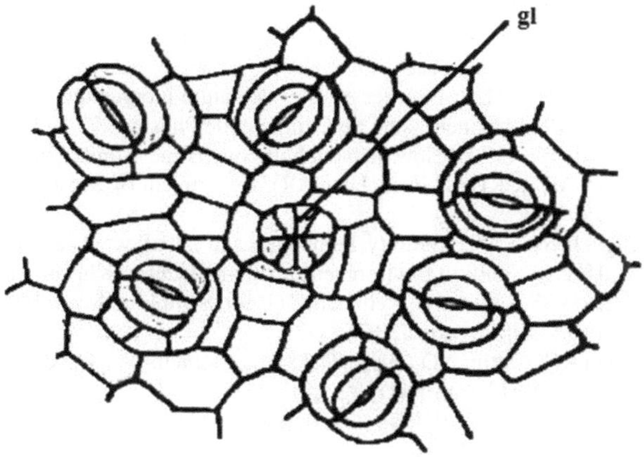

Fig. 5.110 Salt gland (*gl*) of *Ipomoea pes-caprae* (Warming 1897)

References

Amarasinghe LW (1989) Variation in salt secretory activity of microhairs in grasses. Aust J Plant Physiol 16:219–229

Andersson NJ (1849) Atlas öfver Den Skandinaviska Florans naturliga familier. Stockholm, General Stabens Lithografiska Inträttning

APG—Angiosperm Phylogeny Group (2003) An update of the Angiosperm Phylogeny Group Classification for the orders and families of flowering plants: APG II. Bot J Linn Soc 141:399–436

Arcangeli G (1890) Sulla structura delle foglie dell'*Atriplex nummularia* Lind. in relazione alla assimilazione. Nuovo Giorn Bot Ital 22:426–430

Aslam Z, Jeschke WD, Barrett-Lennard EG, Setter TL, Watkin E, Greenway H (1986) Effects of external NaCl on the growth of *Atriplex amnicola* and the ion relations and carbohydrate status of the leaves. Plant Cell Environ 9:571–581

Balasubramanian T, Khan SA (2002) Mangroves of India. State-of-the-art report. ENVIS Publication series 21. Environmental information System Centre, Centre of Advanced Study in marine Biology, Annamalai University, India
Batanouny KH (1973) Kallkdrusen von *Limoniastrum monopetalum.* Naturwiss Rundschau 26:213–214
Batanouny KH (2001) Adaptations of desert organisms. Plants in the desert of the Middle East. Springer, Berlin, Heidelberg
Batanouny KH, Abo Sitta YM (1977) Eco-physiological studies on halophytes in arid and semi-arid zones. I. Autecology of the salt-secreting halophyte *Limoniastrum monopetalum* (L.) Boiss. Acta Bot Acad Sci Hung 23(1–2):13–31
Baumeister W, Kloos G (1974) Über die salzsekretion bei *Halimione portulacoides* (L.) Aellen. Flora 163:310–326
Baylis GT (1940–1941) Leaf anatomy of the New Zealand mangroves. Trans Proc Royal Soc New Zeal 70:164–170
Bennert HW, Schmidt B (1983) Untersuchungen zur Salzabscheidung bei *Atriplex hymenelytra* (Torr.) Wats. (Chenopodiaceae). Flora 174:341–355
Bentham G, Hooker JD (1876) Genera plantarum ad exemplaria imprimis in Herbariis Kewensibus servata definita, vol 2. Reeve & Co.Williams & Norgate, Covent Garden, p 2
Black RF (1954) The leaf anatomy of Australian members of the genus *Atriplex.* I. *Atriplex vesicaria* Heward and *A. nummularia* Lindli. Aust J Bot 2:269–286
Blasco FR (1991) Les mangroves. La Recherche 22(231):445–453
Bonnier G, Du Sablon LL (1905) Cours de Botanique. Phanérogames. Librairie Générale de l' Enseignement, Paris
Braconnot H (1836) Sur les écailles de nature inorganique produites par les plantes de la famille des Plumbaginées. Ann de Chimie et de Physique 73:373–377
Breckle SW (1995) How do halophytes overcome salinity? In: Biology of Salt Tolerant Plants (ed. by M. A. KHAN, I. A. UNGAR), Department of Botany, University of Karachi, Pakistan, p. 199-213
Breckle SW (2002) Salinity, halophytes and salt affected natural ecosystems. In: Läuchli A, Lüttge U (eds) Salinity: Environment-plants-molecules. Kluwer Academic, New York, pp 53–77
Breckle SW, Freitas H, Reimann C (1990) Sampling *Atriplex* bladders: a comparison of methods. Plant Cell Environ 13:871–873
Brian RC, Cattlin ND (1968) The surface structure of leaves of *Chenopodium album* L. Ann Bot 32:609–612. (abstract)
Brunner C (1909) Beiträge zür vergleichenden Anatomie der Tamaricaceae. Jahrb f Wiss Anst Hamburg 27:89–162
Bunge A (1872) Die Gattung *Acantholimon* Boiss. Mém de l'Acad Impér des Sci de St. Pétersbourg, VIIe sér 18(2):1–72
Campbell CJ, Strong JE (1964) Salt gland anatomy in *Tamarix pentandra.* Southwest Nat 9:232–238
Campbell N, Thomson WW (1976) The ultrastructure of *Frankenia* salt glands. Ann Bot 40:681–686
Campbell N, Thomson WM, Platt K (1974) The apoplastic pathway of transport to salt glands. J Exp Bot 25(84):61–69
Carpenter JL (1983) Leaf glands and salt secretion in *Avicennia marina* (Forsk) Vierh. The University of Sydney, Australia, Hons Thesis
Chermezon H (1910) Recherches anatomiques sur les plantes littorales. Ann Sci Nat, sér. 9, Botany 12:117–313
Cronquist A (1981) An integrated system of classification of flowering plants, Second edn. Columbia University Press, New York
Cross BD (1909) Observations on some New Zealand halophytes. Trans Proc New Zeal Inst 42:545–574

D'Ambrogio AS, Fernandez S, Gonzales E, Furlan I, Frayssinet N (2000) Estudios morfoanatomicas y citologicos en *Atriplex sagittifolia* Speg. (Chenopodiaceae). Bol Soc Argent Bot 35(3–4):215–226

Dajic Z (2006) Salt stress. In: Madhava Rao KV, Raghavendra AS, Reddy KJ (eds) Physiology and molecular biology of stress tolerance in plants. Springer, Dordrecht, pp 41–100

Das S (2002) On the ontogeny of stomata and glandular hairs in some Indian mangroves. Acta Bot Croat 61(2):199–205

Das S, Ghose M (1996) Anatomy of leaves of some mangroves and their associates of Sundarbans. West Bengal Phytomorphol 46(2):139–150

Dawes CJ (1998) Marine botany, Second edn. Wiley, New York

de Araujo SAM, Silveira JAG, Almeida TD, Rocha IMA, Morais DL, Viegas RA (2006) Salinity tolrance of halophyte *Atriplex nummularia* grown under increasing NaCl leaves. Rev Bras Engenharia Agrocola Ambiental 10(4):848–854

de Bary A (1877) Vergleich Anat d Vegetationsorgane d Phanerogamen und Farne

Bary A de (1884) Comparative anatomy of the vegetative organs of the Phanerogams and Ferns, engl. ed. Oxford at the Clarendon Press

De Fraine E (1916) The morphology and anatomy of the genus *Statice* as represented at Blakeney Point. I. *Statice binervosa* G.E. Smith and *Statice bellidifolia* D.C. (= *Statice reticulata*). Ann Bot (London) 30:239–282

Decker JP (1961) Salt secretion by *Tamarix pentandra* Pall. Forest Sci 7:214–217

Deschida WJ, Platt-Aloia KA, Thomson WW (1992) Epidermal peels of *Avicennia germinans* (L.) Stearn: a useful system to study the function of salt glands. Ann Bot 70:501–509. (abstract)

Deslongchamps L (1820) Dictionnaire des Sciences naturelles, vol 17. F. G. Levrault, Paris

Dickison WC (2000) Integrative plant anatomy. Harcourt Academic, San Diego

Drennan PM, Berjak P (1982) Degeneration of the salt glands accompanying foliar maturation in *Avicennia marina* (Forsskal) Vierh. New Phytol 90:165–176

Drennan P, Pammenter NW (1982) Physiology of salt excretion in the mangrove *Avicennia marina* (Forsk.) Vierh. New Phytol 91:597–606

Drennan PM, Berjak P, Lawton JR, Panmenter NW (1987) Ultrastructure of the salt glands of the mangrove, *Avicennia marina* (Forssk.) Vierh., as indicated by the use of selective membrane staining. Planta 172:176–183. (abstract)

Endlicher S (1836–1840) Genera plantarum secundum ordines naturales disposita. Beck Universitatis Bibliopolam, Vindobonae apud Fr

Evert RF, Esau K, Eichhorn SE (2006) Esau's plant anatomy, Third edn. Wiley, New York

Fahn A (1967) Plant anatomy. Pergamon, London

Fahn A (1988) Secretory tissues in vascular plants. New Phytol 108:229–257

Fahn A, Shimony C (1977) Development of the glandular and non-glandular leaf hairs of *Avicennia marina* (Forsskal) Vierh. Bot J Linn Soc 74:37–46

Feller IC, Sitnik M (1996a) Mangrove ecology: a manual for a field course. Smithsonian Institution, Washington, DC

Feller IC, Sitnik M (1996b) Mangrove ecology: a manual for a field course. Smithsonian Institution, Washington, DC

Fitting H (1911) Die wasserversorgung und die osmotichen druckverhältnisse der wüstenpflanzen. Ein beitrag zur ökologischen pflanzengeographie. Zeit Bot 3:209–288

Fitzgerald M A, Allaway W G (1991) Appoplastic and symplastic pathways in the leaf of the grey mangrove Avicennia marina (Forsk.) Vierh. New Phytol 119:217-226

Fitzgerald M A, Orlovich D A, Allaway WG (1992) Evidence that abaxial leaf glands are the sites of salt secretion in leaves of the mangrove Avicennia marina (Forssk.) Vierh. New Phytol 120:1-7

Flowers TJ, Flowers SA, Hajibagheri MA, Yeo AR (1990) Salt tolerance in the halophytic wild rice, *Porteresia coarctata* Tateoka. New Phytol 114:675–684

Frayssinet N, Gonzales E, Ambrogio AD, Fernandez S, Furlan I (2007) Estudio citologico, exo y endomorfologico en *Atriplex lampa* (Moq.) D. Dietr. (*Chenopodiaceae*). Polibotanica 24:1–23

Freitas H, Breckle SW (1992) Importance of bladder hairs for salt tolerance of field-grown *Atriplex* species from a Portuguese salt marsh. Flora 187(3–4):283–297

Freitas H, Breckle SW (1993a) Accumulation of nitrate in bladder hairs of *Atriplex* species. Plant Physiol Biochem 31:887–892

Freitas H, Breckle SW (1993b) Progressive cutinization in *Atriplex* bladder stalk cells. Flora 188 (3):287–290

Freycinet de L, Gaudichaud C (1826) Voyage autour du monde fait par ordre du Roi, sur les corvettes de S. M. l'Uranie et la Physicienne, pendant les annés 1817, 1818, 1819 et 1820. Histoire Naturelle: Botanique. Paris, Chez Pillet, Aine, Imprimeur-Libraire, Rue des Grands Augustins no 7

Frey-Wyssling A (1935a) Die Stoffausscheidung der höheren Pflanzen. Springer, Berlin

Frey-Wyssling A (1935b) Ein physiologische System der pflanzlichen Ausscheidungsstoffe. Protoplasma 23:393–409

Gamaley IB (1985) Variaţii kranţ—anatomii u rastenii pustyni Gobi i Karakumi (The variations of the Kranz-anatomy in Gobi and Karakum plants). Bot Jurn SSSR 70:1302–1314

Glenn EP, Brown JJ, Khan MJ (1997) Mechanisms of salt tolerance in higher plants. In: Basra AS, Basra RK (eds) Mechanisms of environmental stress resistance in Plants. Harwood Academic, Amsterdam, pp 83–110

Goodin JR, Mozafar A (1970) Quantitative histochemistry of oxalate in vesiculated hairs of *Atriplex halimus*. Histochem 21(4):366–368. (abstract)

Gorham J (1995) Mechanisms of salt tolerance of halophytes. In: Choukr-Allah R, Malcom CV, Hamdy A (eds) Halophytes and biosaline agriculture. Marcel Dekker Inc, New York, pp 31–54

Griffith W (1854) Notulae ad plantas asiaticas, part IV. In: Dicotyledonous plants, Calcutta. Charles A. Serrao

Griffiths ME, Rotjan RD, Ellmore GS (2008) Differential salt deposition and excretion on leaves of *Avicennia germinans* mangroves. Carrib J Sci 44(2):267–271

Grigore M-N (2008a) Introducere în halofitologie. Elemente de anatomie integrativă. Pim, Iaşi

Grigore M-N (2008b) Cercetări histo-anatomice cu implicaţii ecologice la specii de halofite din Moldova. Teză de doctorat, Universitatea "Alexandru Ioan Cuza" Iaşi, Facultatea de Biologie

Grigore M-N (2012) Romanian salt tolerant plants. Taxonomy and ecology. Tehnopress, Iaşi

Grigore M-N, Toma C (2007) Histo-anatomical strategies of Chenopodiaceae halophytes: adaptive, ecological and evolutionary implications. WSEAS Trans Biol Biomed 12(4):204–218

Grigore M-N, Toma C (2008) Ecological anatomy of halophyte species from the Chenopodiaceae family. Advanced topics on mathematical biology and ecology. In: Proceedings of the 4th WSEAS international conference on mathematical biology and ecology—MABE'08, Acapulco, Mexico, January 25–27. pp 62–67

Grigore M-N, Toma C (2010) Structuri secretoare de săruri la halofite. O abordare integrativă. Edit. Academiei Române, Bucureşti

Grigore MN, Toma C (2016) Structure of salt glands of Plumbaginaceae. Rediscovering old findings from 19th century. 'Mettenius' or 'Licopoli' organs? J Plant Develop 23:37-52

Grigore M-N, Ivanescu L, Toma C (2014) Halophytes. An integrative anatomical study. Springer, Cham

Haberlandt G (1914) Physiological plant anatomy. MacMillan, London

Helder RJ (1956) The loss of substances by cells and tissues (salt glands). In: Ruhland W (ed) Handbuch der Pflanzenphysiologie, vol 2. Springer, Berlin, pp 468–488

Hogarth PJ (2007) The biology of mangroves and seagrasses, Second edn. Oxford University Press, Oxford

Ingrouille M, Eddie B (2006) Plants: Diversity and evolution. Cambridge University Press, Cambridge

Ish-Shalom-Gordon N, Dubinsky Z (1990) Possible modes of salt secretion in *Avicennia marina* in the Sinai. Plant Cell Physiol 31(1):27–32. (abstract)

Jacoby B (1999) Mechanisms involved in salt tolerance of plants. In: Pessarakli M (ed) Handbook of plant and crop stress, Second edn. CRC, Boca Raton, pp 97–124

Jáuregui D, Castro M, Ruiz-Zapata T, Lapp M (2014) Anatomía de los órganos vegetativos de dos especies de *Atriplex* (Chenopodiaceae) de Venezuela. Rev Biol Trop (Int J Trop Biol) 64 (4):1625–1636

Jeschke WD, Stelter W (1983) Ion relations of garden orache, *Atriplex hortensis* L.: growth and ion distribution at moderate salinity and the function of bladders hairs. J Exp Bot 34:795–810

Kabbash AM (2016) Macroscopic and microscopic charachterization of *Atriplex halimus* L. growing in Egypt and in vitro evaluation of its ytotoxic activity. World J Pharm Pharm Sci 5(6):84–100

Kamienski F (1880) Vergleichende Anatomie der Primulaceen. Abhand. Naturforsch Ges Halle 14:141–230

Karimi SH, Ungar IA (1989) Development of epidermal salt hairs in *Atriplex triangularis* Willd., in response to salinity, light intensity and aeration. Bot Gaz 150(1):68–71. (abstract)

Kathiresan K, Bingham BL (2001) Biology of mangroves and mangrove ecosystems. Adv Mar Biol 40:81–251

Khan MA, Gul B (2006) Salt tolerant plants of Coastal Sabkhat of Pakistan. In: Barth HJ, Böer B (eds) Sabkha ecosystems, The Arabian Peninsula and adjacent countries, vol I. Kluwer Academic, Dordrecht, pp 123–140

Khan MA, Qaiser M (2006) Halophytes of Pakistan: characteristics, distribution and potential economic usage. In: Khan MA, Kust GS, Barth HJ, Böer B (eds) Sabkha ecosystems, vol II. Springer Dordrecht, West and Central Asia, pp 129–153

Koyro HW, Lieth H (2008) Global water crisis: the potential of cash crop halophytes to reduce the dilemma. In: Lieth H, Sucre MG, Herzog B (eds) Mangroves and halophytes restoration and utilisation. Springer, Dordrecht, pp 7–20

Kubitzki K (1993) Plumbaginaceae. In: Kubitzki K, Rohwer JG, Bittrich V (eds) The families and genera of vascular plants, vol 2. Springer, Berlin, Germany, pp 523–530

Lal PN (2002) Integrated and adaptive mangrove management framework—an action oriented option for the New Millenium. In: Lacerda LD (ed) Mangrove ecosystems. Function and management. Springer, Berlin, pp 235–254

Lambers H, Chapin FS III, Pons TL (2008) Plant physiological ecology, Second edn. Springer, New York

Lambert M, Turner J (2000) Commercial forest plantations on saline lands. CSIRO Publishing, Collingwood

Levitt J (1972) Response of plants to environmental stresses. Academic, New York

Levering C A, Thomson W W (1971) The ultrastructure of the salt gland of Spartina foliosa. Planta 97:183-196

Licopoli G (1866) Richerche microscopiche sopra alcuni organi particolari della *Statice monopetala*. Sulla formazione di alcune organi nella *Statice monopetala* destinati all'escrezione di sostanza minerale (Extras from) Ann dell' Acad degli Aspiranti Naturalisti di Napoli:1–14

Licopoli G (1879) Gli stomi e le glandole delle piante. Atti dell' R Acad d Sci Fis e Mat 8(5):1–69

Lincevskii IA, Cerniakovskoi EG (1952) Plumbaginaceae. In: Shishkin BK, Bobrov EG (eds) Flora SSSR, vol 18. Izdatel'stvo Akademii Nauk SSSR, Moskva, Leningrad, pp 292–474

Lindley J (1846) The vegetable kingdom: or, the structure, classification and uses of plants, illustrated upon the natural system, second edn. Braddbury and Evans, London

Liphschitz N, Waisel Y (1974) Existence of salt glands in various genera of the Gramineae. New Phytol 73:507–513

Lledo MD, Crespo MB, Cameron KM, Fay MF, Chase MW (1998) Systematics of Plumbaginaceae based upon cladistic analysis of rbcL sequence data. Systematic Bot 23:21–29

Lledo MD, Karis PO, Crespo MB, Fay MF, Chase MW (2001) Phylogenetic position and taxonomic status of the genus *Aegialitis* and subfamilies Staticoideae and Plumbaginoideae (Plumbaginaceae): evidence from plastid DNA sequences and morphology. Plant Systematics Evol 229:107–124

Lledo MD, Crespo MB, Fay MF, Chase MW (2005) Molecular phylogenetics of *Limonium* and related genera (Plumbaginaceae): biogeographical and systematic implications. Am J Bot 92 (7):1189–1198
Loudon J (1842) The ladies' flower-garden of ornamental annuals. William Smith, London
Lüttge U (2002) Mangroves. In: Läuchli A, Lüttge U (eds) Salinity: environment-plants-molecules. Kluwer Academic, New York, pp 113–135
Lüttge U (2008) Physiological ecology of tropical plants, second edn. Springer, Berlin
Mahi Z, Dedaldechamp F, Belkhodja M, Lemoine R (2015) Anatomical features of *Atriplex halimus* L. to saline environments. IJSRST 1(6):69–76
Marcum KB (1999) Salinity tolerance mechanisms of grasses in the subfamily Chloridoideae. Crop Sci 39:1153–1160
Marcum KB (2001) Growth and physiological adaptations of grasses to salinity. In: Pessarakli M, Dekker M (eds) Handbook of plant and crop physiology, Second edn. Basel, New York, pp 623–636
Marcum KB, Murdoch CL (1990) Salt glands in the Zoysieae. Ann Bot 66:1–7
Marloth R (1887) Zur Bedeutung der Salz Drüsen der Tamariscineen. Ber deutsch bot Ges 5:319–324
Martinet MJ (1872) Organes des sécretion des végetaux. Ann Sci Nat, Bot 5(14):91–232
Maury P (1886) Etude sur l'organisation et distribution géographique des Plombaginacées. Ann Sc Nat, sér 7, Bot 4:1–134
Metcalfe CR, Chalk L (1972) Anatomy of the dicotyledons, vol 2. Clarendon, Oxford, pp 1075–1084
Mettenius G (1856) Filices Horti Botanici Lipsiensis. Die Farne Des Botanischen Gartens zu Leipzig. Verlag von Leopold Voss, Leipzig
Mohr H, Schopfer P, Lawlor G, Lawlor DW (1995) Plant physiology. Springer, Berlin
Monteil P (1906) Anatomie comparée de la feuille des Chénopodiacées, Thèse no. 9, Ecole Supérieure de Paris
Moore DM (1972) Plumbaginaceae. In: Tutin TG, Heywood VH, Burges NA, Moore DM, Valentine DH, Walters SM, Webb DA (eds) Flora Europaea, vol 3. Cambridge University Press, Cambridge, pp 29–51
Moţiu T, Toma C, Tiron A, Niţă M (1987) Contribuţii la cunoaşterea structurii organelor vegetative de *Limonium gmelini* (Willd). O. Ktze. An Şt Univ "Al I Cuza" Iaşi, s. II a (Biol) 33:11–14
Mozafar A, Goodin JR (1970) Vesiculated hairs: a mechanism for salt tolerance in *Atriplex halimus* L. Plant Physiol 45:62–65
Mullan DP (1931) On the occurrence of glandular hairs (salt glands) on the leaves of some Indian halophytes. J Ind Bot Soc 10:184–209
Nandy (Datta) P, Das S, Ghose M (2005) Relation of leaf micromorphology with photosynthesis and water afflux in some Indian mangroves. Acta Bot Croat 64(2):331–340
Orcutt DM, Nilsen KT (2000) The physiology of plants under stress: soil and biotic factors. Wiley, Hoboken, NJ
Osmond CB (1974) Leaf anatomy of Australian saltbushes in relation to photosynthetic pathways. Austral J Bot 22(1):39–44. (abstract)
Osmond C B Lüttge U, West K R, Pallaghy C K, Shacher-Hill B (1969) Ion absorption in Atriplex leaf tissue. II. Secretion of ions to epidermal bladders. Austral J Biol Sci 22:797–814
Packham JR, Willis AJ (1997) Ecology of dunes, salt marsh and shingle. Chapman & Hall, London
Pallaghy CK (1970) Salt relation of *Atriplex* leaves. In: Jones R (ed) The biology of *Atriplex*. Div Plant Industry, CSIRO, Canberra, pp 57–61
Pandey SN, Sinha BK (2005) Plant physiology, Fourth edn. Vikas Publishing House Pvt Ltd, Tamil Nadu, India
Paulsen O (1912) Studies on the vegetation of the Transcaspian lowlands. In: The second Danish Pamir expedition (conducted by Olufsen O). Gyldendalske Boghandel, Nordisk Forlag, Copenhagen

Pax F (1897) Plumbaginaceae. In: Engler A, Prantl K (eds) Die natürlichen Pflanzenfamilien 4 (1–2). von Wilhelm Engelmann, Leipzig, pp 116–125

Pax F, Knuth R (1905) Primulaceae (IV. 237). In: Engler A (ed) Das Pflanzenreich. Regni vegetabilis conspectus. Verlag von Wilhelm Engelmann, Leipzig, pp 1–386

Pollak G, Waisel Y (1979) Ecophysiology of salt secretion in *Aeluropus littoralis* (Gramineae). Physiol Plant 47:177–184

Pora E (1969) Mangrovele. Natura 21(3):45–53

Pouchet FA (1883) The Universe; or, the wonders of creation. The infinitely great and the infinitely little, Seventh edn. H. Hallett and Company, Portland, Me

Pyykkő M (1966) The leaf anatomy of East Patagonian xeromorphic plants. Ann Bot Fennici 3 (3):453–622

Răvăruţ M (1960) Plumbaginaceae. In: Săvulescu T (ed) Flora RPR, vol 7. Academiei RPR, Bucureşti, pp 21–40

Redondo-Gomez S, Mateos-Naranjo E, Davy AJ, Fernandez-Munoz F, Castellanos EM, Luque T, Figueroa ME (2007) Growth and photosynthetic responses to salinity of the salt-marsh shrub *Atriplex portulacoides*. Ann Bot 100(3):555–563

Reyes SA (1997) Plumbaginaceae. In: Sosa V (ed) Flora de Veracruz, Instituto de Ecologia, AC. Xalapa, Ver, vol 97. University of California, Riverside, CA, pp 1–11

Rozema J, Gude H (1981) An ecophysiological study of the salt secretion of four halophytes. New Phytol 89:201–217

Rozema J, Riphagen J (1977) Physiology and ecological relevance of salt secretion by the salt gland of *Glaux maritima* L. Oecologia (Berl) 29:349. (abstract)

Rozema J, Riphagen I, Sminia T (1977) A ligt and electron-microscopical study on the structure and function of the salt gland of *Glaux maritima*. New Phytol 79:665–671

Ruhland W (1915) Untersuchungen über die Hautdrüsen der Plumbaginaccen. Ein Beitrag zur Biologie der Halophyten Jahrb f Wiss Bot 55:409–498

Saenger P (2002) Mangrove ecology, silviculture, and conservation. Kluwer Academic, Dordrecht

Salama FM, El-Naggar SM, Ramadan T (1999) Salt glands of some halophytes in Egypt. Phyton (Horn, Austria) 39(1):91–105

Schimper AFW (1903) Plant geography upon a physiological basis. Clarendon, Oxford

Scholander PF (1968) How mangroves desalinate water. Physiol Plant 21:251–261

Scholander PF, Hammel HT, Hemmingsen EA, Garry W (1962) Salt balance in mangroves. Plant Physiol 37(6):722–729

Scholander PF, Hammel HT, Bradstreet ED, Hemmingsen EA (1965) Sap pressure in vascular plants. Science 148:339–345

Scholander PF, Bradstreet ED, Hammel HT, Hemmingsen EA (1966) Sap concentration in halophytes and some other plants. Plant Physiol 41:529–532

Schtscherback J (1910) Über die Salzausscheidung durch die Blätter von *Statice gmelini*. (Vorläufige Mitteilung). Ber deutsch bot Ges 28:30–34

Schmidt J (1905) Bidrag til kundskab om skuddene hos den gamle verdens mangrovetraeer. Bot Tidsskr 26:1-113

Short PS, Wightman GM (2011) Plumbaginaceae. In: Short PS, Cowie ID (eds) Flora of the Darwin Region, vol 1. Northern Territory Herbarium, Palmerston, Australia, pp 1–4

Shimony C, Fahn A (1968) Light and electron microscopical studies on the structure of salt glands of Tamarix aphylla L. J Linn Soc Bot London 60:283-288

Shimony C, Fahn A, Reinhold L (1973) Ultrastructure and ion gradient in the salt glands of Avicennia marina (Forssk) Vierh. New Phytol 72:27-36

Singh G (2004) Plant systematics: an integrated approach (second ed). Science Publishers, New Hampshire

Skelding AD, Winterbotham J (1939) The structure and development of the hydathodes of *Spartina townsendii* Groves. New Phytol 39:69–79

Smith JAC, Popp M, Lüttge U, Cram WJ, Diaz M, Griffiths H, Lee HSJ, Medina E, Schäfer Stimmel KH, Thonke B (1989) Ecophysiology of xerophytic and halophytic vegetation of a

coastal alluvial plain in northern Venezuela. VI. Water relations and gas exchange of mangrove. New Phytol 111:293–307
Solereder H (1908) Systematic anatomy of the Dicotyledons. In: A handbook for laboratories of pure and applied Botany, vol 1. Clarendon, Oxford
Stenlid G (1958) Salt losses and redistribution of salts in higher plants. In: Ruhland W (ed) Handbuch der Pflanzenphysiologie, vol 4. Springer, Berlin, pp 615–637
Stocking C (1956) Guttation and bleeding. In: Ruhland W (ed) Handbuch der Pflanzenphysiologie, vol 3. Springer, Berlin, pp 489–502
Strasburger E, Noll F, Schenck H, Schimper AFW (1894) Lehrbuch der Botanik für Hochschulen. Jena, von Gustav Fischer Verlag
Sutherland G H, Eastwood A (1916) The physiological anatomy of Spartina townsendii. Ann Bot 30:333-351
Smaoui M A (1971) Différenciation des trichomes chez Atriplex halimus L. C R Acad Sci Paris sér D 273:1268-1271
Takhtajan A (2009) Flowering plants, Second edn. Springer, New York
Taleisnik EL, Anton AM (1988) Salt glands in *Pappophorum* (Poaceae). Ann Bot 62:383–388. (abstract)
Thomson W W (1975) The structure and function of salt glands. In. Plants in saline environments (ed, by. Poljakoff-Mayber A., Gale J.), Springer Verlag, Berlin, Heidelberg, New York, p. 118-146
Thomson W W, Liu L L (1967) Ultrastructural features of the salt gland of Tamarix aphylla L. Planta, 73:201-220
Thomson WW, Platt-Aloia K (1979) Ultrastructural transitions associated with the development of the bladder cells of the trichomes of *Atriplex*. Cytobios 25:105–114
Thomson WW, Berry WL, Liu LL (1969) Localization and secretion of salt by the salt glands of *Tamarix aphylla*. Proc Nat Acad Sci USA 63:310–317
Toderich KN, Tsukatani T, Black CC, Takabe K, Katayama Y (2002) Adaptation of plants to metal/salt contamined environments: glandular structure and salt excretion. Kyoto Institute of Economic Research, Kyoto University, Discussion Paper no. 552, 18p
Tomlinson PB (1994) The botany of mangroves. Cambridge University Press, Cambridge
Tomlinson P B (1995) The Botany of mangroves. Cambridge University Press
Troughton JH, Card KA (1974) Leaf anatomy of *Atriplex buchananii*. New Zeal Bot J 12:167–177
Ungar IA (1991) Ecophysiology of vascular halophytes. CRC, Boca Raton
van der Bakhuizen Brink RC (1921) Revisio generis *Avicenniae*. Bull Jard Bot Btzg, sér 3 (3):199–223
van Tieghem MPH (1898) *Avicenniacées* et *Symphorémacées*. J Bot 12:345–365
Vesque J (1883) Contributions a l'histologie systématique de la feuille des *Caryophyllinées*. Ann Sci Nat, sér 6 Bot 15:105–148
Volkens G (1884) Die Kalkdrüsen der Plumbagineen. Ber deutsch bot Ges 2:334–342
Volkens G (1887) Die Flora der Aegyptisch-Arabischen Wüste. Gebrüder Borntraeger, Berlin
von Minden M (1899) Über die Aussonderung wässeriger Lösungen bei den *Nicotiana*-Arten und *Glaux maritima*. Die Bedeutung des Epithems bei der Sekretion, part V—Beiträge zur anatomischen und physiologischen Kenntnis Wasser-secernierender Organe. Bibliotheca Botanica 46:56–76
Vuillemin P (1887) Recherches sur quelques glandes épidermiques. Ann Sci Nat, sér 7, Bot 5:153–177
Wahid A (2003) Physiological significance of morpho-anatomical features of halophytes with particular reference to Cholistan Flora. Int J Agric Biol 5(2):207–212
Waisel Y (1972) Biology of halophytes. Academic, New York
Waisel Y, Neumann R, Eshel Y (1966) Mineral uptake of plants. Mada 10:273–279
Walter H, Breckle SW (2002) Walter's vegetation of the Earth: the ecological systems of the geobiosphere, Fourth edn. Springer, Berlin

Warming E (1890) Fra Vesterhavskystens Marskegne. Vidensk Meddel Fra D naturh Foren Kjøben V(1):206–239
Warming E (1897) Halophyt-studier. D Kgl Danske Vidensk Selsk Skr, 6, Raekke, naturvidenskabeling og mathematisk Afd. VIII 4:173–272
Warming E (1909) Oecology of plants: an introduction to the study of plant-communities. Clarendon, Oxford
Warming E, Graebner P (1914) Eug. Warming's Lehrbuch der Ökologischen Plflanzengeographie. Gebrüder Borntraeger, Berlin
West KR (1970) The anatomy ox *Atriplex* leaves. In: Jones R (ed) The biology of *Atriplex*. Div. Plant Industry, CSIRO, Canberra, pp 11–15
Wickens GE (1998) Adaptations of desert organisms. Ecophysiology of economic plants in arid and semi-arid lands. Springer, Berlin
Wight R (1850) Icones plantarum Indiae Orientalis, vol IV. P.R. Hunt, American Mission Press, Madras
Wilie N (1883) On stamens og bladets bygning hos *Avicennia nitida* L. Bot Tidsskr 13:33–44
Wood A (1861) Class-book of Botany: being outlines of the structure, physiology, and classification of plants, with a flora of the United States and Colorado. A. S. Barnes & Burr, Troy: Moore & Nims, New York
Wood JG (1925) The selective absorption of chlorine ions and the absorption of water in the genus *Atriplex*. Aust J Exp Biol Med Sci 2:45–56. (abstract)
Woronin M (1885) Notiz über die Structur der Bläter von *Statice monopetala* L. Bot Zeit 43:177–191
Zemke E (1938–1939) Anatomische untersuchungen an Pflanzen der Namibwüste (Deutsch-Südwestafrika). In: Renner O (ed) Flora oder Allgemeine Botanische Zeitung, neue folge, vol 33. Verlag von Gustav Fischer, Jena, pp 365–416
Ziegler H, Lüttge U (1967) Die Salzdrüsen von *Limonium vulgare*. II Mitteilung: Die Lokalisierung des Chloride Planta 74:1–17

Chapter 6
Kranz Anatomy

Generally, the plant leaf is considered the organ with highest plasticity in terms of structural and adaptive value. Indeed, without insisting on the well-known role and functions that leaf plays in the life plant, it can be stated that the leaf represents the physical support for functional processes emerging from the long process of evolution and adaptation of plant facing many harmful environmental factors (Grigore 2008).

In several cases, it was observed that leaves of many species that grows in difficult conditions (salinity, aridity) present peculiar features related to special differential tissue functionality, in the direction of a better adaptation to the environment.

From all of these adaptations, Kranz anatomy is very interesting as a perfect example of connection between structure and functional processes in C_4 photosynthetic plants. It has been noticed a long time ago that nervures from *Atriplex* species are surrounded by a sheath of cubic cells containing chloroplasts, higher than other mesophyll cells (Laetsch 1968).

Generally, it is considered that the botanist Moser (1934) used first time the expression containing the word "Kranz," in relation to the foliar anatomy of *Atriplex tataricum*, where he observed that the nervures are surrounded by chlorenchymatic sheaths (*Der Kranztypus in der Gattung Atriplex*); he also depicted this type of structure. Literally, Kranz means "wreath, corona" that explicitly illustrates the way in which these tissues are arranged around vascular bundles. Thereafter, it was discovered that monocots can also present this kind of leaf structure. In fact, Moser only systematized and deepened this issue, explicitly referring to "*Kranztypus in der Gattung Atriplex*" (p. 380), but Volkens in the two papers (1887—mentioned by Moser and another from 1893—consulted by us) mentioned about the arrangement of palisade tissue around nervures in *Atriplex* species (*A. halimus*, *A. roseum*, and *A. sibiricum*): "*Nervenbündel (. . .), um die sich Palissaden im Kranze herumlegen.*" (Volkens 1893, p. 64—see also Fig. 6.1).

However, since Moser mentioned Volkens' paper from 1887 and Volkens himself uses this description in the work from 1893, where he introduces drawings

M.-N. Grigore, C. Toma, *Anatomical Adaptations of Halophytes*,
DOI 10.1007/978-3-319-66480-4_6

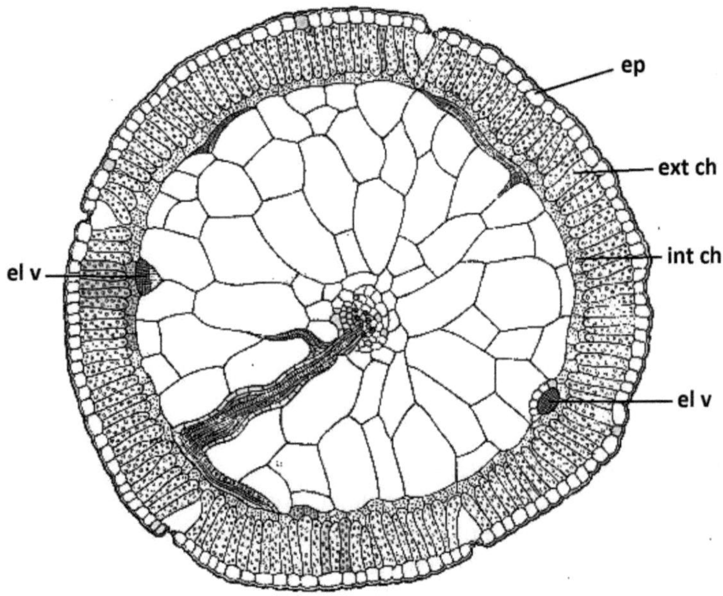

Fig. 6.1 Cross section through the lamina of *Zygophyllum simplex* (*ep* epidermis, *ext ch* external chlorenchyma = palisade tissue, *int ch* internal chlorenchyma = bundle sheath, *el v* vascular elements? If yes, then the structure would fit with *salsoloid* sybtype) (Volkens 1887)

from 1887 monograph, we should consider, based on this data, that the first mention of Kranz structure in *Chenopodiaceae* was made in 1887. Close to this period of time, Arcangeli (1890), Italian botanist, also described a similar structure in *Atriplex nummularia.*

A close connection was made between this type of structure and C_4 plants.

It is well known that in the great majority of C_4 plants, functioning of the C_4 pathway requires metabolic cooperation of two closed and distinct chlorenchyma tissues: an external one (or photosynthetic carbon assimilative—PCA) and an inner bundle sheath (or photosynthetic carbon reductive—PCR) tissue. These tissues are arranged concentrically with respect to vascular tissues, forming a structural pattern known as Kranz anatomy (Muhaidat et al. 2007). This structural type provides one of the best examples of the intimate connection between plant form and function and represents a suite of structural characters that have evolved repeatedly from C_3 ancestors (Dengler and Nelson 1999; Kellog 1999; Sage 2001, 2004).

However, in order to simplify the strictly anatomical language used in this chapter, we will refer to photosynthetic carbon assimilative (PCA) tissue as external chlorenchyma and to photosynthetic carbon reductive (PCR) as internal chlorenchyma, respectively.

This internal structure physically partitions the biochemical events of the C_4 pathway into two major phases. In the first step, atmospheric CO_2 is initially assimilated into C_4 acids by PCA tissue-specific phosphoenolpyruvate carboxylase. In the second phase, these acids diffuse into the PCR compartment, where they are decarboxylated, and the released CO_2 is refixed by PCR tissue-specific Rubisco. This biphasic C_4 system enhances CO_2 levels around Rubisco, suppressing photorespiration and improving plant carbon balance (Kanai and Edwards 1999).

In fact, the biochemical and physiological events associated with C_4 photosynthesis are indefinitely much complicated, and in this chapter, attention will be maintained only to its structural support, the Kranz anatomy.

Consequently, three subtypes of C_4 biochemical pathways have been described, as follows (Kanai and Edwards 1999):

1. NADP-ME (NADP-malic enzyme)
2. NAD-ME (NAD-malic enzyme)
3. PEP-CK (phosphoenolpyruvate-carboxykinase)

These three types occur in *Poaceae*, each of which is characterized by a set of "classical" sequence of anatomical and ultrastructural features, which include the number of vascular bundle sheath layers, the position of chloroplasts within internal chlorenchyma cells, the development of chloroplast grana, the size and number of mitochondria, and the occurrence of a suberin lamella within the internal chlorenchyma cell walls (Gutierrez et al. 1974; Hattersley and Browning 1981; Yoshimura et al. 2004; Ueno et al. 2005).

In dicots, only NADP-ME and NAD-ME subtypes can be found. These two subtypes cannot be obviously distinguished on the basis of anatomical characteristics; for example, PCR cell chloroplasts have a centripetal position in both NADP-ME and NAD-ME subtypes, with a few exceptions, where chloroplasts are centrifugally placed. As in C_4 grasses, the two subtypes are distinguished by a pronounced cytological dimorphism between cells of the two types of chlorenchyma tissues (Voznesenskaya et al. 1999). Thus, PCR cell chloroplasts in NADP-ME subtype eudicots have greatly reduced grana (associated with a high ratio between photosynthetic systems I and II), while those of PCA cells have well-developed grana (high levels of photosynthetic systems I and II). The higher photosynthetic system I: II ratio in PCR of NADP-ME species reflects a higher proportion of cyclic electron flow, related to linear electron flow. The opposite is true for PCR chloroplasts of NAD-ME C_4 plants (Kanai and Edwards 1999; Voznesenskaya et al. 1999; Takabayashi et al. 2005).

Some time ago, anatomical characteristics were considered sufficient to consider a plant as C_4 type, i.e., having or not a Kranz anatomy structure type. However, in time, it was demonstrated that these characters should not be generalized, and they are insufficient to make absolute classifications. In addition, it was observed that the presence of these structures is not compulsory for a plant to be considered as C_4 (Shomer-Ilan et al. 1975).

Moreover, such correlations made between photosynthetic pathway and anatomical support of processes taking place within it led to classifications of different types of foliar anatomy in the *Chenopodiaceae*.

Carolin et al. (1975) have conducted the first ultrastructural study in the *Chenopodiaceae*, identifying and describing different anatomical structures;

interestingly, some taxonomic correlations between different groups were done, thus suggesting that certain taxonomic reorganizations might be useful.

After 25 years from Carolin's study, Jacobs (2001) has reviewed in detail the anatomical types of species from *Chenopodiaceae*, on structural and ultrastructural considerations. In the next paragraphs, only the structural configurations related to C_4 photosynthesis will be described (Jacobs 2001):

1. *Atriplicoid* (C_4) (Carolin et al. 1975). The Kranz cells form a parenchymatous sheath around the bundles, except for a gap on the abaxial side, and the mesophyll is arranged radially. The Kranz cell walls are thicker than those of the mesophyll. The Kranz cell chloroplasts generally have more starch grains than the mesophyll chloroplasts, although chloroplasts in both cell types have well-formed grana. The Kranz cells have larger densities of mitochondria between the elongated plastids.
2. *Kochioid* (C_4) (Carolin et al. 1975). The Kranz cells form arcs along the xylem of peripheral bundles. Most leaves have extensive central aqueous tissue. There is a main bundle and several peripheral bundles with the Kranz cells forming a partial parenchymatic sheath. On the main bundle, this parenchymatic sheath is on the adaxial side but often is not present toward the base of the leaf. In *Kochia scoparia*, the central aqueous tissue is reduced and the lateral bundles opposite to each other are pressed together and the Kranz cells form a partial parenchymatic sheath interrupted laterally. The Kranz cell walls are thicker than those of the mesophyll, and there is no clear increase in the density of mitochondria. The Kranz cell chloroplasts tend to be centripetal and have ill-developed grana (mostly one or two appressions) and well-developed starch grains. The mesophyll cell chloroplasts have well-defined grana and fewer and smaller starch grains than those of the Kranz chloroplasts. There is usually only one layer of mesophyll cells, and in transection, most of these appear to be in contact with a Kranz cell.
3. *Salsoloid* (C_4). Similar to the Kochioid type except that both the Kranz (inner layer) and mesophyll (outer layer) cells tend to form complete layers around the leaf. There is still a peripheral network of vascular tissue associated with the Kranz cells, but the main bundle is more clearly central in position for most of the length of the leaf. The mesophyll is generally one layered, but there is sometimes a hypodermis. The ultrastructure of both cells is much the same as in the Kochioid type with the exception that the Kranz cells have smaller vacuoles in the Salsoloid type than in the Kochioid type.
4. *Kranz-Suaedoid* (C_4). (Carolin et al. 1975). The Kranz cells form a more or less complete layer between the aqueous tissue and the mesophyll cell layer. The vascular tissue forms a network in the lateral longitudinal plane, and there is no peripheral network. The plastids of the Kranz cell tend to be centripetal and have larger starch grains, and larger and more grana, than those of the mesophyll. The Kranz cells also tend to have a large vacuole and a higher density of mitochondria than the mesophyll cells.

5. *Kranz-Halosarcoid* (C_4). (Carolin et al. 1982). There are two distinct layers of chloroplast-containing cells. The cells of the outer layer have small intercellular spaces except close to the stomata where these spaces are considerably larger. The chloroplasts have well-defined but small grana and rarely hold starch grains. There may be isolated groups of non-photosynthetic cells embedded among the chloroplast-containing cells. The cells of the inner layer tend to be more isodiametric, have thicker walls, and have denser mitochondria than the cells of the outer layer. The chloroplasts of the inner cells tend to be centrifugal, with small grana and usually large starch grains. The peripheral vascular bundles are often close to this inner layer of Kranz cells.

In order to simplify things and to make a more operational classification, this classification should be reformulated in fewer words. Actually, four types of anatomical configuration can be maintained: *atriplicoid*, *kochioid*, *suaedoid*, and *salsoloid* (Gamaley 1985; Fisher et al. 1997; Kadereit et al. 2003; Voznesenskaya et al. 1999; Pyankov et al. 2001; Muhaidat et al. 2007). The four anatomical types are readily discerned on the basis of photosynthetic tissue arrangement with respect to the vascular and other tissues. In the *atriplicoid* type, the internal chlorenchyma forms a complete (or nearly complete) sheath around vascular bundles. In the *kochioid* type, internal chlorenchyma cells are confined to the exterior of peripherally positioned veins and do not form a continuous layer. Both the *salsoloid* and the *suaedoid* types are characterized by having a continuous stratum of internal chlorenchyma tissue at the periphery of leaves and photosynthetic stems. These two types are distinguished by the position of the vascular bundles in relation to the chlorenchyma. In the *salsoloid* type, minor vascular bundles are located adjacent to the internal chlorenchyma, while larger vascular bundles are more deeply embedded in water storage parenchyma. In the *suaedoid* type, all vascular bundles are centrally placed in water storage tissue and none in direct contact with cells of internal chlorenchyma (Gamaley 1985; Muhaidat et al. 2007; Grigore et al. 2012a, b, 2014).

However, it should be mentioned in respect of *salsoloid* and *suaedoid* types that sometimes, when analyzing a permanent slide with leaves of halophytic species, these "continuous" chlorenchymatic layers may present interruptions in some points of the cross section (Grigore et al. 2014). Perhaps, this is due rather to a technical procedure or selected level to be sectioned from leaf.

As already stated, Kranz anatomy types were evidenced in halophytes a long time ago by early botanists; of course, they do not nominate them as species related to C_4 photosynthesis or belonging to various types of Kranz anatomy, because of limitations imposed by historic period.

For instance, Volkens in his monograph on flora from Egyptian desert mentioned several halophytic species and gave for them respective drawings of microscopic cross sections: *Zygophyllum simplex* (Fig. 6.1), *Anabasis articulata* (Fig. 6.2), *Atriplex halimus* (Fig. 6.3), *Tribulus alatus* (Fig. 6.4), *Bassia muricata* (Fig. 6.5), *Halogeton alopecuroides* (Fig. 6.6), *Salsola longifolia* (Fig. 6.7), and *Haloxylon schweinfurthii* (Fig. 6.8). Drawings with cross sections from chenopods species are

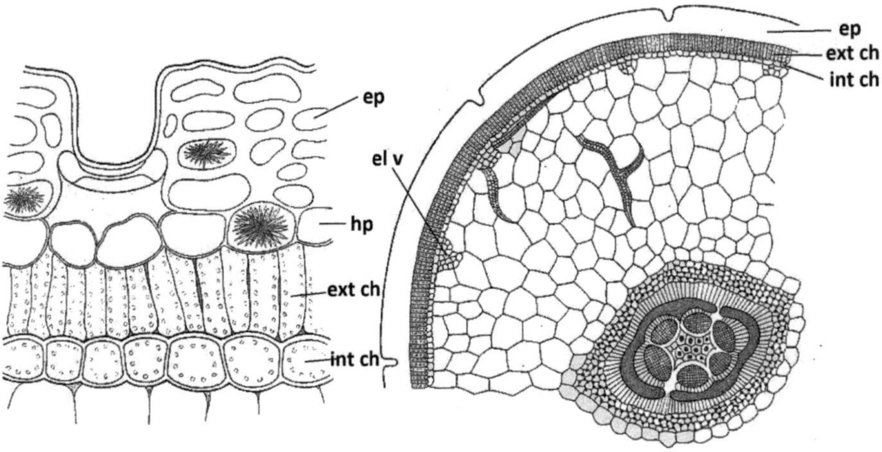

Fig. 6.2 Cross section through the shoot of *Anabasis articulata* (*ep* epidermis, *ext ch* external chlorenchyma, *int ch* internal chlorenchyma, *el v* vascular elements? If yes, then the structure would fit with *salsoloid* sybtype, otherwise to *suaedoid*) (Volkens 1887)

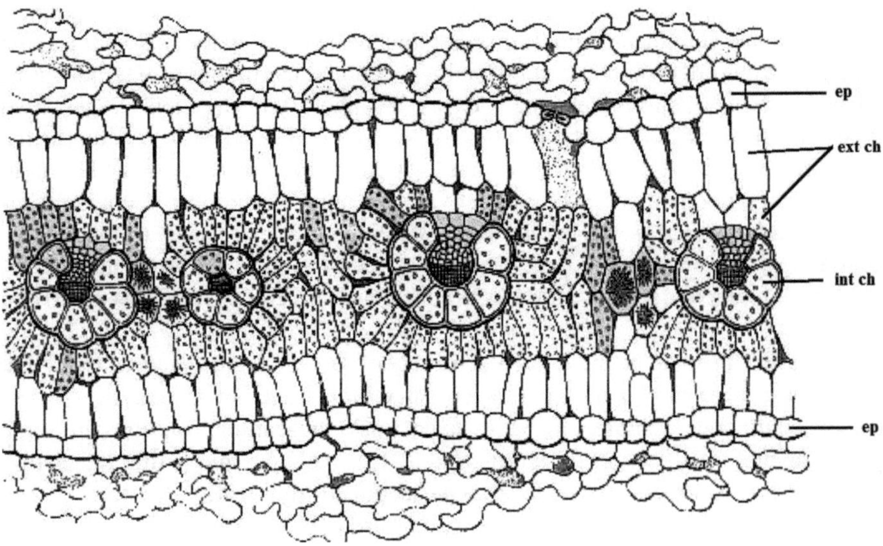

Fig. 6.3 Cross section through the lamina of *Atriplex halimus* (*ep* epidermis, *ext ch* external chlorenchyma, *int ch* internal chlorenchyma; *atriplicoid* sybtype) (Volkens 1887, 1893)

then included in the chapter with general characters of *Chenopodiaceae* (1893). However, he did not explain in detail the nature of pictured tissues; on its figures (and the authors' following) we included the corresponding explanations. When something was questionable, the "?" is given in the figure explanations. It is included especially in respect of possible vascular elements in direct connection

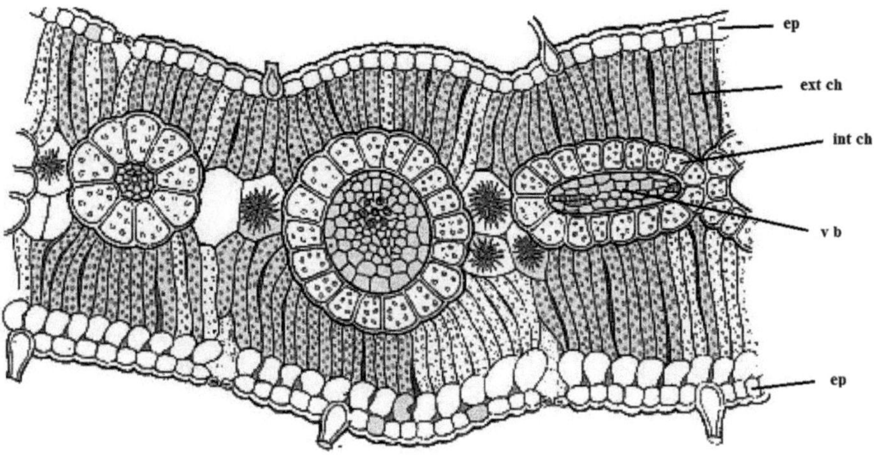

Fig. 6.4 Cross section through the lamina of *Tribulus alatus* (*ep* epidermis, *ext ch* external chlorenchyma, *int ch* internal chlorenchyma, *v b* vascular bundle; *atriplicoid* sybtype) (Volkens 1887)

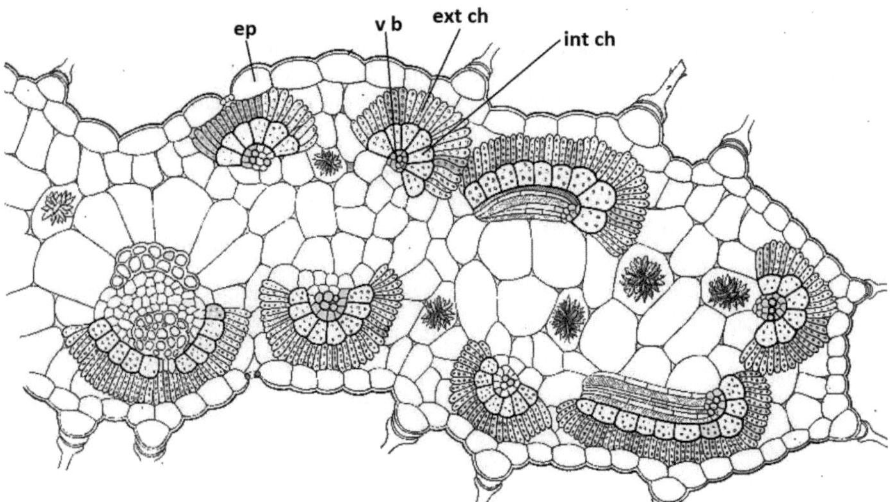

Fig. 6.5 Cross section through the lamina of *Bassia muricata* (*ep* epidermis, *ext ch* external chlorenchyma, *int ch* internal chlorenchyma, *v b* vascular bundle; *kochioid* sybtype) (Volkens 1887)

with inner chlorenchyma, which would change between *salsoloid* and *suaedoid* subtypes (see above paragraphs).

However, Muhaidat et al. (2007) include *Zygophyllum simplex* in the *kochioid* subtype, and their micrograph (and not drawing) is obviously clear. On their image, the vascular bundles are surrounded incompletely by an internal sheath and support the inclusion of this species in the *kochioid* subtype.

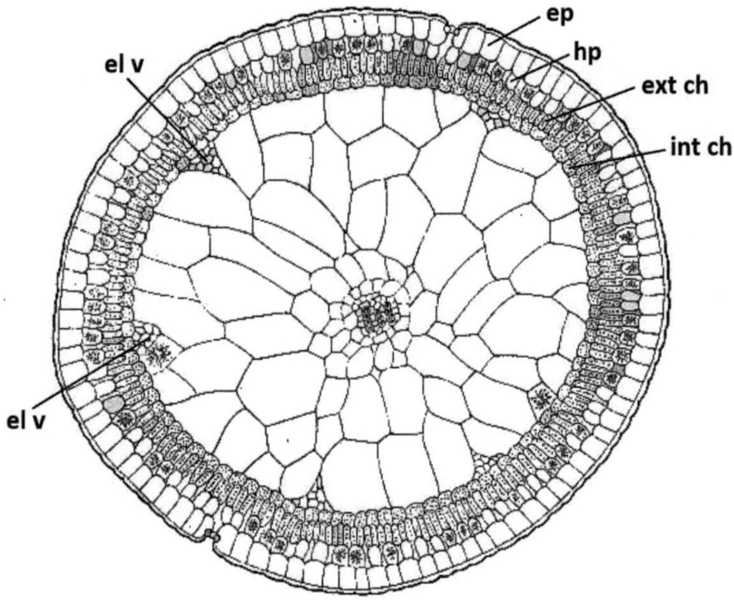

Fig. 6.6 Cross section through the shoot of *Halogeton alopecuroides* (*ep* epidermis, *ext ch* external chlorenchyma, *int ch* internal chlorenchyma, *hp* hypodermis?, *el v* vascular elements? If yes, then the structure would fit with *salsoloid* sybtype, otherwise to *suaedoid*) (Volkens 1887)

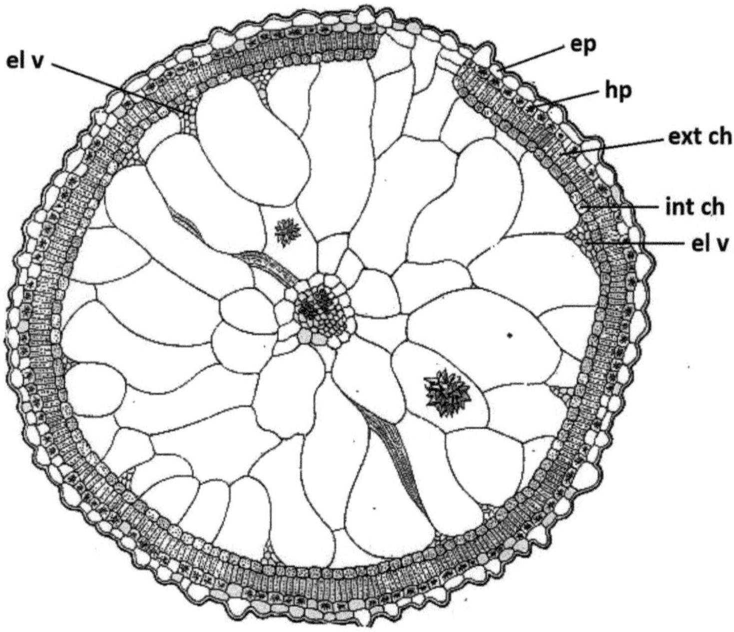

Fig. 6.7 Cross section through the lamina of *Salsola longifolia* (*ep* epidermis, *ext ch* external chlorenchyma, *int ch* internal chlorenchyma, *hp* hypodermis?, *el v* vascular elements? If yes, then the structure would fit with *salsoloid* sybtype, otherwise to *suaedoid*) (Volkens 1887, 1893)

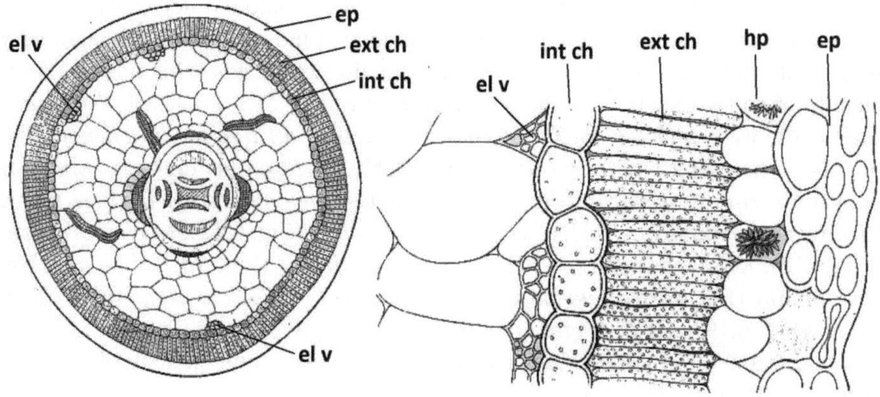

Fig. 6.8 Cross section through the shoot of *Haloxylon schweinfurthii*, general view—*left side*; detail—*right side* (*ep* epidermis, *ext ch* external chlorenchyma, *int ch* internal chlorenchyma, *hp* hypodermis?, *el v* vascular elements? If yes, then the structure would fit with *salsoloid* sybtype, otherwise to *suaedoid*) (Volkens 1887)

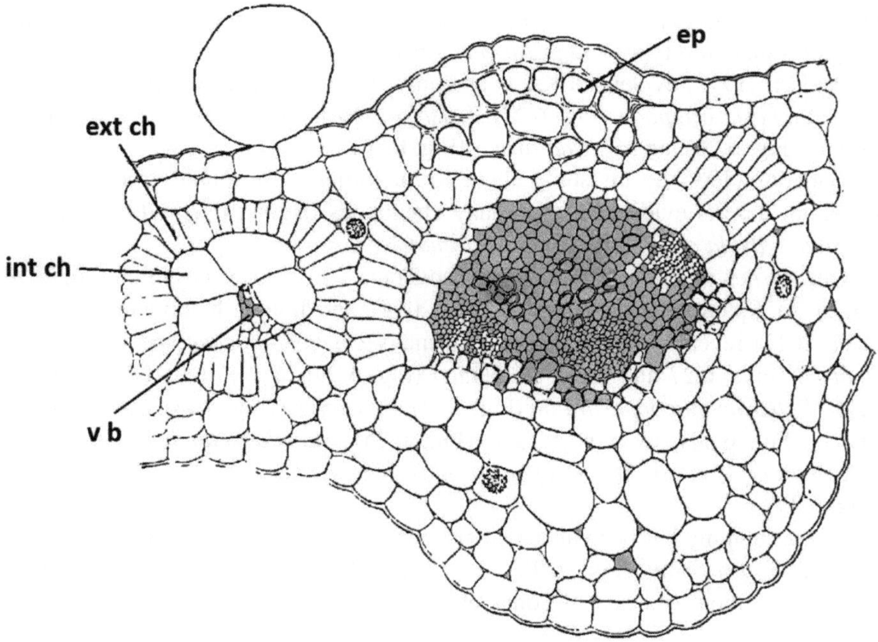

Fig. 6.9 Cross section through the lamina of *Atriplex arenaria* (*ep* epidermis, *ext ch* external chlorenchyma, *int ch* internal chlorenchyma, *v b* vascular bundle; *atriplicoid* sybtype) (Monteil 1906)

Monteil (1906) in his consistent study on leaf structure in *Chenopodiaceae* species evidenced Kranz anatomy in many halophytic taxa. Of course, as in the case of botanists of that time, the expression Kranz anatomy was not used yet.

However, he clearly speaks about palisade tissue and "endodermic" sheath, tissues that are clearly identified on drawings he included in the study. In addition, a very subtle but interesting observation is made with allusion to this foliar sheath (nominated not so rigorously with the word "endodermic"): *"(...) beneath these palisade cells, we found the sheath already reported in* ***all halophytic*** *chenopods."* (p. 114). Therefore, Kranz anatomy is noticeable based on his drawings observations in *Atriplex arenaria* (Fig. 6.9), *Camphorosma monspeliaca* (Fig. 6.10), *Kirilovia eriantha* (Fig. 6.11), *Corispermum orientale* (Fig. 6.12), *Kochia arenaria* (Fig. 6.13), *K. scoparia* (Fig. 6.14), *Chenolea muricata*[1] (Fig. 6.15), *Echinopsilon hyssopifolia* (Fig. 6.16), *Suaeda altissima* (Fig. 6.17), *Salsola soda* (Fig. 6.18), *S. kali* (Fig. 6.19), and *S. tragus* (Fig. 6.20).

Chermezon (1910) identified a Kranz anatomy structure in leaf of *Atriplex crassifolia* (Fig. 6.21); he also made a very important observation, related to the disposition of chlorenchymatic tissues that later will be explicitly designated as Kranz anatomy. Thus, he refers to its foliar anatomy:"*(leaf) is different especially by the clearly* ***radial*** *disposition of palisade tissue* ***around chlorophyll sheath*** *of nervures*" (p. 236).

In Warming's (1897) drawings about anatomical structure of halophytes, Kranz anatomy can be noticeable in *Haloxylon ammodendron* (Fig. 3.39—Succulence chapter—*salsoloid* type) and in *Atriplex farinosa* (Fig. 6.22, *atriplicoid* subtype).

Paulsen (1912) evidenced Kranz anatomy structures in several halophytic species: *Anabasis eriopoda* (Fig. 6.23), *Salsola arbuscula* (Fig. 6.24), *Horaninowia ulicina* (Fig. 6.25), and *Suaeda lipskii* (Fig. 6.26).

Gamaley (1985) in his study about Kranz variations in plants from Gobi and Karakum deserts evidenced it in many chenopods (*Bassia hyssopifolia, Atriplex sibirica, Salsola collina,* and *Suaeda arcuata*).

Our studies in Romanian (Grigore and Toma 2007, 2008; Grigore et al. 2011, 2012a, b), Spanish (Grigore et al. 2011; 2014), and Iranian halophytes (Safiallah et al. 2017) revealed Kranz anatomy in many *Chenopodiaceae* species: *Atriplex tatarica* (Figs. 6.27 and 6.28), *A. glauca* (Fig. 6.29), *A. halimus* (Figs. 6.30 and 6.31), *Petrosimonia oppositifolia* (Fig. 6.32), *P. triandra* (Fig. 6.33), *Camphorosma annua* (Fig. 6.34), *C. monspeliaca* (Fig. 6.35), *Suaeda splendens* (Fig. 6.36), *Salsola kali* (Fig. 6.37), *S. oppositifolia* (Fig. 6.38), and *Bassia hyssopifolia* (Fig. 6.39).

Kranz anatomy has been also evidenced in several halophytic Iranian species of *Bassia: B. prostrata* (Fig. 6.40), *B. pilosa* (Fig. 6.41), and *B. turkestanica* (Fig. 6.42) (photo courtesy of Somayeh Safiallah).

[1]A serious problem is that related to taxonomical nomenclature. Chenopods species have many synonyms, used by various botanists. For instance, this is synonymous with *Bassia muricata*, and *Echinopsilon hyssopifolia*, with *Bassia hyssopifolia*. However, sometimes slight differences occur between drawings of the same species given by different botanists (see next paragraphs).

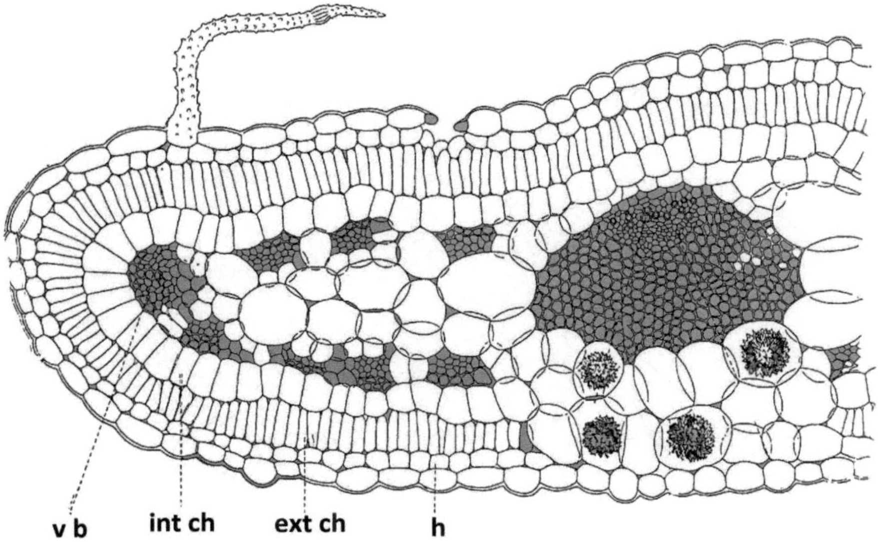

Fig. 6.10 Cross section through the lamina of *Camphorosma monspeliaca* (*h* hypodermis, *ext ch* external chlorenchyma, *int ch* internal chlorenchyma, *v b* vascular bundle; *salsoloid* sybtype) (Monteil 1906)

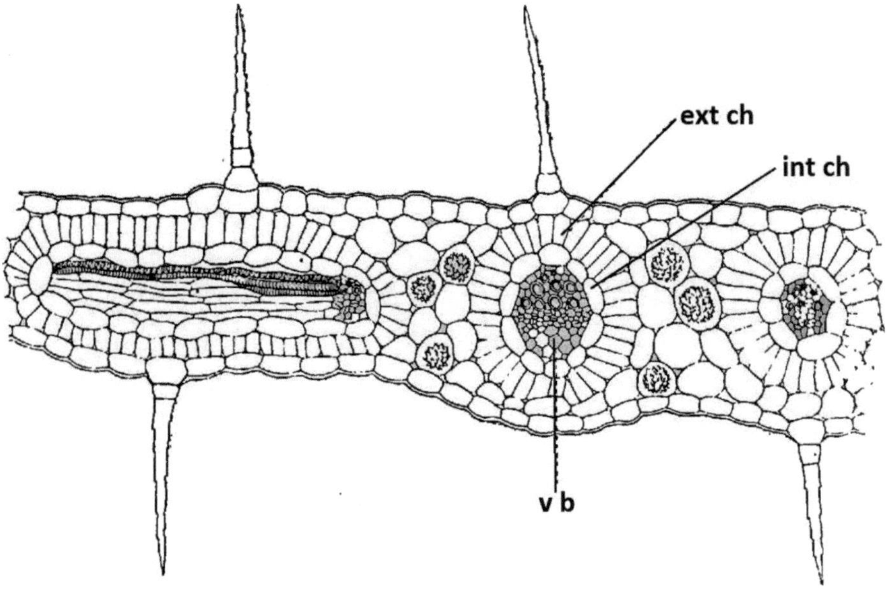

Fig. 6.11 Cross section through the lamina of *Kirilovia eriantha* (*ext ch* external chlorenchyma, *int ch* internal chlorenchyma, *v b* vascular bundle; *atriplicoid* sybtype) (Monteil 1906)

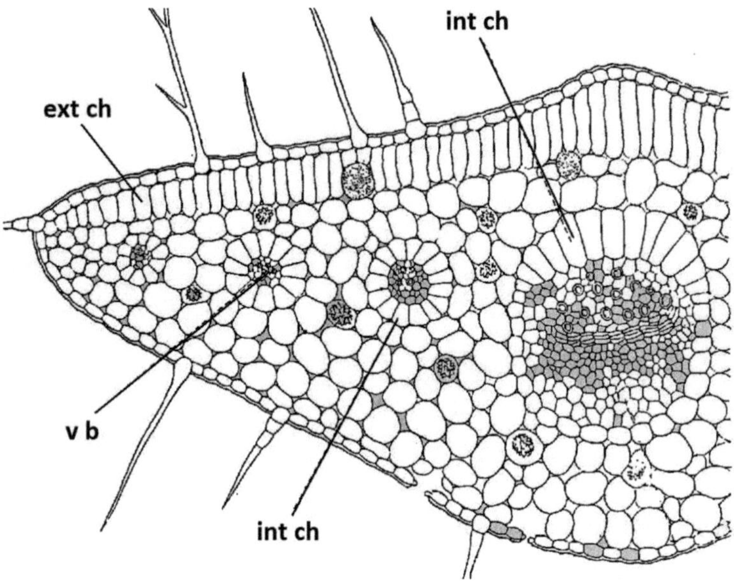

Fig. 6.12 Cross section through the lamina of *Corispermum orientale* (*ext ch* external chlorenchyma, *int ch* internal chlorenchyma, *v b* vascular bundle; *atriplicoid* sybtype? The internal and external ch. seem not to be in direct contact) (Monteil 1906)

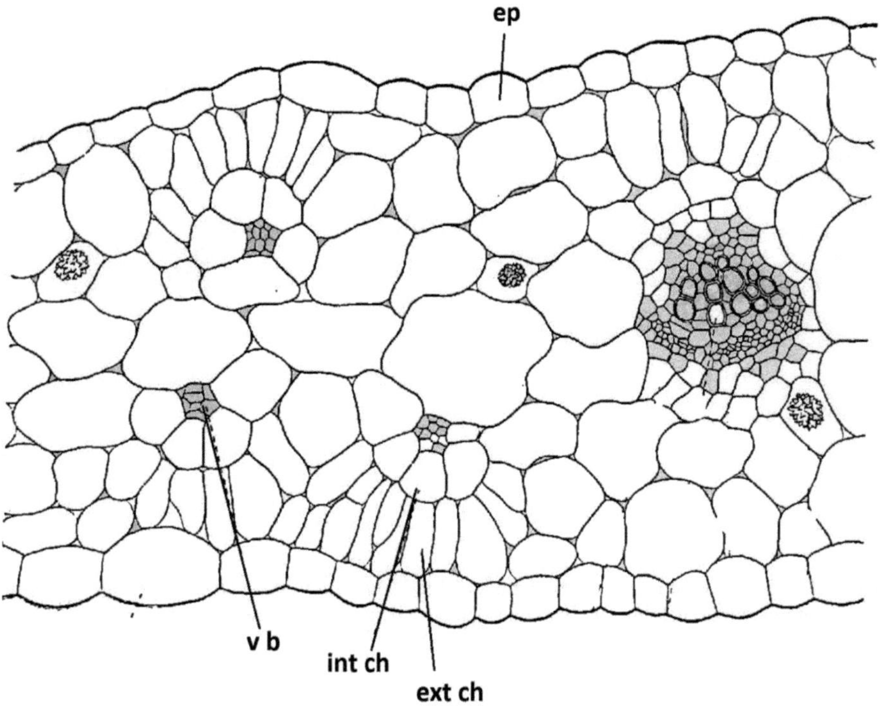

Fig. 6.13 Cross section through the lamina of *Kochia arenaria* (*ep* epidermis, *ext ch* external chlorenchyma, *int ch* internal chlorenchyma, *v b* vascular bundle; *kochioid* sybtype) (Monteil 1906)

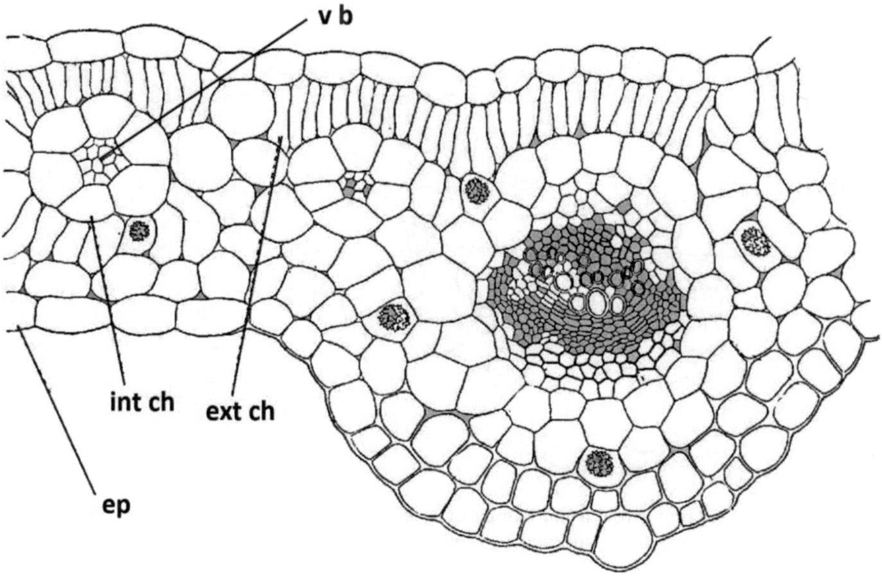

Fig. 6.14 Cross section through the lamina of *Kochia scoparia* (*ep* epidermis, *ext ch* external chlorenchyma, *int ch* internal chlorenchyma, *v b* vascular bundle; *atriplicoid*) (Monteil 1906)

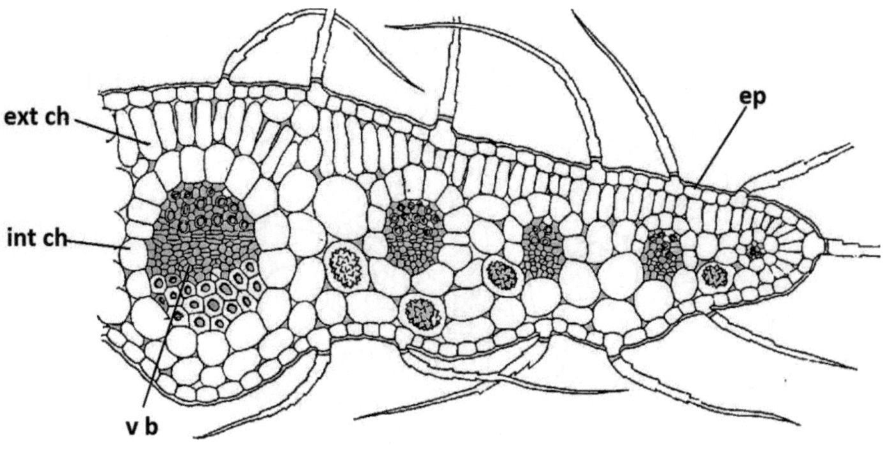

Fig. 6.15 Cross section through the lamina of *Chenolea muricata* (*ep* epidermis, *ext ch* external chlorenchyma, *int ch* internal chlorenchyma, *v b* vascular bundle; *kochioid* subtype—see Fig. 6.6 for comparisons) (Monteil 1906)

The *atriplicoid* subtype found by us in *Atriplex tatarica* was also evidenced by Jacobs (2001) and Muhaidat et al. (2007) and can be found in other species of *Atriplex*: *A. lampa* (Pyykkö 1966) and *A. buchananii* (Troughton and Card 1974).

As we already emphasized, the cross sections should be analyzed carefully, because the continuity/discontinuity of chlorenchyma layers imposes the appurtenance to a subtype or other. For this reason, what we previously considered in

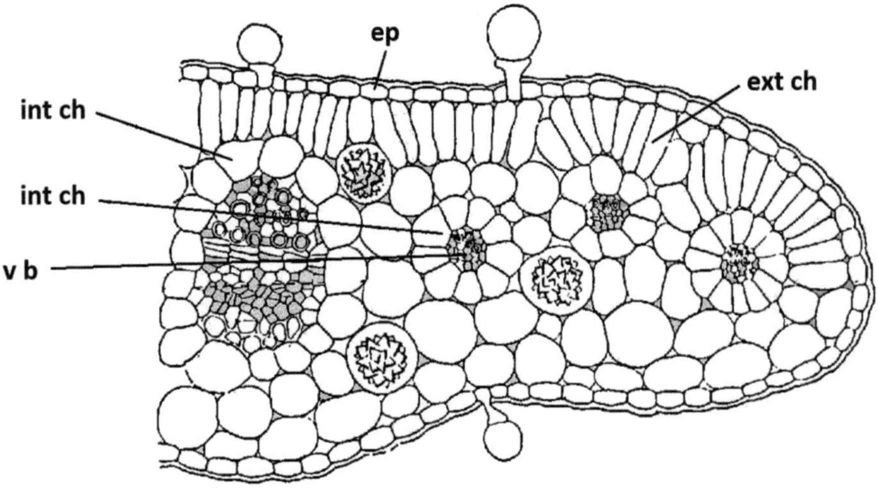

Fig. 6.16 Cross section through the lamina of *Echinopsilon hyssopifolia* (*ep* epidermis, *ext ch* external chlorenchyma, *int ch* internal chlorenchyma, *v b* vascular bundle; *kochioid* subtype) (Monteil 1906)

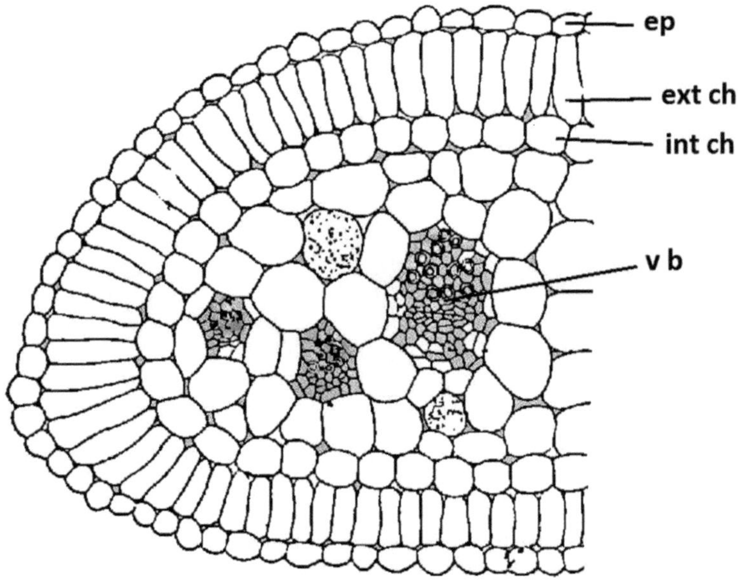

Fig. 6.17 Cross section through the lamina of *Suaeda altissima* (*ep* epidermis, *ext ch* external chlorenchyma, *int ch* internal chlorenchyma, *v b* vascular bundle; *suaedoid* subtype) (Monteil 1906)

Petrosimonia species as belonging to *kochioid* sybtype—because the chlorenchyma layers seemed on analyzed slides as discontinuous—belong in fact to *salsoloid* sybtype (Grigore et al. 2014).

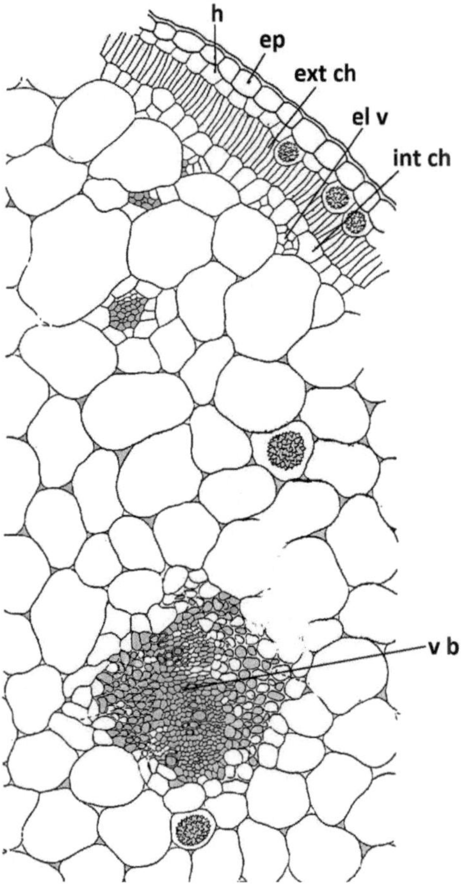

Fig. 6.18 Cross section through the lamina of *Salsola soda* (*ep* epidermis, *h* hypodermis, *el v* vascular elements, *ext ch* external chlorenchyma; *int ch* internal chlorenchyma; v (**b**) vascular bundle; *salsoloid* subtype) (Monteil 1906)

The same is true for *Camphorosma* species, which belongs to salsoloid subtype too.

Muhaidat et al. (2007) reviewed the structural diversity of Kranz anatomy in C_4 eudicots. The *atriplicoid* subtype species has been found in *A. rosea* and *Atriplex polycarpa*; *kochioid* subtype was found in *Kochia scoparia* and *Zygophyllum simplex*. *Salsoloid* subtype was evidenced in *Salsola komarovii* and *suaedoid* subtype, in *Suaeda vermiculata*.

It has been shown that aridity and salinity are important factors promoting stomatal closure and thus reduce intercellular CO_2 levels, stimulating photorespiration and aggravating a CO_2 substrate deficiency (Guy et al. 1980; Adam 1990). Together, the combination of drought, increased salinity, low humidity, and high temperature produces the greatest potential for photorespiration and CO_2 deficiency (Ehleringer and Monson 1993). In addition, drought or salinity stresses further increase CO_2 compensation points, because lower stomatal conductance and

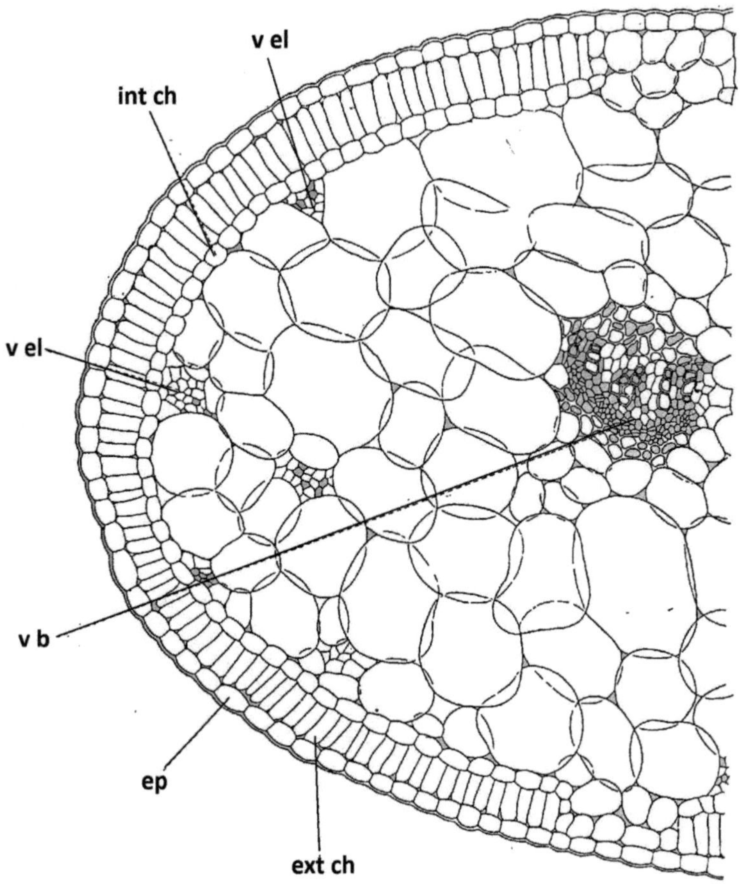

Fig. 6.19 Cross section through the lamina of *Salsola kali* (*ep* epidermis, *el v* vascular elements, *ext ch* external chlorenchyma, *int ch* internal chlorenchyma, *v b* vascular bundle; *salsoloid* subtype) (Monteil 1906)

photosynthetic capacity reduce carbon income, allowing respiration to consume proportionally more of carbon acquired by the plant (Sage 2004).

Evolutionarily speaking, it seems like anatomical modifications (Kranz type) represented a preconditioning step in occurrence of this photosynthetic type (Sage 2004); to evolve an effective CO_2 concentration mechanism, the distance between mesophyll and bundle sheath cells has to decline to allow for rapid diffusion of metabolites (Raghavendra 1980; Ehleringer et al. 1997).

Even with all exposed data at our disposal, it is still difficult to find a direct correlation between salinity factor and Kranz anatomy structures. All investigated species by us are xero-halophytes and obligatory halophytes, excepting *A. tatarica*.

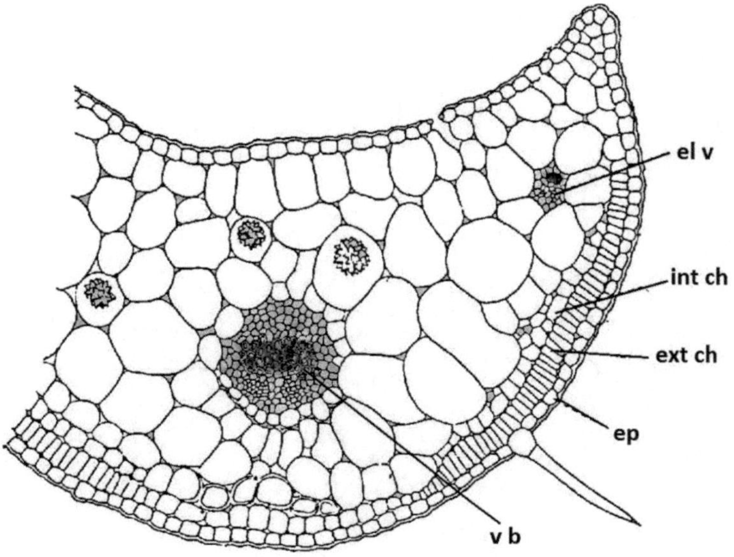

Fig. 6.20 Cross section through the lamina of *Salsola tragus* (*ep* epidermis, *el v* vascular elements, *ext ch* external chlorenchyma, *int ch* internal chlorenchyma, *v b* vascular bundle; *salsoloid* subtype) (Monteil 1906)

In its native distribution area of Middle and western part of Central Asia, this species occupies solonetz sandy and clayey banks of rivers and lakes, coastal solonchaks, and solonetz alluvial trails and is frequently found as a weed in roadside ditches and in villages (Kochánková and Mandák 2008).

C_4 species form a particularly high proportion of the herbaceous flora of saline environments, even in cool temperate regions (Long and Mason 1983). Apparently, the inherently higher water use efficiency of C_4 species would have two theoretical advantages in saline environments (Long 1999). First, saline soils have a soil water potential of around—2.5 MPa; to extract water, the halophytes must generate a lower water potential, even though this exceeds limits that can apparently be tolerated by many mesophytic vascular plants. Transpiration must be minimal, and the higher water use efficiency of C_4 species would confer the advantage of maximizing carbon gain per unit of water lost. Second, plant mineral content is inversely correlated to water use efficiency as an assumed result of increased passive uptake with increased transpiration. For a halophyte, increased transpiration increases the energy needed to exclude Na^+ and Cl^- (Long and Mason 1983).

It has been suggested that halophytes are, in fact, a special case among xerophytes (Wiessner 1899; Henslow 1895; Schimper 1903; Kearney 1904; Warming 1909; Clements 1920; McDougall 1941; Grigore and Toma 2010). This implies the occurrence of some mechanisms serving to protect the water reserves of the plant in periods of drought or high potential evapotranspiration when soil water potential

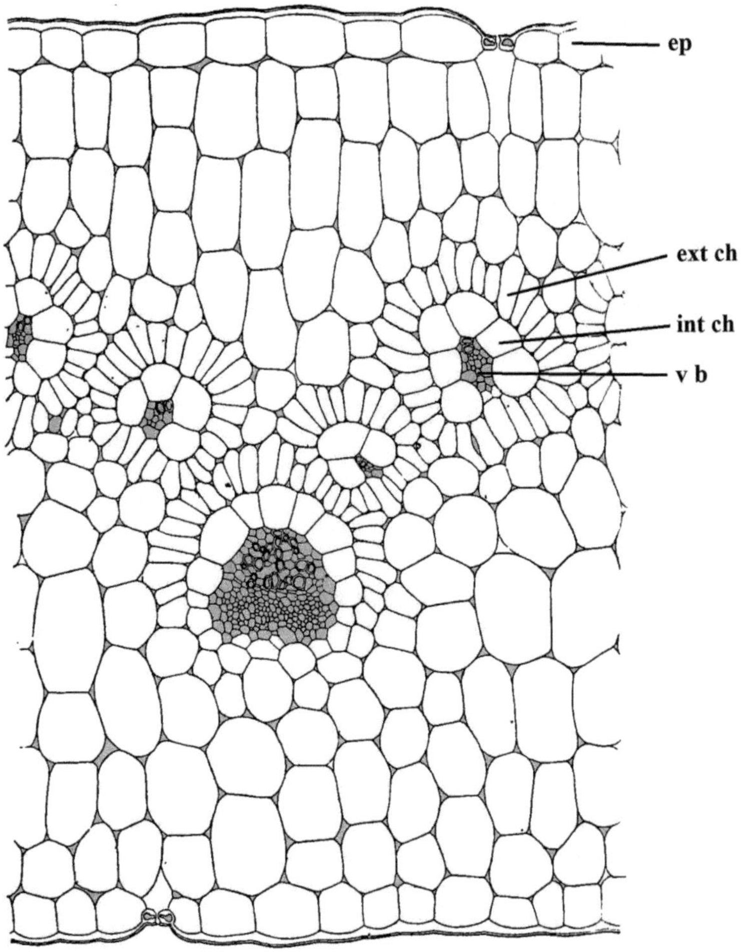

Fig. 6.21 Cross section through the lamina of *Atriplex crassifolia* (*ep* epidermis, *ext ch* external chlorenchyma, *int ch* internal chlorenchyma, *v b* vascular bundle; *atriplicoid* subtype) (Chermezon 1910)

falls. A cost of xeromorphy is increased resistance to diffusion of CO_2 to the mesophyll; because of the low leaf intercellular pressure, necessary to saturate C_4 photosynthesis, this cost is minimized in C_4 species.

Despite the fact that C_4 species represent only about 8000 of the estimated 250,000–300,000 land plants species (Sage et al. 1999), they are major components of biomes that cover more than 35% of the earth's land surface area. These species are dominant in tropical and subtropical grassland and savanna, warm temperate grassland and savanna, arid steppe, beach dunes, salt marshes, salt desert, hot deserts, and semideserts.

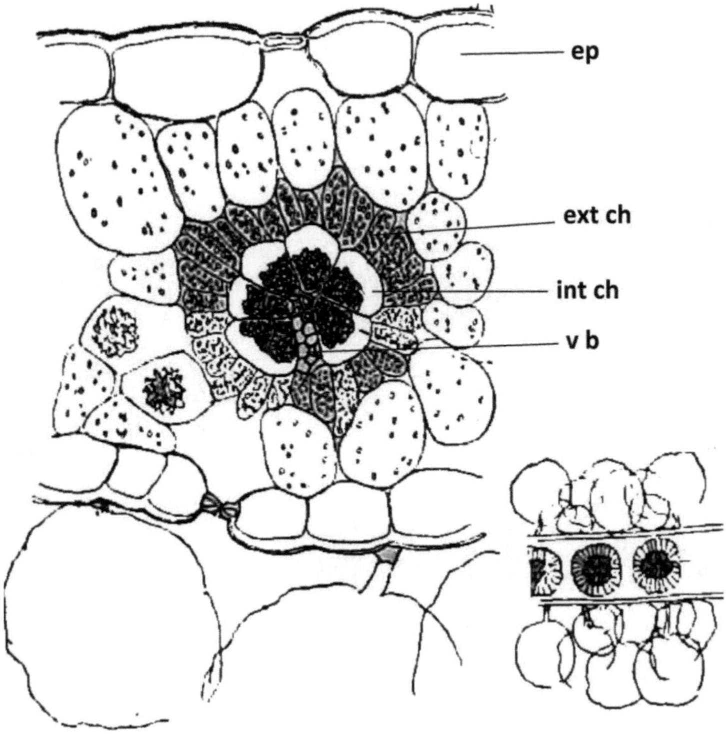

Fig. 6.22 Cross section through the lamina of *Atriplex farinosa* (*ep* epidermis, *ext ch* external chlorenchyma, *int ch* internal chlorenchyma, *v b* vascular bundle; *atriplicoid* subtype; *right* below—general view of Kranz anatomy) (Warming 1897)

C_4 also represents an important ecological strategy in certain desert shrubs, most notably species of *Atriplex*, particularly in saline soils (Keeley and Rundel 2003). In these species, the key adaptation is the ability to maintain growth under high summer temperatures and drought conditions at a time when C_3 species are dormant. The maximal rates of photosynthesis in these desert C_4 species are generally no higher than that of concurring C_3 species, but the water use efficiency is far greater. In addition, C_4 plants have higher nitrogen use efficiency.

Some studies certify the close relationship between C_4 photosynthesis and extreme habitats, such as deserts and salinized areas. Thus, Wang (2007) identified among species vegetating in the deserts of China that 36.5% of the *Chenopodiaceae* species were found with C_4 photosynthesis, which was about 48% of the total C_4 species. These taxa were predominantly members of the genera *Anabasis, Atriplex, Kochia, Salsola,* and *Suaeda.*

Other studies sustain the facts mentioned above: there is a close relationship between some special morphotypes and respective photosynthetic type. In an

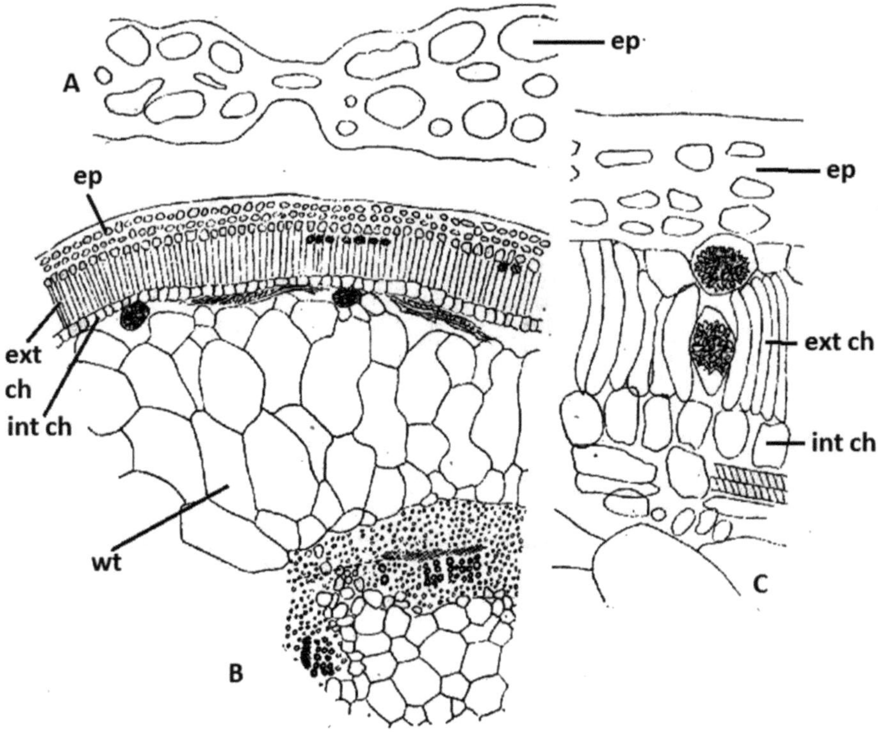

Fig. 6.23 *Anabasis eriopoda*—Cross section through the stem (**b**, **c**) (*ep* epidermis, *ext ch* external chlorenchyma, *int ch* internal chlorenchyma, *wt t* water storage tissue, *C* epidermis) (Paulsen 1912)

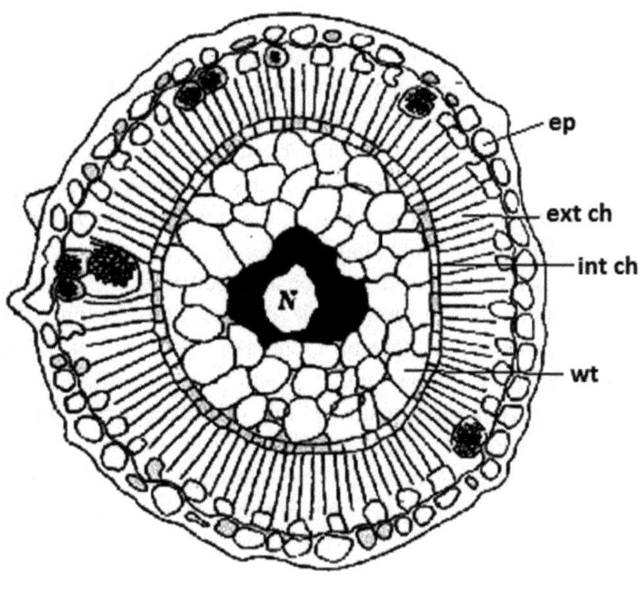

Fig. 6.24 *Salsola arbuscula*—Cross section through the lamina (*ep* epidermis, *ext ch* external chlorenchyma, *int ch* internal chlorenchyma, *wt t* water storage tissue) (Paulsen 1912)

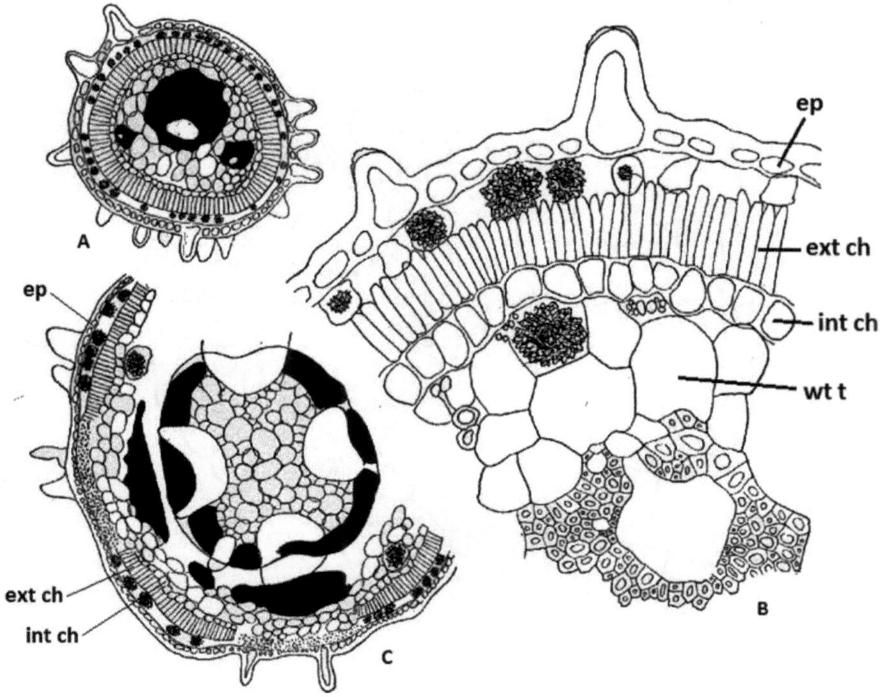

Fig. 6.25 *Horaninowia ulicina*—Cross section through the lamina (**a**, general view and **b**, detail) and stem (**c**) (*ep* epidermis, *ext ch* external chlorenchyma, *int ch* internal chlorenchyma, *wt t* water storage tissue) (Paulsen 1912)

ecological work, it was observed that halophytes and xerophytes with articulated stems and stem succulents of *Anabasis* type are exclusively C_4. Leaf succulent halophytes and xerophytes are also predominantly C_4 (Akhani et al. 1997).

Additional results obtained by Pyankov et al. (2000) referring to C_4 plants from Mongolia also suggest the relevance of this photosynthetic pathway on plants growing in extreme environmental conditions. The Chenopodiaceae comprises the greatest number of C_4 plants (about 41 species). Additionally, the C_4 Chenopodiaceae make up 45% of the total chenopods and are very important ecologically in saline areas and cold arid deserts. NADP-ME tree-like species with a salsoloid type of Kranz anatomy, such as *Haloxylon ammodendron* and *Iljinia regelii*, plus shrubby *Salsola* and *Anabasis* species, were the plant most resistant to environmental stresses. Most of the annual C_4 chenopods species are halophytes, are succulent, and occur in saline and arid habitats.

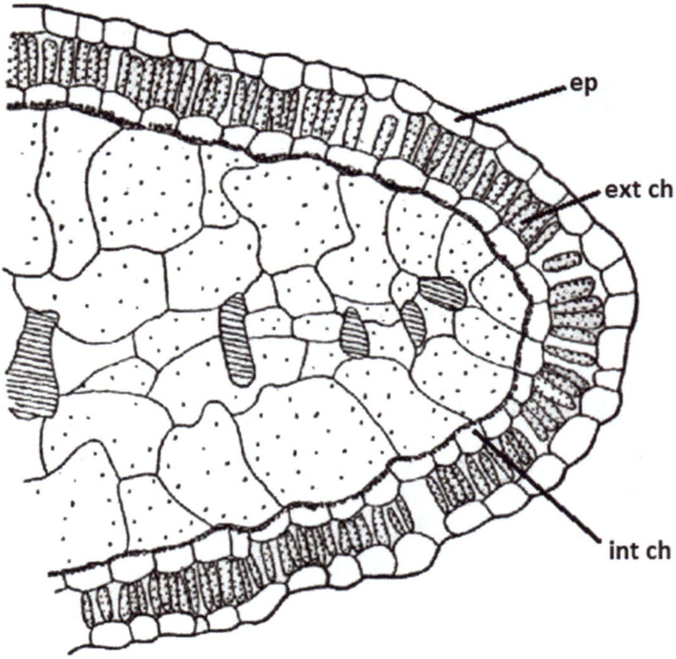

Fig. 6.26 *Suaeda lipskii*—Cross section through the lamina (*ep* epidermis; *ext ch* external chlorenchyma, *int ch* internal chlorenchyma) (Paulsen 1912)

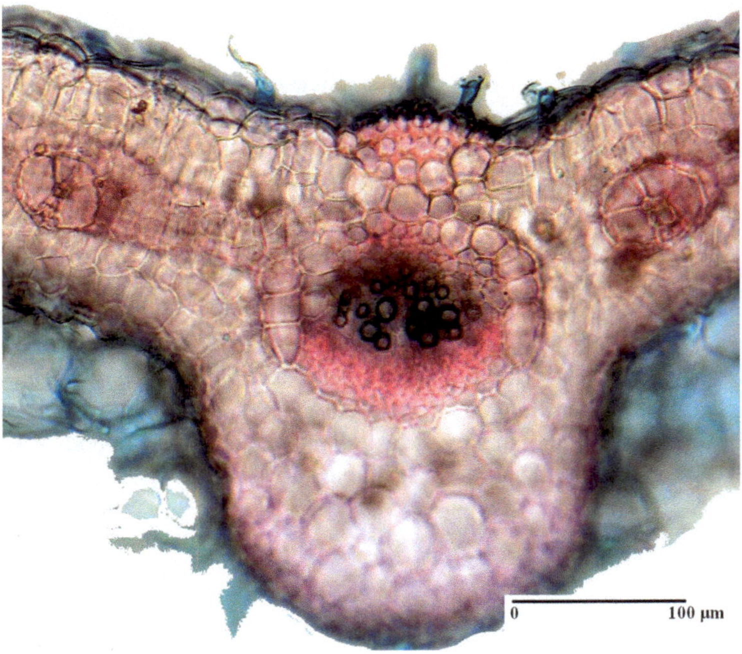

Fig. 6.27 Cross sections through the lamina of *Atriplex tatarica* (RO)

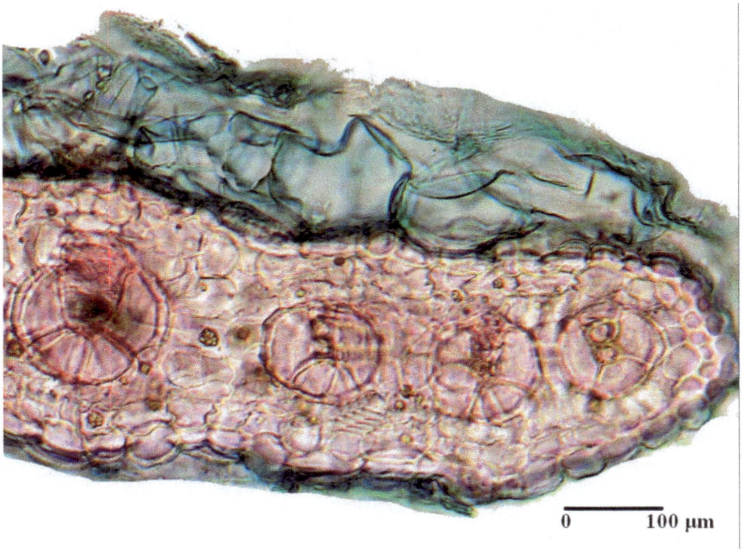

Fig. 6.28 Cross sections through the lamina of *Atriplex tatarica* (RO)

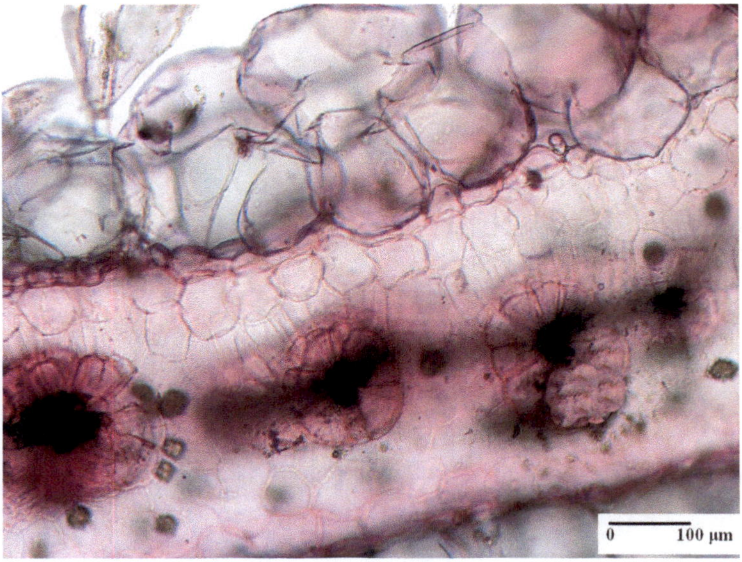

Fig. 6.29 Cross sections through the lamina of *A. glauca* (ESP)

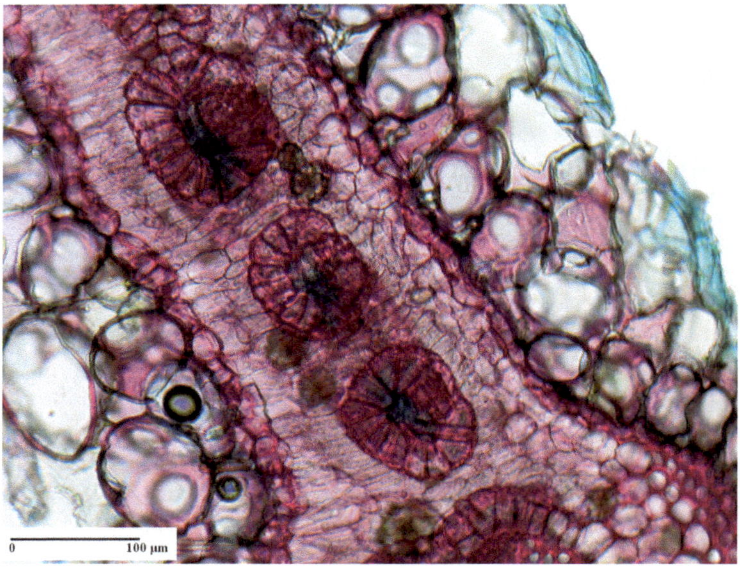

Fig. 6.30 Cross sections through the lamina of *A. halimus* (ESP)

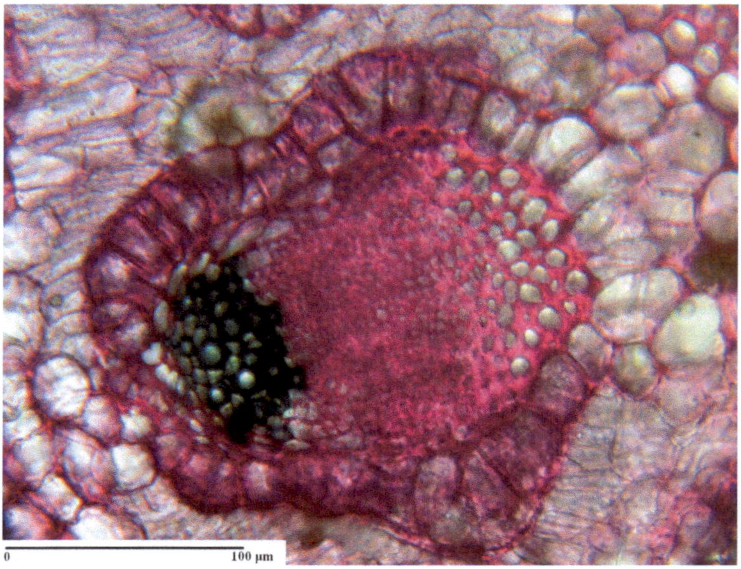

Fig. 6.31 Cross sections through the lamina of *Atriplex halimus* (ESP)

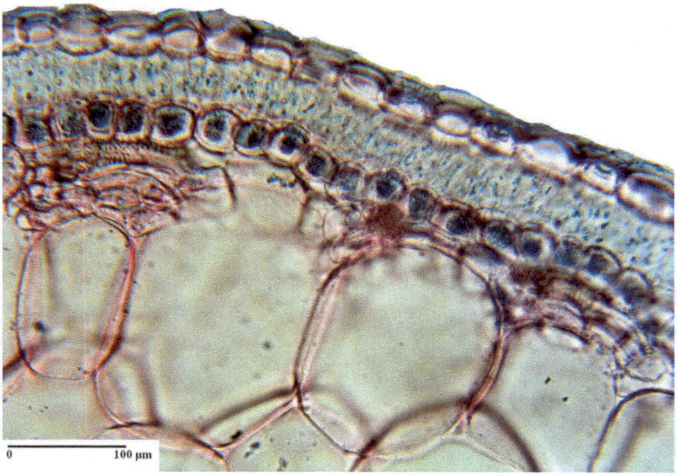

Fig. 6.32 Cross sections through the lamina of *Petrosimonia oppositifolia* (RO)

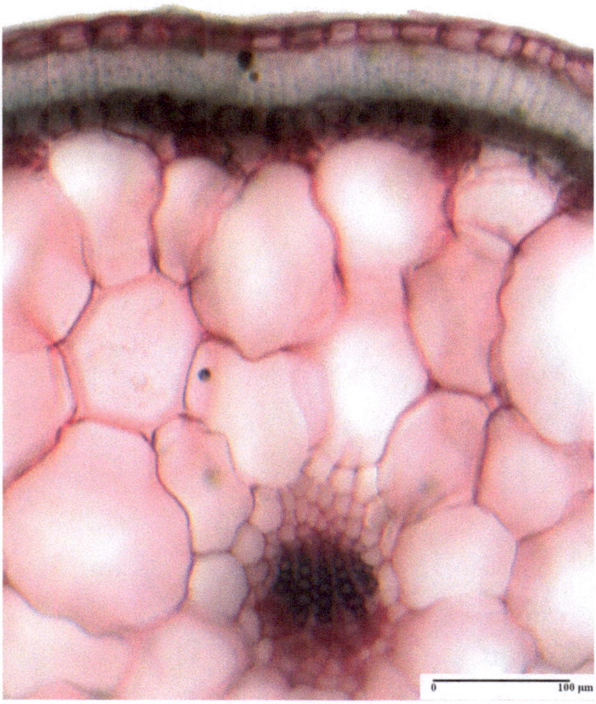

Fig. 6.33 Cross sections through the lamina of *P. triandra* (RO)

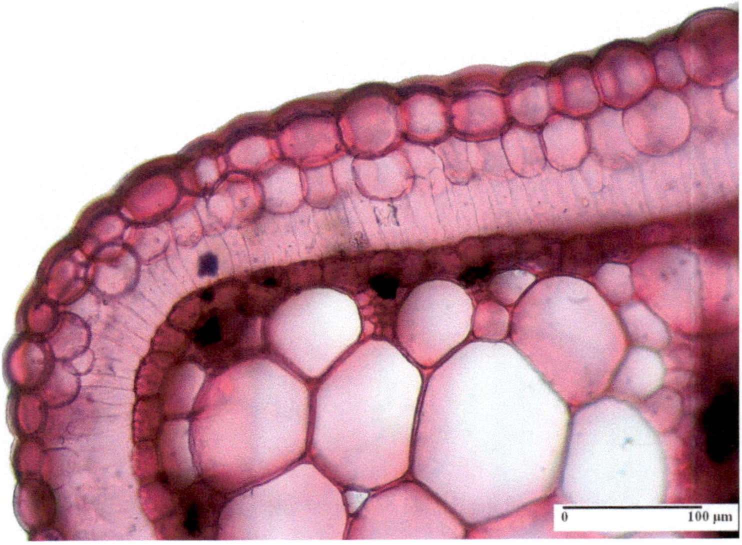

Fig. 6.34 Cross sections through the lamina of *Camphorosma annua* (RO)

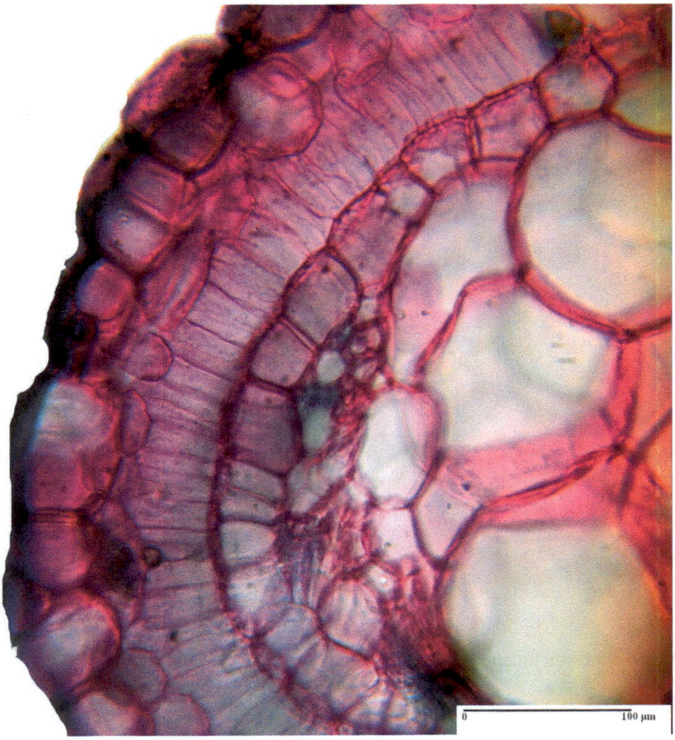

Fig. 6.35 Cross sections through the lamina of *C. monspeliaca* (RO)

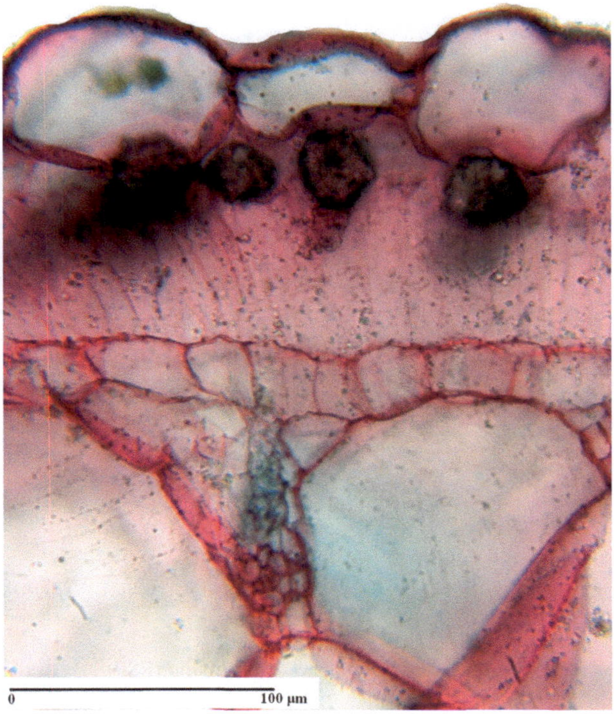

Fig. 6.36 Cross sections through the lamina of *Suaeda splendens* (ESP)

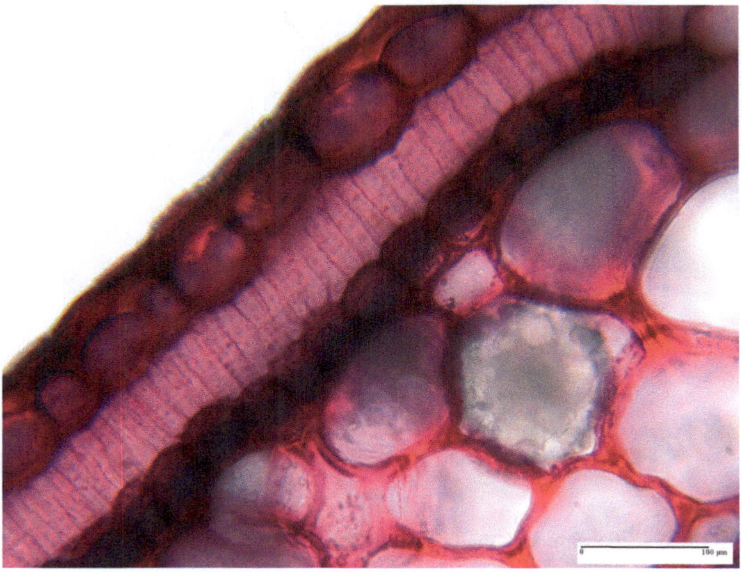

Fig. 6.37 Cross sections through the lamina of *Salsola kali* (ESP)

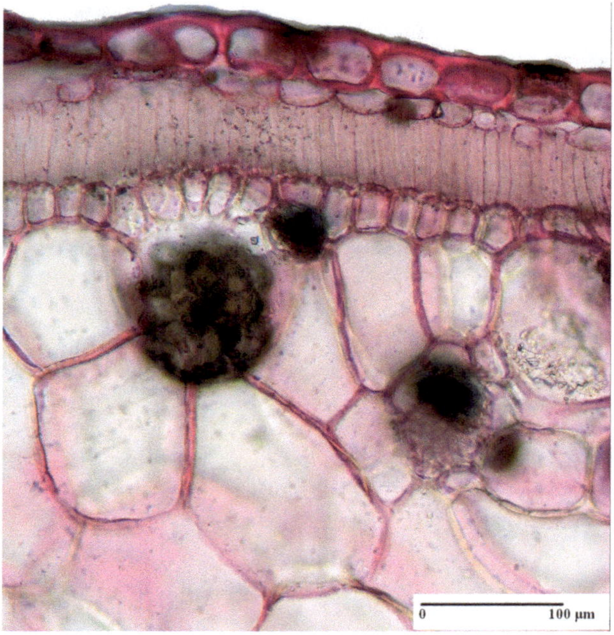

Fig. 6.38 Cross sections through the lamina of *S. oppositifolia* (ESP)

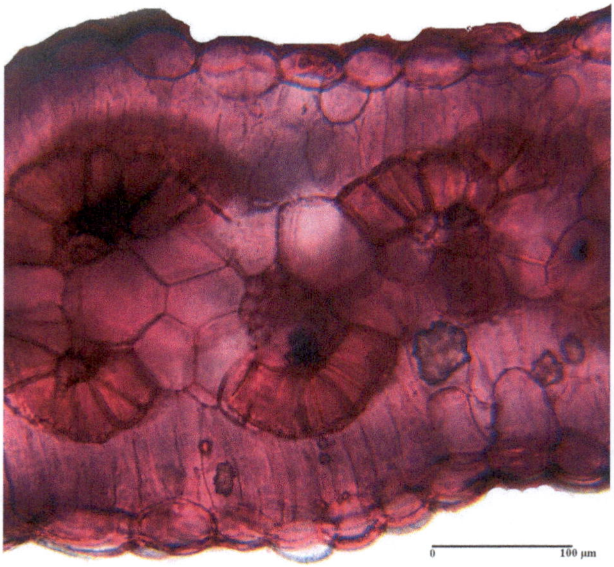

Fig. 6.39 Cross sections through the lamina of *Bassia hyssopifolia* (ESP)

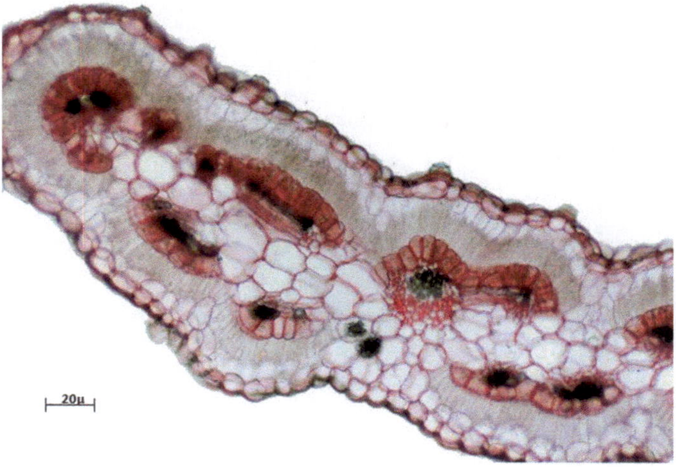

Fig. 6.40 Cross sections through the lamina of *Bassia prostrata* (photo courtesy of Somayeh Safiallah)

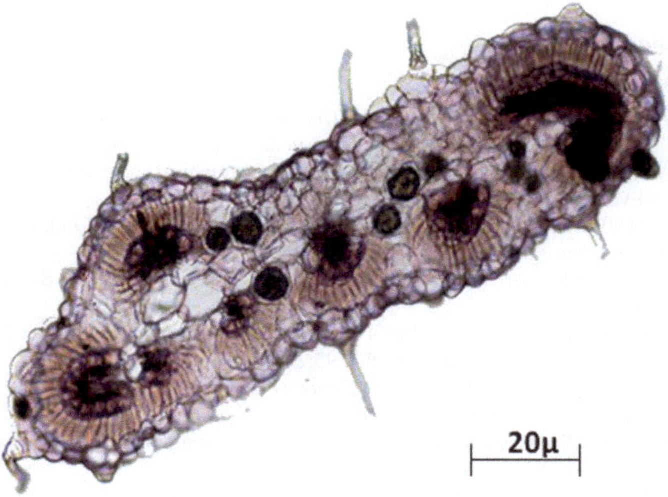

Fig. 6.41 Cross sections through the lamina of *B. pilosa* (photo courtesy of Somayeh Safiallah)

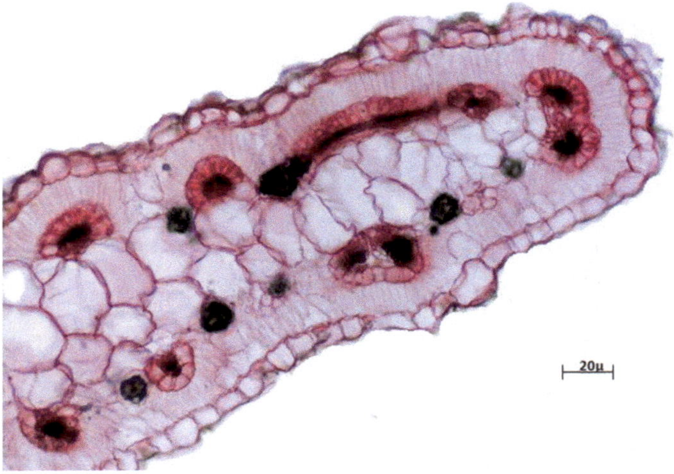

Fig. 6.42 Cross sections through the lamina of *B. turkestanica* (photo courtesy of Somayeh Safiallah)

References

Adam P (1990) Saltmarsh ecology. Cambridge University Press, Cambridge

Akhani H, Trimborn P, Ziegler H (1997) Photosynthetic pathways in *Chenopodiaceae* from Africa, Asia and Europe with their ecological, phytogeographical and taxonomical importance. Plant Syst Evol 206(1–4):187–221

Arcanheli G (1890) Sulla struttura delle foglie dell' *Atriplex nummularia* Lind. in relazione alla assimilazione. Nuova giorn Ital 22:426–430

Chermezon H (1910) Recherches anatomiques sur les plantes littorales. Ann Sci Nat sér 9 Bot 12:117–313

Carolin RC, Jacobs SWL, Vesk M (1975) Leaf structure in *Chenopodiaceae*. Bot Jahr Syst Pflanzengeschichte and Pflanyengeographie 95:226–255

Carolin RC, Jacobs SWL, Vesk M (1982) The chlorenchyma of some members of the *Salicornieae* (*Chenopodiaceae*). Aust J Bot 30:387–392

Clements FE (1920) Plant indicators: the relation of plant communities to process and practice. Carnegie Institution, Washington

Dengler NG, Nelson T (1999) Leaf structure and development in C_4 plants. In: Sage RF, Monson RK (eds) C_4 plant biology. Academic, San Diego, pp 133–172

Ehleringer JR, Monson RK (1993) Evolutionary and ecological aspects of photosynthetic pathway variation. Ann Rev Ecol Syst 24:411–439

Ehleringer JR, Cerling TE, Helliker BR (1997) C_4 photosynthesis, atmospheric CO_2 and climate. Oecologia 112:285–299

Fisher DD, Schenk HJ, Thorsch JA, Ferren WR (1997) Leaf anatomy and subgeneric affiliations of C3 and C4 species of *Suaeda* (*Chenopodiaceae*) in North America. Am J Bot 84:1198–1210

Gamaley IB (1985) Variaţii kranţ—anatomii u rastenii pustyni Gobi i Karakumi (The variations of the Kranz-anatomy in Gobi and Karakum plants). Bot Jurn SSSR 70:1302–1314

Grigore M-N (2008) Introducere în Halofitologie. Elemente de Anatomie Integrativă. PIM, Iaşi

Grigore M-N, Toma C (2007) Histo—anatomical strategies of Chenopodiaceae halophytes: adaptive, ecological and evolutionary implications. WSEAS Trans on Biol and Biomed 4:204–218

Grigore M-N, Toma C (2008) Ecological anatomy of halophyte species from the *Chenopodiaceae* family. Advanced topics on mathematical biology and ecology (Proceedings of the 4th WSEAS International Conference on Mathematical Biology and Ecology—MABE '08, Acapulco, Mexico, January 25–27, 2008), pp 62–67

Grigore M-N, Toma C (2010) Halofitele. Aspecte de anatomie ecologică. Edit. Univ. "Al. I. Cuza", Iaşi

Grigore M-N, Toma C, Ivănescu L (2011) Anatomical and ecological observations on Mediterranean halophytes: *Suaeda* Forssk. ex Scop. genus. Lucr. Şt. (Horticultură). USAMV "Ion Ionescu de la Brad", Iaşi 54(1):23–28

Grigore M-N, Toma C, Zamfirache M-M, Boscaiu M, Olteanu Z, Cojocaru D (2012a) Ecological anatomy in halophytes with C_4 photosynthesis: discussing adaptative features in endangered ecosystems. Carpathian J of Earth and Environ Sci 7(2):13–21

Grigore M-N, Toma C, Zamfirache M-M, Ivănescu L (2012b) A survey of anatomical adaptations in Romanian halophytes. Towards an ecological interpretation. Fres Environ Bull 21 (11b):3370–3375

Grigore M-N, Ivănescu L, Toma C (2014) Halophytes: an integrative anatomical study. Springer, Cham, Heidelberg

Guttierez M, Gracen VF, Edwards GE (1974) Biochemical and cytological relationships in C_4 plants. Planta 119:279–300

Guy RD, Reid DM, Krouse HR (1980) Shifts in carbon isotope ratios of two C_3 halophytes under natural and artificial conditions. Oecologia 44:241–247

Hattersley PW, Browning AJ (1981) Occurrence of the suberized lamella in leaves of grasses of different photosynthetic types. I. In parenchymatous bundle sheats and PCR ("Kranz") sheaths. Protoplasma 109:371–401

Henslow G (1895) The origin of plant-structures by self-adaptation to the environment. Kegan Paul, Trench, Trübner & Co, Ltd, Paternoster House, Charing Cross Road, London

Jacobs SWL (2001) Review of leaf anatomy and ultrastructure in the *Chenopodiaceae (Caryophyllales)*. J Torrey Bot Soc 128:236–253

Kadereit G, Borsch T, Weising K, Freitag H (2003) Phylogeny of *Amaranthaceae* and *Chenopodiaceae* and the evolution of C_4 photosynthesis. Int J Plant Sci 164(6):959–986

Kanai R, Edwards GE (1999) The biochemistry of C_4 photosynthesis. In: Sage RF, Monson RK (eds) C_4 plant biology. Academic, San Diego, pp 59–87

Kearney TH (1904) Are plants of sea and dunes true halophytes? Bot Gaz 37:424–436

Keeley JE, Rundel OW (2003) Evolution of CAM and C_4 carbon-concentrating mechanisms. Int J Plant Sci 164(3 Suppl):55–77

Kellog EA (1999) Phylogenetic aspects of the evolution of C_4 photosynthesis. In: Sage RF, Monson RK (eds) C_4 Plant biology. Academic, San Diego, pp 411–444

Kochánková J, Mandák B (2008) Biological flora of Central Europe: *Atriplex tatarica* L. Perspect Plant Ecol Evol System 10:217–229

Laetsch WM (1968) Chloroplast specialization in dicotyledons possessing the C_4—dicarboxylic acid pathway of photosynthetic CO_2 fixation. Am J Bot 55:875–883

Long SP (1999) Environmental responses. In: Sage RF, Monson RK (eds) C_4 plant biology. Academic, San Diego, pp 215–249

Long SP, Mason CF (1983) Saltmarsh ecology. Blackie, Glasgow

Mcdougall WB (1941) Plant ecology, 3rd edn. Lea & Febiger, Philadelphia

Monteil P (1906) Anatomie compareé de la feuille des *Chenopodiacees*. Travaux de Laboratoire de Matiere Medicale de l'École Supérieure de Pharmacie de Paris 4:5–156

Moser H (1934) Untersuchungen űber die Blattstruktur von *Atriplex* Arten und ihre Beziehungen zur Systematic. Beih Bot Centralbl 52:378–388

Muhaidat R, Sage RF, Dengler NG (2007) Diversity of Kranz anatomy and biochemistry in C_4 eudicots. Am J Bot 94(3):362–381

Paulsen O (1912) Studies on the vegetation of the Transcaspian lowlands. The second Danish Pamir expedition conducted by Olufsen O, Copenhagen, Gyldendalske Boghandel, Nordisk Forlag

Pyankov V, Artyusheva EG, Edwards GE, Black CC Jr, Soltis PI (2001) Phylogenetic analysis of tribe *Salsoleae* (*Chenopodiaceae*), based on ribosomal ITS sequences: implications for the evolution of photosynthesis types. Am J Bot 88(7):1189–1198

Pyankov VI, Gunin PD, Tsoog S, Black CC (2000) C_4 plants in the vegetation of Mongolia: their natural occurrence and geographical distribution in relation to climate. Oecologia 123 (1):15–31

Pyykkö M (1966) The leaf anatomy of East Patagonian xeromorphic plants. Ann Bot Fennici 3 (3):453–622

Raghavendra AS (1980) Characteristics of plant species intermediate between C_3 and C_4 pathways of photosynthesis: their focus of mechanism and evolution of C_4 syndrome. Photosynthetica 14:271–173

Sage RF (2001) Environmental and evolutionary preconditions for the origin and diversification of C_4 photosynthesis syndrome. Plant Biol 3:202–213

Sage RF (2004) The evolution of C_4 photosynthesis. New Phytol 161:341–370

Sage RF, Wedin DA, Li M (1999) The biogeography of C_4 photosynthesis: patterns and controlling factors. In: Sage RF, Monson RK (eds) C_4 plant biology. Academic, San Diego, pp 313–373

Schimper AFW (1903) Plant geography upon a physiological basis. Clarendon, Oxford

Shomer-Ilan A, Beer S, Waisel Y (1975) *Suaeda monoica,* a C_4 plant without typical bundle sheats. Plant Physiol 56:676–679

Safiallah S, Hamdi SMM, Grigore M-N, Sara J (2017) Micromorphology and leaf ecological anatomy of *Bassia* halophyte species (*Amaranthaceae*) from Iran. Acta Biologica Szegediensis 61(1):85–93

Takabayashi A, Kishine M, Asada K, Endo T, Sato F (2005) Differential use of two cyclic electron flows around photosystem I for driving CO_2-concentration mechanism in C4 photosynthesis. Proc Natl Acad Sci USA 102(46):16898–16903

Troughton JH, Card KA (1974) Leaf anatomy of *Atriplex buchananii*. New Zeal J Bot 12:167–177

Ueno O, Yoshimura Y, Sentoku N (2005) Variation in the activity of some enzymes of photorespiratory metabolism in C_4 grases. Ann Bot 96:863–869

Volkens G (1887) Die Flora der aegyptisch-arabischen Wüste auf Grundlage anatomisch-physiologischer Forschungen. Gebrüder, Borntraeger, Berlin

Volkens G (1893) Chenopodiaceae. In: Engler A, Prantl K (eds) Die Natürlichen Pflanzenfamilien, 3(1a): 36-91

Voznesenskaya EV, Franceschi VR, Pyankov VI, Edwards GE (1999) Anatomy, chloroplast structure and compartmentation of enzymes relative to photosynthetic mechanisms in leaves and cotyledons of species in the tribe *Salsoleae (Chenopodiaceae)*. J Exp Bot 50 (341):1779–1795

Wang RZ (2007) C_4 plants in the deserts of China: occurrence of C_4 photosynthesis and its morphological functional types. Photosynthetica 45(2):167–171

Warming E (1897) Halophyt-studier. D Kgl Danske Vidensk Selsk Skr 6, Raekke, naturvidenskabeling og mathematisk Afd. VIII 4:173–272

Warming E (1909) Oecology of Plants. An introduction to the study of plant-communities. Clarendon, Oxford

Wiessner J (1899) Uber die Formen der Anpassung der Blatter an die Lichtstarke. Biol Centralbl 19:1–14

Yoshimura Y, Kubota F, Ueno O (2004) Structural and biochemical bases of photorespiration in C_4 plants: quantification of organelles and glycine decarboxylase. Planta 220:307–317

Chapter 7
Successive (Additional) Cambia

At a glance, discussions about this phenomenon in a work dealing with anatomical adaptations in halophytes might be regarded as inappropriate. First of all, we treat it as a reality found in many halophytic species that we have investigated or that have been investigated elsewhere. Second, we tried to question if this phenomenon could have an ecological and adaptive significance in relation to salinity factor (Grigore and Toma 2006; Grigore 2008). Not in the last, it should be mentioned that this phenomenon largely occurs also in species that are not halophytic.

This phenomenon is primarily known due to successive concentric rings from *Beta vulgaris*. It appears in roots and stems of other genera, such as *Amaranthus, Atriplex, Bougainvillea, Chenopodium, Cycas, Mirabilis, Phytolacca, Welwitschia, Petrosimonia, Halimione, Camphorosma, Suaeda,* and *Salicornia*, and many species of *Chenopodiaceae* that includes the most representative halophytes among this ecological group of plants (Grigore 2008, 2012; Grigore and Toma 2007, 2008, 2010; Grigore et al. 2012, 2014).

Successive cambia phenomenon (also known as supernumerary or additional cambia) is being considered by some authors (Hérail 1885; Metcalfe and Chalk 1972; Şerbănescu-Jitariu and Toma 1980; Fahn and Zimmermann 1982) as a structural anomaly. Supernumerary cambia refer to vegetative axial organs (the root and the stem), and it consists shortly of the following succession of histological events: the typical, general structure is generated by a normal cambium that generally produces a few secondary phloemic and xylemic vessels. Afterward, on behalf of pericycle, an additional (supernumerary) cambium is born, this one generating a ring of fundamental cellulosic parenchyma, where the vascular bundles are placed circularly, with the phloem outside and the xylem inside. Each normal cambium is born after that from the phloemic parenchyma produced by the former cambium (Esau and Cheadle 1969; Rajput and Rao 1999).

However, there is no uniform method governing the formation of the supernumerary cambia. The first supernumerary cambium may arise in the area of primary phloemic parenchyma, between pericycle and secondary phloem (Hayward 1938). In ring formation, the activation of the successively formed secondary cambia is

M.-N. Grigore, C. Toma, *Anatomical Adaptations of Halophytes*,
DOI 10.1007/978-3-319-66480-4_7

centrifugally progressive. The mentioned author states that the manner in which the vascular elements are distributed indicates that the secondary cambia do not occur as complete rings but consist of discrete sectors derived from the pericycle.

It is obvious that these products of successive cambia appear as bands or strands of secondary phloem and xylem (vascular increments) that are embedded in a "background" of parenchyma or fibers (conjunctive tissue). In relation to conjunctive tissue, Romanian literature dealing with plant anatomy has not used this term during time; in recent works (Grigore 2008; Grigore and Toma 2010), it has been introduced and adapted. In the Anglo-Saxon literature, the term conjunctive tissue refers to a fundamental tissue, parenchymatic or not, where the products of successive cambia are noticeable. Schenck (1893) and Pfeiffer (1926) used this term; the latter accurately applied it to many of the genera in which the phenomenon is known today. Other authors have referred to successive cambia under the vague terms "anomalous secondary thickening" or "included phloem," despite that the secondary phloem of successive cambia is not included within xylem at all (Stevenson and Popham 1973). Carlquist (2007) underlined that maintenance of such terms may have been furthered by those who are involved with xylem identification and who therefore seek for simple terms. The distinctive appearance of successive cambia and their products can easily be learned by xylem anatomists. The desire to use such a term as "included phloem" probably indicates a desire to consider the background tissue of plants with successive cambia as "xylem"; however, although often xylemic in texture, this background tissue is not xylem in the ordinary sense. Xylem anatomists who have dealt in detail with plants with successive cambia have used the term conjunctive tissue for the background of fibers and/or parenchyma in which vascular increments are embedded (Carlquist 2007).

Although various authors described the histology of investigated species with decent accuracy, understanding of the ontogeny of successive cambia has been problematic. One cause is that soft and hard tissues are intermixed in stems and roots with successive cambia. Soft tissues and stages in their development are frequently damaged and uninterruptable when such axes are sectioned untreated on a sliding microtome.

A cause of misinterpretation is due to the fact that xylem anatomists who highlighted this phenomenon often studied only a single genus and sometimes even a single species. Although generally accepted as a unique plan that explains the manifestation of this phenomenon in various cases, some species express its "extreme" (*Stayneria, Gnetum africanum, Mendoncia*), which can lead to problems of interpretation (Carlquist 2007).

Successive cambia are found in 34 families of dicotyledons (Carlquist 2001) in which the species of the order *Caryophyllales* predominate; they also occur in *Welwitschia, Gnetum* species (Carlquist 1996), *Cycas* (Terrazas 1991), *Encephalartos,* and *Macrozamia* (Greguss 1968) (Carlquist 2007).

However, the number of species that present this phenomenon is only a small part of the total number of species. Nevertheless, a logical question arises: why this "pattern" repeatedly evolved and, especially, what is its adaptive value (if any)?

Attention should be paid that the concept of successive cambia differs from that called "interxylary phloem formed from a single cambium" (Carlquist 2001). These instances occur in groups such as *Combretaceae* (van Vliet 1979), *Onagraceae* (Carlquist 1975), and *Strychnos* and allied genera of *Loganiaceae* (Cockrell 1941; Mennega 1980). In this case, interxylary phloem strands or bands formed from a single cambium do not form pairwise relationship to strands or bands of vessels or vessel groups; the vessels are distributed within the secondary xylem. In the case of successive cambia, the strands or bands of secondary phloem occur external to strands or bands of secondary xylem, respectively. These strands or bands (considered vascular increments here) are separated from each other by conjunctive tissue. Because conjunctive tissue is not secondary xylem, the term "interxylary" is not appropriate in plants with successive cambia. This fact was noted by Stevenson and Popham (1973) and others.

Carlquist (2007) discusses the ontogeny and consequences of successive cambia activity and provides a hypothetical generalized plan for describing the functioning of successive cambia. He promotes the idea that there is a master cambium that produces externally secondary cortex, rays, conjunctive tissue, vascular cambium, and secondary phloem and xylem, internally.

Secondary phloem and secondary xylem are born from the vascular cambium in each vascular increment. Vascular cambia function indefinitely so that a master cambium and a series of vascular cambia (each in a vascular increment) function indefinitely. The master cambium either remains active as long as an axis is actively growing (although it may become quiescent following the initiation of each vascular increment and associated conjunctive tissue) or, less commonly, may be reinvented in the secondary cortex.

As far as it is known, the first vascular cambium in species with successive cambia produces secondary xylem and phloem in the same way as does the vascular cambium in plants that have a single vascular cambium that ensures the secondary growing of roots and stems. The master cambium forms from a periclinal division in cortical cells of stem or pericycle in the root (Fig. 7.1). These divisions form a band of the indefinite circumference or a cylinder around the entire axis (Kirchoff and Fahn 1984). In stems, the parenchyma between the secondary phloem of the first vascular cylinder and the master cambium is parenchyma from the primary cortex. The conjunctive tissue is produced from the master cambium; it is not formed adjacent to the first vascular cylinder but is formed inside to each subsequent vascular increment.

Conjunctive tissue differs from cortical parenchyma by the fact that its cells are arranged in radial rows, while cortical cells are larger in diameter. After the origin of the master cambium, each vascular cambium continues to add secondary xylem and secondary phloem to its own vascular increment. Therefore, secondary phloem and xylem continue to add to the original vascular cylinder; therefore, a master cambium and an indefinite number of vascular cambia coexist.

There is no a consensus on master cambium concept. As Carlquist (2007) underlined, several terms are used in parallel with "master cambium." Thus, anomalous cambium, a lateral meristem, primary thickening meristem, second

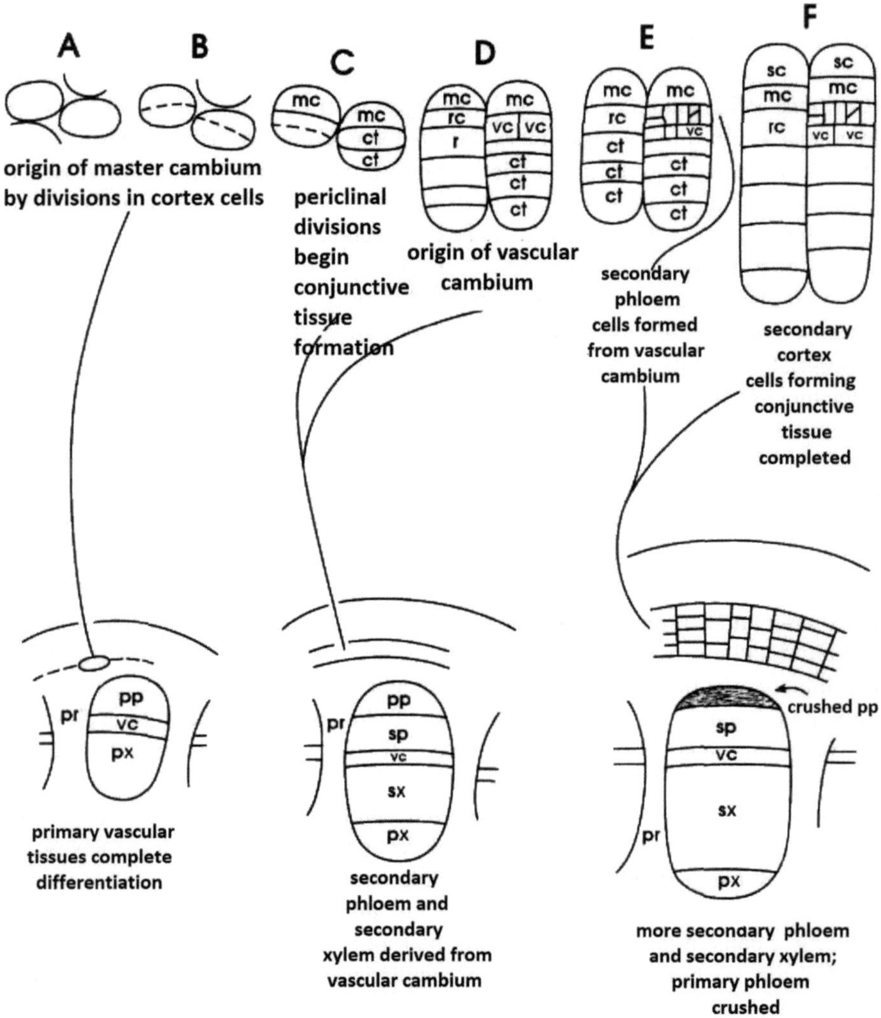

Fig. 7.1 Origin of the master cambium (in a hypothetical stem cross section) and stages in the production of cell types (*c* cortex, *ct* conjunctive tissue, *mc* master cambium, *pd* periderm, *pp* primary phloem, *pr* primary ray, *r* ray, *rc* ray cambium; *sc* secondary cortex; *sp* secondary phloem, *sx* secondary xylem, *vc* vascular cambium) (Carlquist 2007)

cambium, and supernumerary cambium include some examples of terms used to describe this phenomenon. For instance, Esau and Cheadle (1969) use the terms "old cambium" and "new cambium" in *Bougainvillea*; there is still a certain inconsistency in this case because sometimes they claim that vascular cambium (in plants wits successive cambia) produces phloem outside and xylem to the inside, while later they state that each of the cambia produces "xylem and conjunctive tissue to the inside and phloem and conjunctive tissue to the outside."

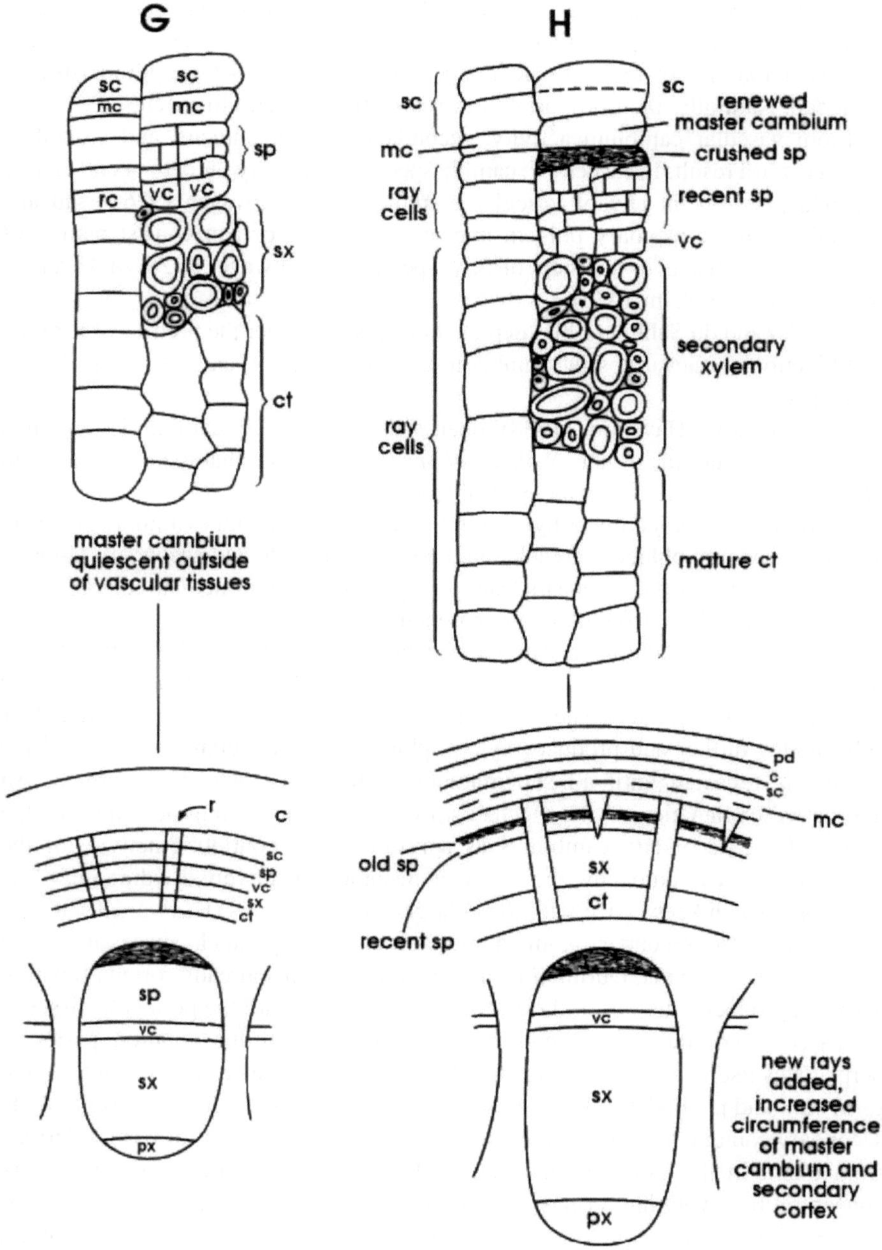

Fig. 7.1 (continued)

The term "master cambium," although used with caution by Carlquist (2007), is considered the most suitable to explain the formation of conjunctive tissue and vascular cambia internally, a situation observed in many taxa of the order *Caryophyllales*, but also in many other situations.

As mentioned, conjunctive tissue from *Beta* is entirely parenchymatic. The large familiarity of this example has led some authors to regard this tissue as being always a parenchyma. In fact, to this tissue, fibers, sclereids, and sometimes idioblasts may be added. Naturally, their proportion is variable from species to species.

From vascular cambium activity, secondary phloem outward and secondary xylem inward result. In successive cambia species, secondary phloem occurs before secondary xylem, in any of vascular increments (Artschwager 1926; Esau and Cheadle 1969). Secondary phloem may contain fibers or can consist mainly of parenchyma cells, as in the roots of *Beta* species (Artschwager 1926) or *Mirabilis* (Mikesell and Popham 1976).

Bonnier and du Sablon (1905) depicted in a good manner the process of forming and functioning the successive cambia in the root (Fig. 7.2) and stem (Fig. 7.3) of *Beta* species.

Subsequently, Hayward (1938) dealt with successive cambia issue when described in detail the organs of *Beta vulgaris*. He believed that the point of origin of these cambia varies with the level of axis considered.

Thus, in the root and lower hypocotyl the first supernumerary cambium arises in a zone of primary phloem parenchyma, between pericycle and secondary phloem. In the upper hypocotyl, it appears from pericycle, in the intermediate region, either from the pericycle or from phloemic parenchyma.

The forming of additional cambia is explained in detail by Hayward (1938): when the cambium initial undergoes the first division, the outer of the two daughter cells becomes the initial of a new supernumerary cambium while the inner daughter cell divides further and produces xylem, phloem, and medullary ray tissue. This process is repeated until all supernumerary cambiums have been formed. An alternative explanation of the mechanism of tertiary thickening accounts for the origin of the secondary cambiums as a result of the continued activity of the pericycle. Early in ontogeny, the pericycle becomes an actively dividing multilayered zone which keeps pace with the enlargement of the axis (Hayward 1938).

Likely, these secondary cambia may occur in very quick succession from pericyclic parenchyma. During this interval, these cambia already formed function actively, producing tissues of their corresponding rings until the pericyclic tissue is increased by continued radial and tangential divisions. In this situation, pericycle perpetuates itself as the external zone of the axis and produces phellogen that will form cork and phelloderm. Regardless of the hypothesis regarding the origin of the secondary cambia, there is an agreement in terms of subsequent of tissues derived from them. In the situation analyzed by Hayward (1938), there are five to six concentric relatively large rings, outside of which there may be several narrower; internal rings are not equal in width. This might suggest that in a ring formation, activation of successive cambia is centrifugally progressive and that some of these function simultaneously. Because of this way of development, it is possible to determine the ontogeny of each ring in a centripetal direction; the outermost, next to periderm, is entirely meristematic consisting of cambial cells (derived from pericycle), parenchyma, and undifferentiated vascular elements. The first of these to differentiate are sieve tubes and companion cells.

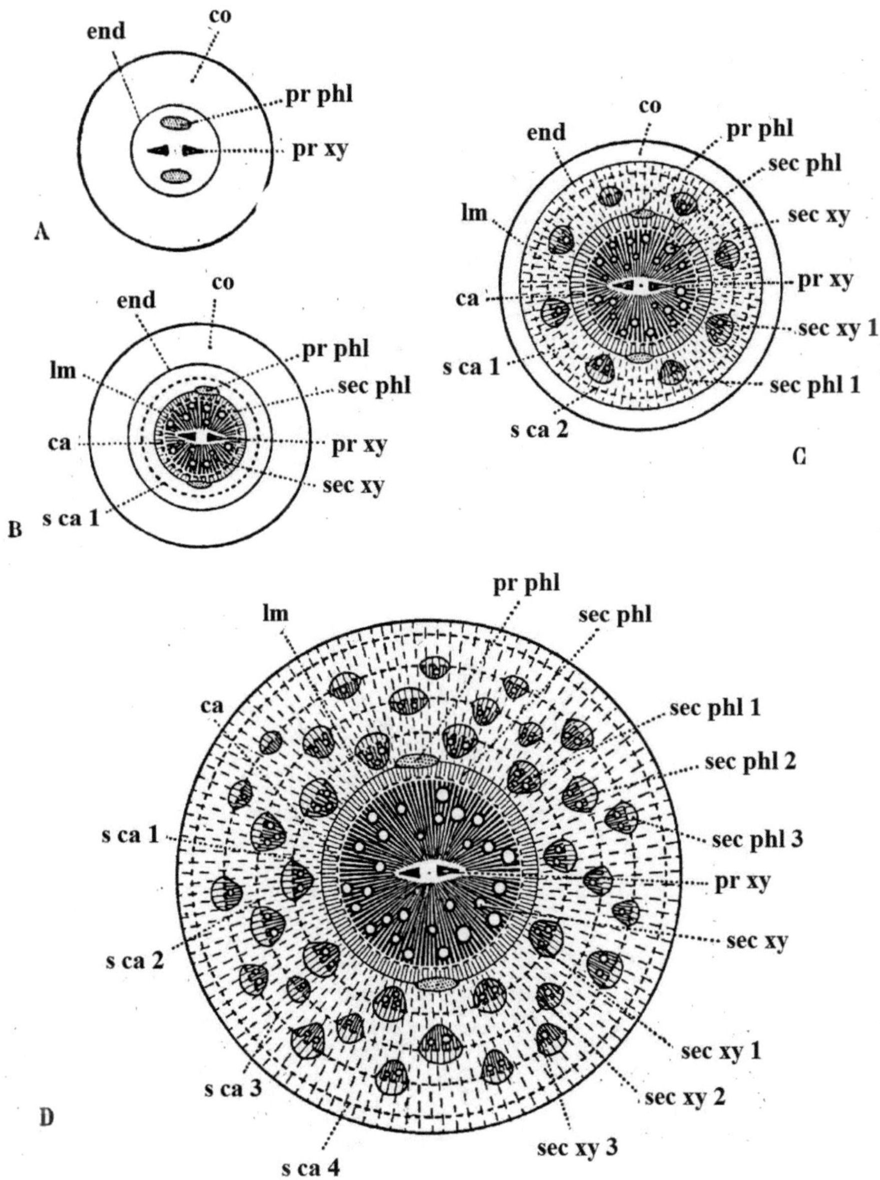

Fig. 7.2 Schema of functioning of the successive cambia in the root of *Beta* (*ca* normal cambium, *co* cortex, *end* endodermis, *pr xy* primary xylem, *pr phl* primary phloem, *sec phl* secondary phloem—1, 2, 3 indicate progressive stages, *sec xy* secondary xylem—1, 2, 3 indicate progressive stages, *s ca* 1, 2, 3—supernumerary cambia) (Bonnier and Du Sablon 1905)

In most cases, the xylem elements are arranged in narrow radial bands and are separated by the ray parenchyma. Usually, the production of phloemic elements precedes the differentiation of xylemic cells. The manner in which vascular

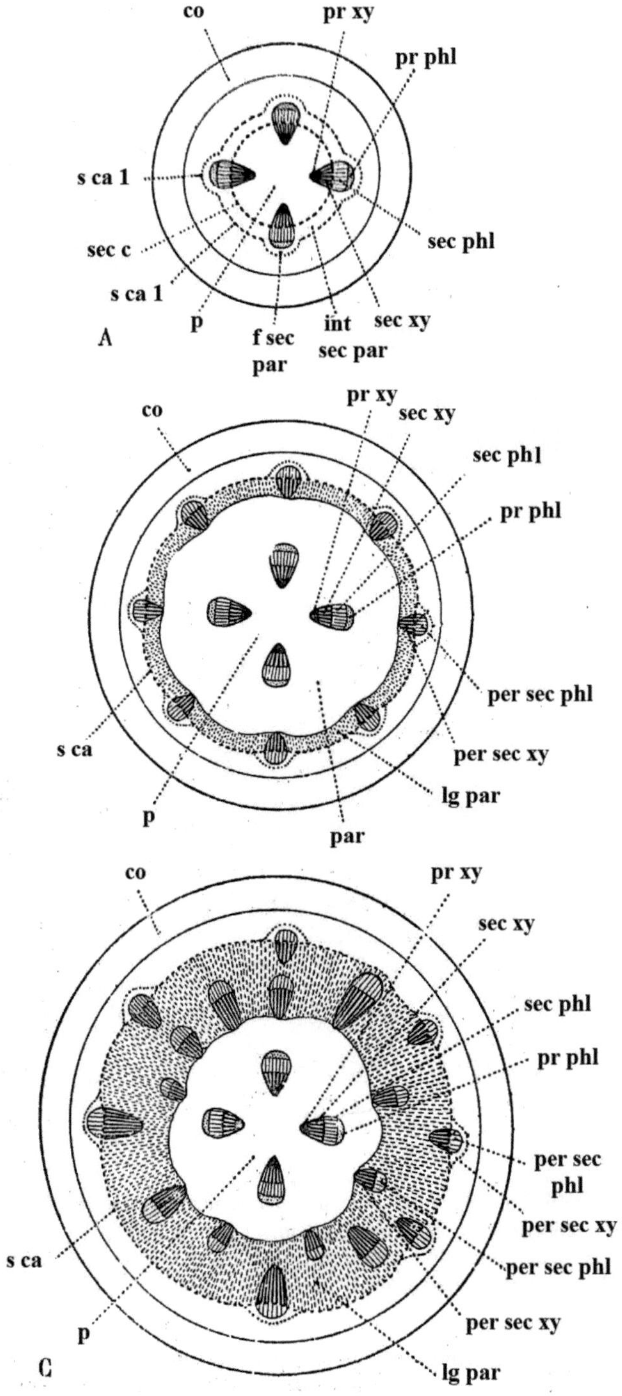

Fig. 7.3 (**a–c**) Schema of functioning of the successive cambia in the stem of *Beta* (*ca* normal cambium, *co* cortex, *end* endodermis, *f sec par* fascicular secondary parenchyma, *int sec par* interfascicular secondary parenchyma, *lg par* lignified parenchyma, *pith* pith, *per sec xy* pericyclic secondary xylem, *per sec phl* pericyclic secondary phloem, *pr xy* primary xylem, *pr phl* primary phloem, *sec phl* secondary phloem, *sec xy* secondary xylem, *s ca* supernumerary cambia) (Bonnier and Du Sablon 1905)

elements are distributed indicates that these cambia do not appear as complete rings but consist of discrete sectors derived from pericycle. Parenchymatous cells between radial rows of vascular elements are of pericyclic origin and should, according to Hayward, to be regarded as pericyclic rays.

However, the history anatomical observations of successive cambia phenomenon dates back to the middle of the nineteenth century. Many botanists revealed the production of additional cambia, including mainly observations on species from *Chenopodiaceae*, including the beet and many halophytic species (Gernet 1859; Regnault 1860; Wiessner 1867; De Bary 1877; Droysen 1877; Prillieux 1877; Gheorhgieff 1887; Sanio 1863; van Tieghem 1870–1871; Weiss 1883; Hérail 1885; Morot 1885; Fron 1899; Volkens 1893).

Interestingly, they were able to correlate this structural "anomaly" with the role of pericycle. For instance, Hérail (1885) discusses successive cambia in *Chenopodiaceae* in the chapter dedicated to anomalies of pericycle. Despite such botanists offering admirable examples of very accurate descriptions of investigated structures, the language they used is not so precise because of the limitations of époque; many terms, in the sense we know today, were not yet present in the botanical language of that time. For example, the term cambium and other meristematic tissue-related terms were not used in their descriptions. Nevertheless, the significance of them may be foreseen or may be deducted from the expressions of many French botanists. For instance, Hérail (1885) wrote about pericycle fragmentation (perhaps with the meaning of differentiation), conjunctive tissue, generating zones (*zones génératrices*), consecutive meristems (*méristèmes consecutifs*), and successive meristems (*méristèmes successifs*). In addition, he observed that there are several exceptions, in the sense that chenopods species may not present successive cambia, such as *Camphorosma monspeliaca*.

However, successive cambia and their increments were depicted since the late seventeenth century, by Grew's iconic work in plant anatomy (Grew 1682) (Fig. 7.4). Nehemiah Grew and Marcello Malpighi are considered the two cofounders of modern plant anatomy. Grew (1682) gave in the plates from his book a drawing of a cross section through the root of beet (Fig. 7.5), where the numerous rings produced by successive cambia are easily noticeable. However, he did not use in his book terms such as cambia or successive cambia; the limitations from his book are logical, as we are talking about one of the first monographs in plant anatomy. From text analysis, evidence from a greater number of (wood) rings can be found. For instance, in the Book II chapter dealing with wood root, he suggests several times that in some species (including the beet) the number and size of rings may differ: "*the number and size of which **rings** differs (. . .) in **beet**, they (rings) are narrower, but **more***" (p. 70). In the Book II of the Roots, dealing with pith (p. 75), he said: "*and in many others there are parenchymous parts, of the same substantial nature with the pith, distributed betwixt the **several** rings of vessels, and every where visible, from the top to the bottom, as in Beet, Fenil.*"

THE
ANATOMY
OF
PLANTS.
WITH AN
IDEA
OF A
Philoſophical Hiſtory of Plants,
And ſeveral other
LECTURES,
Read before the
ROYAL SOCIETY.

By *NEHEMIAH GREW* M.D. Fellow of the *ROYAL SOCIETY*, and of the *COLLEGE* of *PHYSICIANS.*

Printed by *W. Rawlins*, for the Author, 1682.

Fig. 7.4 Nehemiah Grew's Book page of Plant anatomy from 1682

But perhaps the earliest rigorous anatomical observations of successive cambia—together with ink drawings—are those of Gernet (1859), for stems of halophytes *Salsola kali* (Fig. 7.6) and *Haloxylon ammodendron* (Fig. 7.7), and that of Regnault (1860), for stem of *Eurotia ceratoides* (Fig. 7.8).

Morot (1885) dealt with successive cambia phenomenon, which is discussed in the section "Production of vascular bundles in/from pericycle" of his work. Describing the structure of *Atriplex nitens*, he used terms such as partial meristems (*méristèmes partiels*); he concluded that the stem structure of *A. nitens* is similar to that of *Phytolacca*, where a "succession of alternating hard and soft concentric

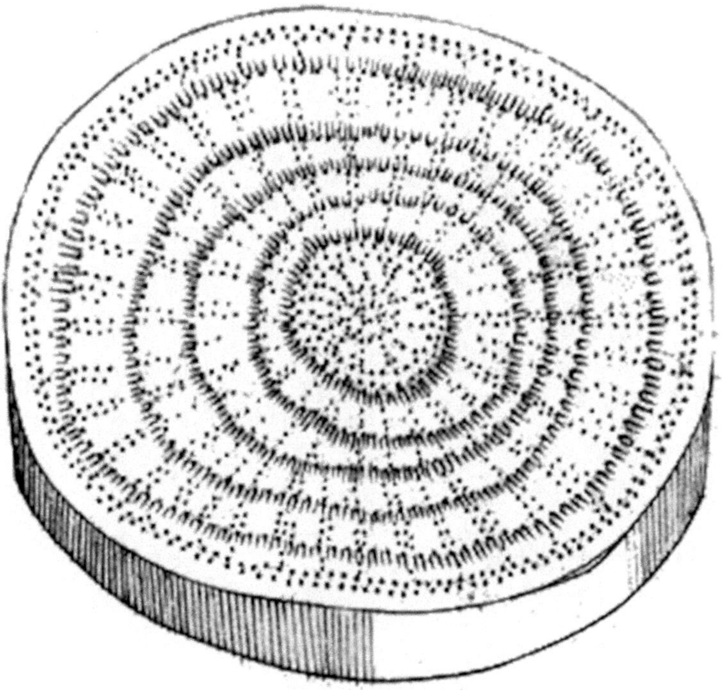

Fig. 7.5 Cross section through the root of beet (*Beta* spp) as depicted by Grew (1682). Numerous concentric rings produced by successive cambia can be easily distinguished

layers, ***all*** of pericyclic origin" is noticeable. Moreover, he explicitly uses the term "supernumerary vascular bundles."

Fron (1899) made a consistent study about the structure of root and stem in *Chenopodiaceae* species. The successive cambia are described in detail in two distinct situations: when the root structure is symmetric (and the thickening of the root derives from the activity of additional cambia—*cercles générateurs successifs*, such as in *Chenopodium murale* (Fig. 7.9), or asymmetric, when it arises from the unequal development of successive increments (?) corresponding to each of the two primary vascular bundles (*formations libéro-ligneuses successive de chaque côté des deux faisceaux ligneux primaires*), such as in *Salsola kali* (Fig. 7.10).

In *Salsola kali* (Fig. 7.10), successive cambia are not concentrically arranged, but rather in a spiral, having a normal cambium as a starting point toward the center of the root; this cambium was formed on the internal side of each of the two vascular bundles. This cambium has a spiral-like form and extends through its external extremity, thus increasing the number of secondary bundles and allowing root thickening. Bonnier and du Sablon (1905) explain this arrangement in a spiral of abnormal cambia by the fact that the top of seed radicle is compressed between the cotyledons.

Accordingly, the radicle of seed in *Chenopodiaceae* species is not compressed between the cotyledons and presents this succession of secondary formations in a symmetrical way.

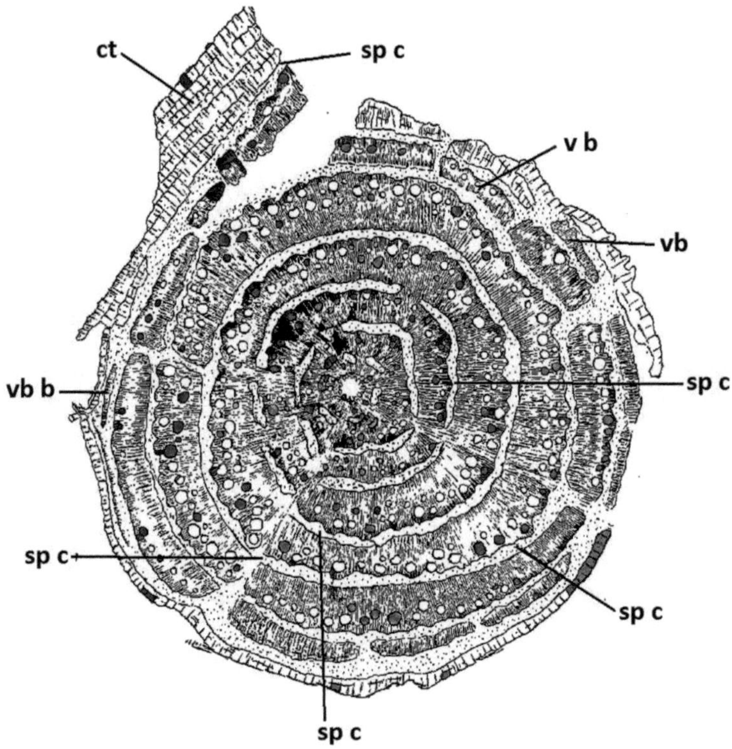

Fig. 7.6 Cross section through the stem of *Salsola kali* (*ct* cortex, *v b* vascular bundle, *sp c* cambia disposed in spiral) (Gernet 1859)

Salsola kali was also anatomically studied and successive cambia were evidenced at stem level (Toma et al 1991, Fig. 7.11).

A special situation was described for the root of *Salicornia macrostachya* (Fron 1899; Bonnier and Du Sablon 1905): in this species, "normal" cambia (*l'assise géné ratrice libéro-ligneuse*) do never act. In the root which is still young, a cambium is produced from pericycle and acts further as the cambium of *Beta*, producing a secondary tissue (conjunctive tissue?), where vascular bundles are embedded (Fig. 7.12).

An "abnormal" situation was also described (Bonnier and Du Sablon 1905) for a species of genus *Camphorosma* (unfortunately, not specified in their text), a "small shrub from the Mediterranean region," where in the first two years, the cambium acts normally and then ceases its activity. In the third year, another layer (cambium) differentiates from pericycle and produces a new circle of secondary vascular bundles. A few years later, the third circle of vascular bundles is formed on the outside of the second one, and the process continues this way.

Artschwager (1920) shows that the anomalous stem structure of *Chenopodium album* derives from a periodically active cambium; it produces xylem centripetally throughout its extent and phloem centrifugally in limited regions. Where phloem is formed, the cambium is "used up." The continuity of the cambium ring is maintained by the formation of new portions outside the phloem groups. The

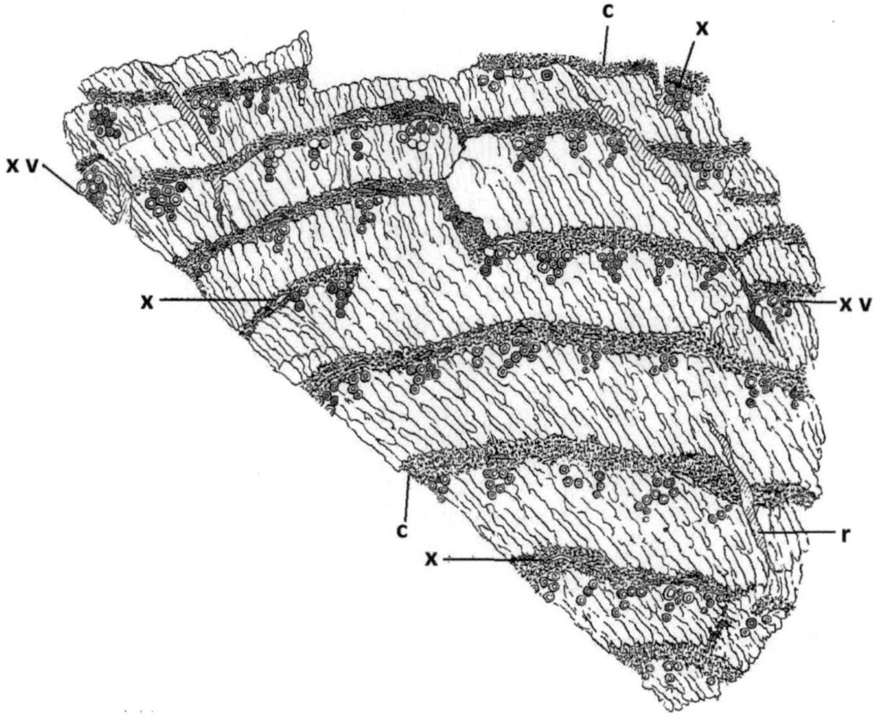

Fig. 7.7 Cross section through the stem of *Haloxylon ammodendron* (*c* cambium, *x* xylem formation, *x v* xylemic vessel, *r* medullary ray?) (Gernet 1859)

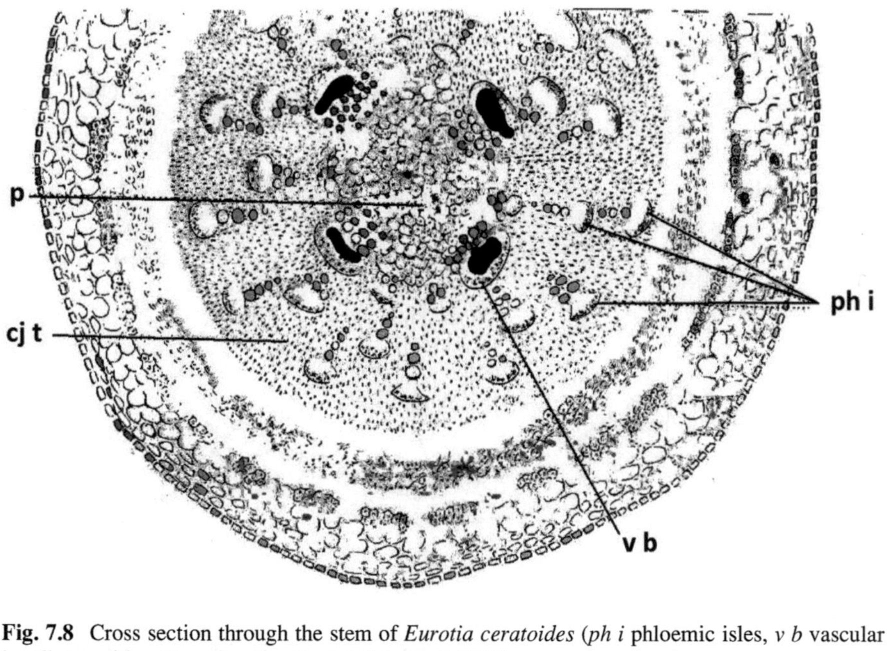

Fig. 7.8 Cross section through the stem of *Eurotia ceratoides* (*ph i* phloemic isles, *v b* vascular bundles, *p* pith, *cj t* conjunctive tissue) (Regnault 1860)

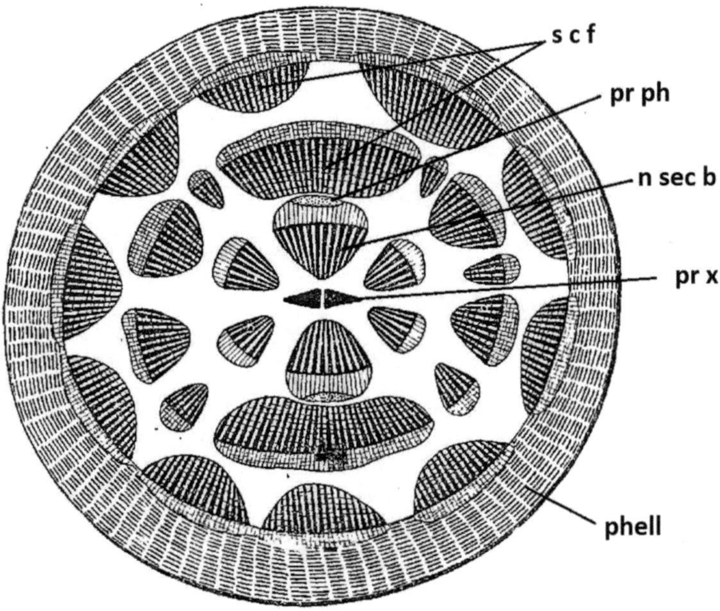

Fig. 7.9 Cross section through the root of *Chenopodium murale*, affected by successive cambia activity (*pr x* primary xylem, *pr ph* primary phloem, *n sec b* normal secondary bundles, *s c f* successive cambia formations, *phell* phelloderm) (Fron 1899)

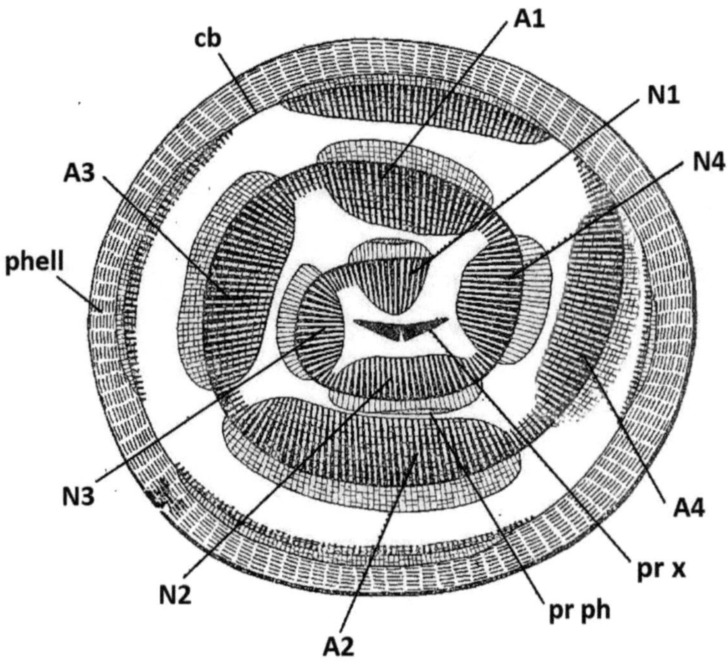

Fig. 7.10 Cross section through the root of *Salsola kali*, affected by successive cambia activity (*pr x* primary xylem, *pr ph* primary phloem, *cb* cambium, *phell* phelloderm, N1, N2, N3, N4—secondary normal productions, A1, A2, A3, A4—successive cambia formations) (Fron 1899)

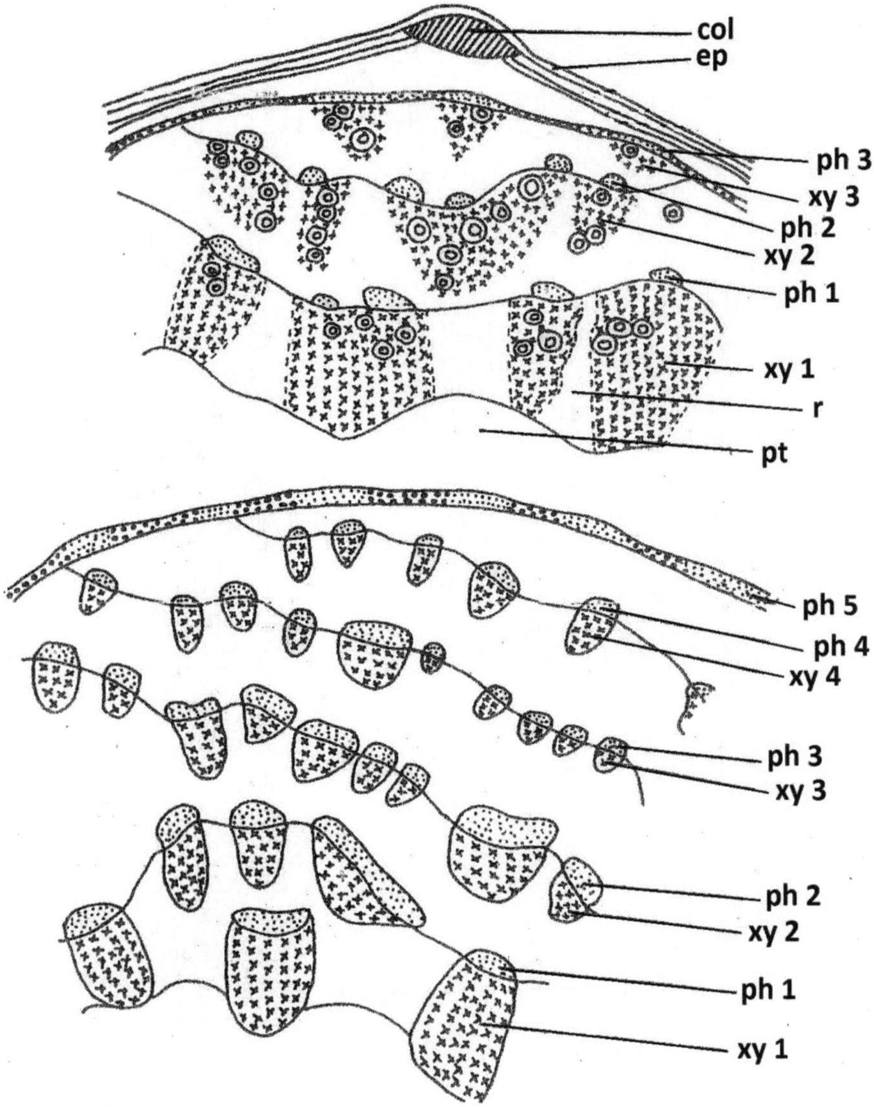

Fig. 7.11 Cross section through the stem (at the base) of *Salsola kali*—sectors with three (above) and five (below) additional cambia (*ep* epidermis, *col* collenchyma, *r* medullary ray, *pt* pith, xy 1,2,3,4,5; ph 1,2,3,4,5—xylem and phloem formed by additional cambia) (schematic view) (Toma et al. 1991). Cross section through a stem (at the base) of *Salsola kali*—sectors with three (*left*) and five additional cambia (*right*) (*m r* medullary ray; xy 1,2,3,4,5; ph 1,2,3,4,5—xylem and phloem formed by additional cambia) (detail) (Toma et al. 1991)

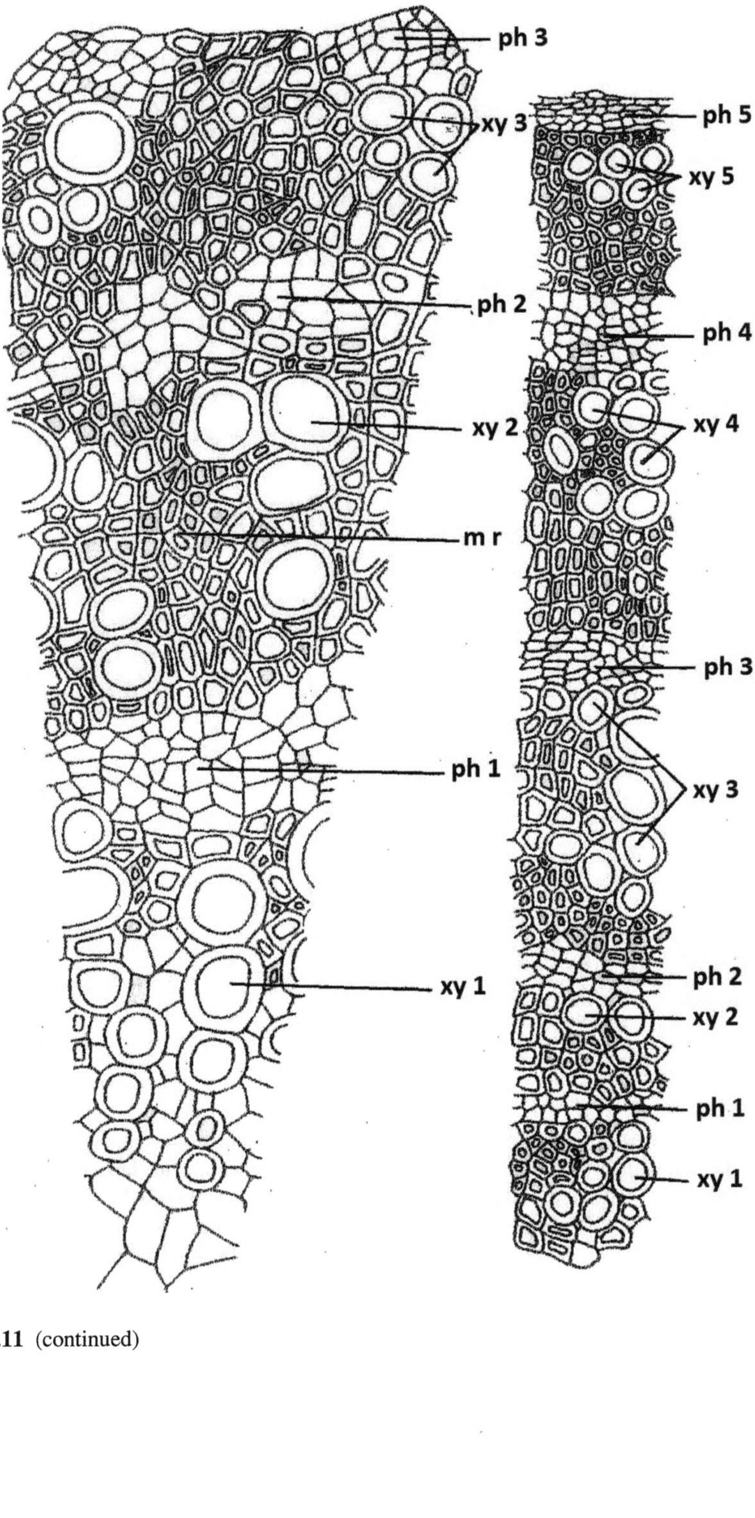

Fig. 7.11 (continued)

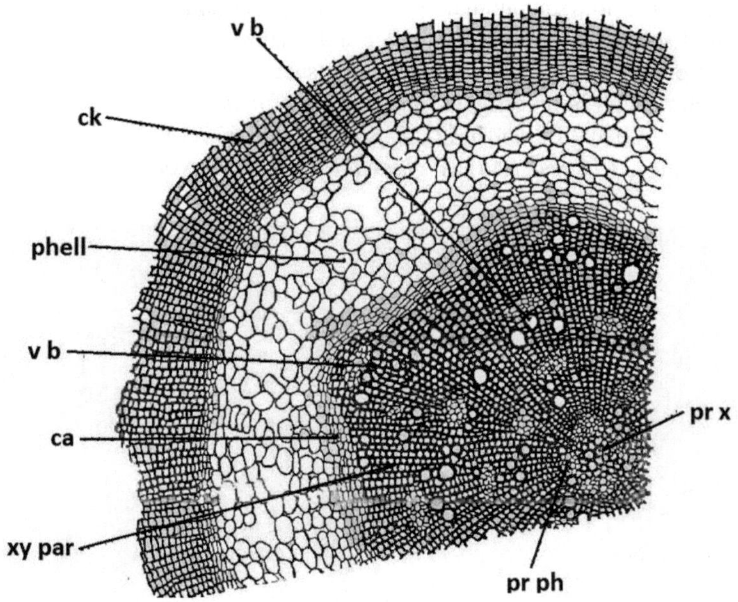

Fig. 7.12 Cross section through the root of *Salicornia macrostachya* (*ca* cambium, *ck* cork, *pr x* primary xylem, *pr ph* primary phloem, *xy par* xylem parenchyma [conjunctive tissue?], *v b* vascular bundle, *phell* phelloderm) (Fron 1899)

phloem of a secondary zone of growth is produced after all or most of the xylem has been formed. It is the normal product of the cambium and only belated in its development. According to him, the conjunctive tissue is not ray tissue in the morphological sense, though it may function as such. However, the issue of rays in chenopods with successive cambia is problematic but very important in clarifying several taxonomical difficulties (Carlquist 2003).

In two species from *Molluginaceae* (*Glinus lotoides* and *G. oppositifolius*), the first cambium ring is active for a short period, being followed by the development of a second cambial ring, formed in the cortical parenchyma (Rao and Rajput 2003).

The successive cambia were observed in many genera from *Amaranthaceae*: *Achyranthes, Acnida, Aerva, Alternanthera, Amaranthus, Bosea, Celosia, Chamissoa, Deeringia, Froelichia, Gomphrena, Hermbstaedtia, Iresine, Pfaffia, Psilotrichum*, and *Pupalia* (Metcalfe and Chalk 1972). According to Joshi (quoted by Metcalfe and Chalk 1972), this type of anomalous secondary thickening from *Amaranthaceae* and *Chenopodiaceae* may be present in the root or base of the stem when it is absent from the rest of the stem. The parenchymatous conjunctive tissue is more abundant in the root than in stem and decreases upwards. For example, in *Achyranthes*, while the secondary cambia form large arcs or complete rings in the root, they are limited in the stem to small arcs. Joshi concluded that formation of a succession of cambia is an ancestral character in these families, the primitive forms having several zones of vascular bundles embedded in parenchymatous ground tissue and formed from a similar number of secondary cambial rings. Evolution has led either to the loss of anomalous thickening from the stem alone or to the reduction of secondary cambia to smaller and smaller segments.

Grigore et al. (2013, 2014) found out that successive (additional) cambia were present in almost all of investigated species from *Chenopodiaceae* family: *Atriplex littoralis* (Figs. 7.13, 7.14, and 7.42), *A. prostrata* (Figs. 7.15, 7.16, 7.17, and 7.43), *A. tatarica* (Figs. 7.18 and 7.44), *A. glauca* (Fig. 7.19), *Bassia sedoides* (Fig. 7.20),

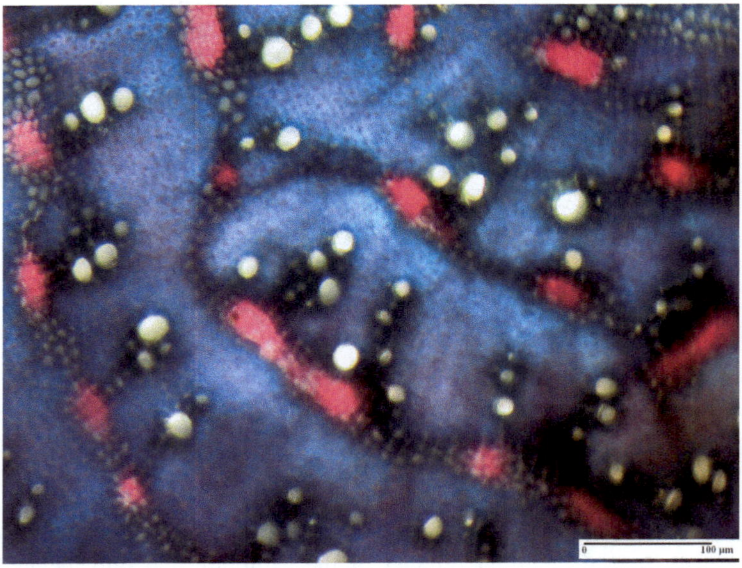

Fig. 7.13 Successive cambia in *Atriplex littoralis* (root, RO)

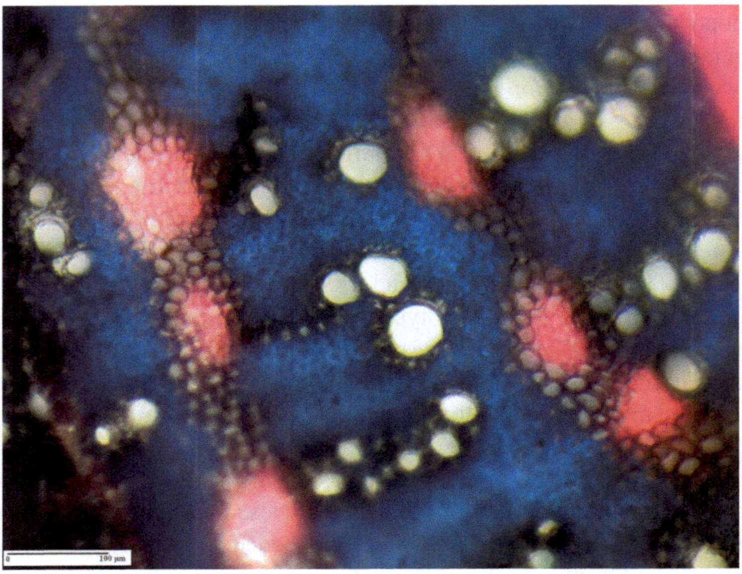

Fig. 7.14 Successive cambia in *Atriplex littoralis* (stem, RO)

Fig. 7.15 Successive cambia in *A. prostrata* (root, basal level, RO)

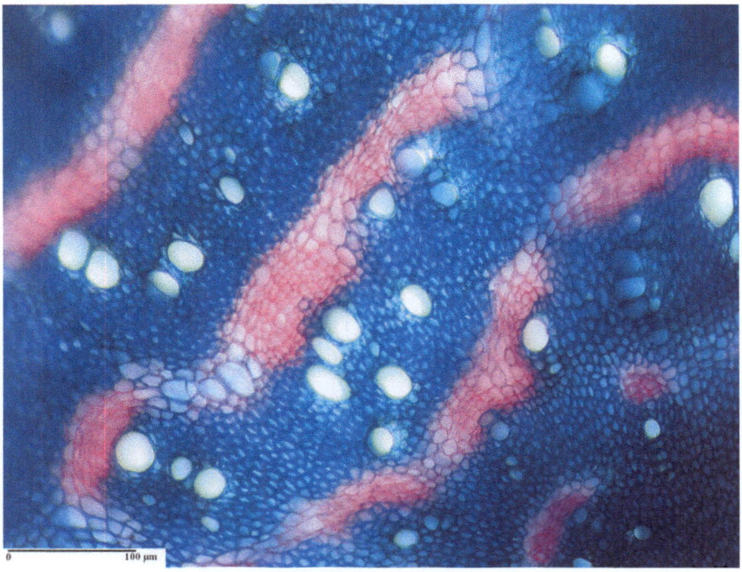

Fig. 7.16 Successive cambia in *A. prostrata* (root, basal level, RO)

Fig. 7.17 Successive cambia in *A. prostrata* (middle level, RO)

Fig. 7.18 Successive cambia in *A. tatarica* (root, RO)

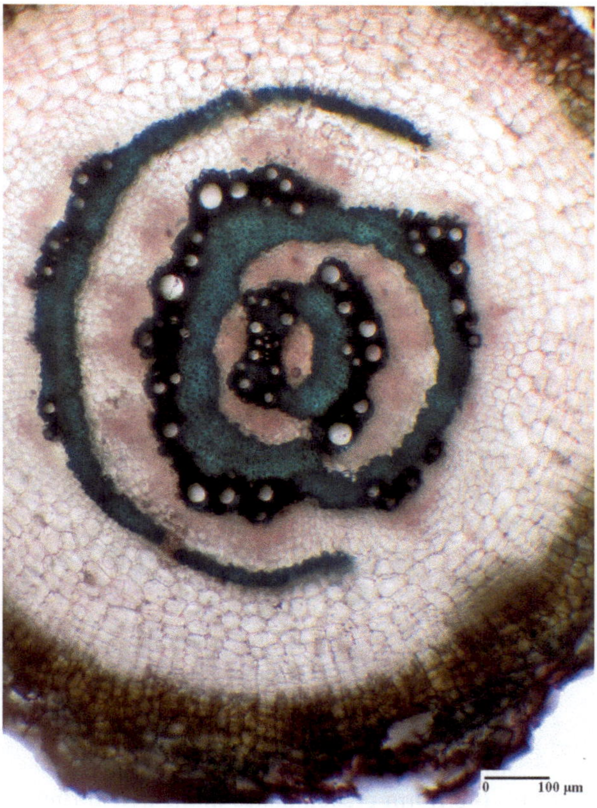

Fig. 7.19 Successive cambia in *Atriplex glauca* (root, ESP)

Camphorosma annua (Figs. 7.21, 7.22, and 7.45), *Chenopodium album* (Figs. 7.23 and 7.24), *Halimione verrucifera* (Figs. 7.25, 7.26, 7.46, and 7.47), *Petrosimonia oppositifolia* (Figs. 7.27, 7.28, and 7.48), *P. triandra* (Figs. 7.29, 7.30, and 7.49), *Salicornia europaea* (Figs. 7.31, 7.32, and 7.50), *S. ramosissima* (Fig. 7.33), *Suaeda maritima* (Fig. 7.34), *S. spicata* (Figs. 7.35 and 7.36), *S. splendens* (Fig. 7.37), *Salsola kali* (Figs. 7.38 and 7.39), and *Sarcocornia fruticosa* (Figs. 7.40 and 7.41). In all mentioned species, successive cambia have been observed in the structure of axial vegetative organs (root and stem).

Descriptions for several Romanian halophytes are given in detail, in the next paragraphs, for a better understanding of successive cambia functioning.

For instance, in *Atriplex prostrata* (Grigore and Toma 2005), the root (Fig. 7.43) presents successive cambia in all sectioned levels. At a lower level, the primary structure is often of diarch type, and in the first ring produced by the additional cambium, vessels are dispersed irregularly, having a large diameter and moderately thickened and lignified walls. Between the vessels, lignified parenchyma cells have cellulosic walls, and the libriform is represented through fibers with moderately thickened and lignified walls. Gradually, a second additional cambium is formed, which produces a relatively compact xylem ring inward and another one of phloem

Fig. 7.20 Successive cambia in *Bassia sedoides* (RO)

outward. Libriform predominates in the xylem ring, parenchyma cells being very rare, as well as the vessels; the latter has a different diameter and is irregularly dispersed. The phloem ring includes sieved tubes, companion cells, and phloem parenchyma cells.

At the middle level, there can be seen four rings of xylem and four of phloem, the overall structure presenting a visible asymmetry and on one side of the section only two rings of xylem and two of phloem being visible. All xylem rings are heavily lignified, in these predominating libriform fibers, with vessels irregularly dispersed. The phloem rings are totally cellulosic, including vascular phloem areas (sieved tubes and companion cells) separated by areas of phloem parenchyma. Here and there, xylem rings are pierced by parenchymatic-cellulosic rays, and in the phloem rings thin layers of elements with lignified walls enter.

Toward root base, there can be noticed six to seven concentric rings of xylem, fully lignified, separated by as many phloem rings, fully cellulosic. Xylem and phloem rings are more sinuous, of different thickness, sometimes fragmented or in contact with each other. In the central portion the xylem tissue predominated, the phloem forming tiny isles completely surrounded by elements with thickened and lignified walls; such situation can be noticed also toward the periphery of the root, where the cellulosic phloem rings are fragmented by radial layers of lignified tissue.

Fig. 7.21 Successive cambia in *Camphorosma annua* (RO)

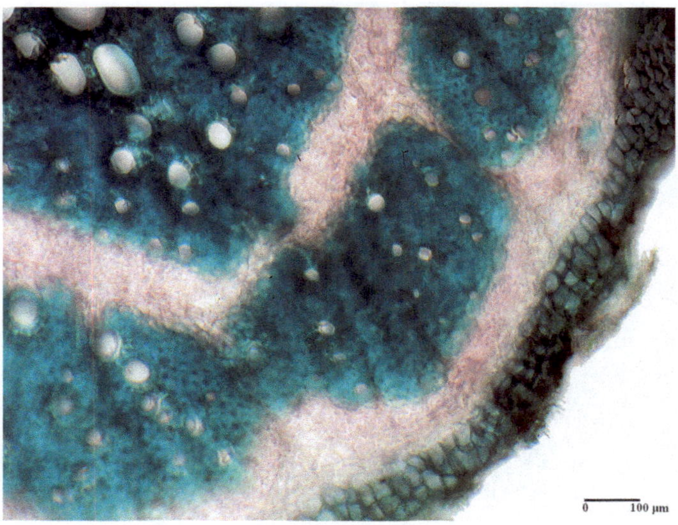

Fig. 7.22 Successive cambia in *Camphorosma annua* (RO)

Fig. 7.23 Successive cambia in *Chenopodium album* (root, ESP)

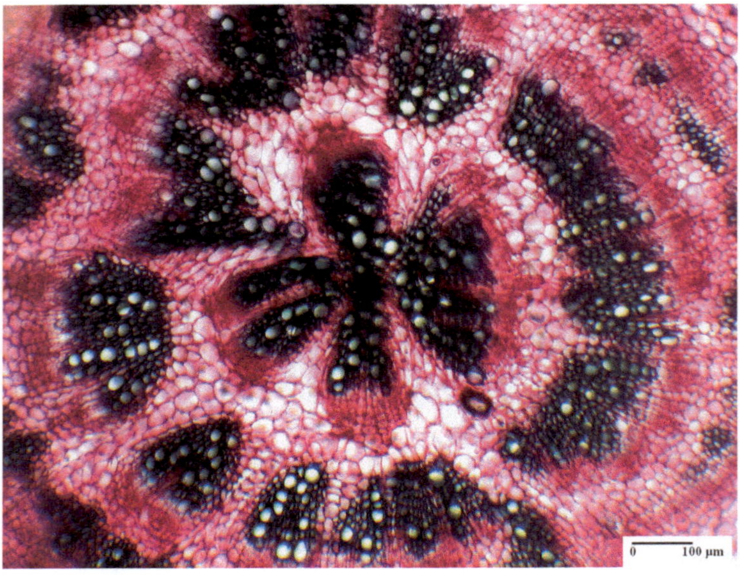

Fig. 7.24 Successive cambia in *Chenopodium album* (root, ESP)

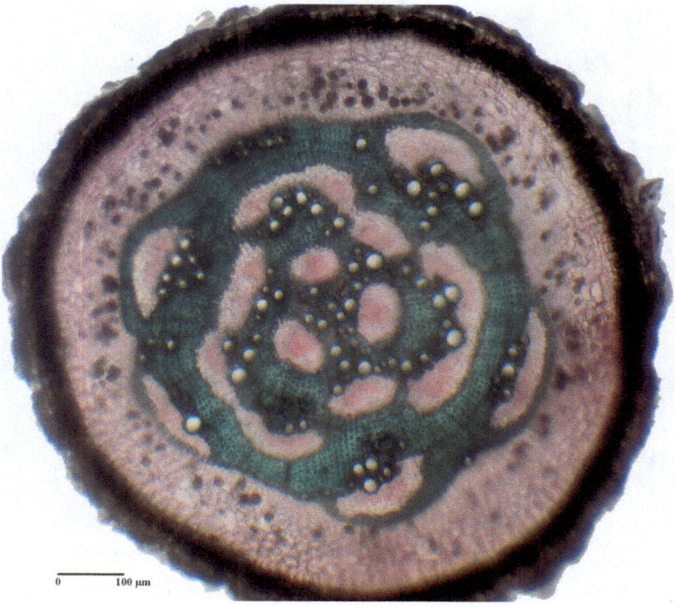

Fig. 7.25 Successive cambia in *Halimione verrucifera* (root, RO)

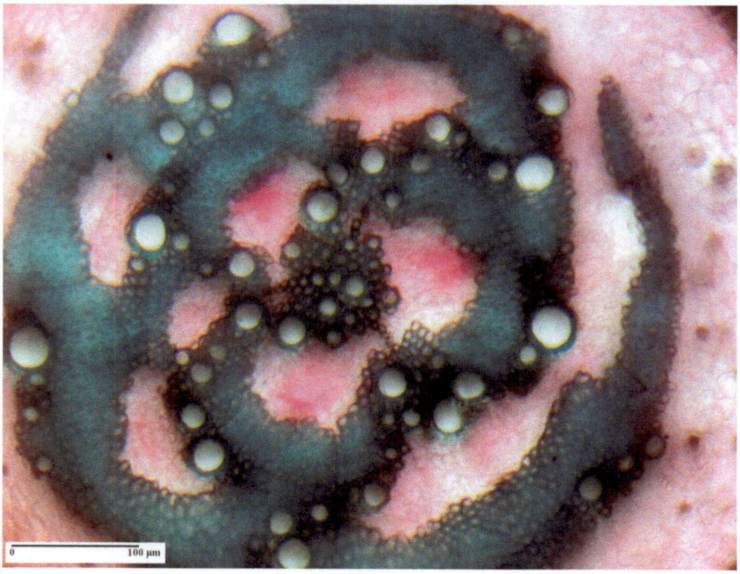

Fig. 7.26 Successive cambia in *Halimione verrucifera* (root, RO)

The center of the organ is occupied by a solid compact xylem body, intensely lignified, which has on one side as well as on the other one two phloem bundles visibly collenchymatous.

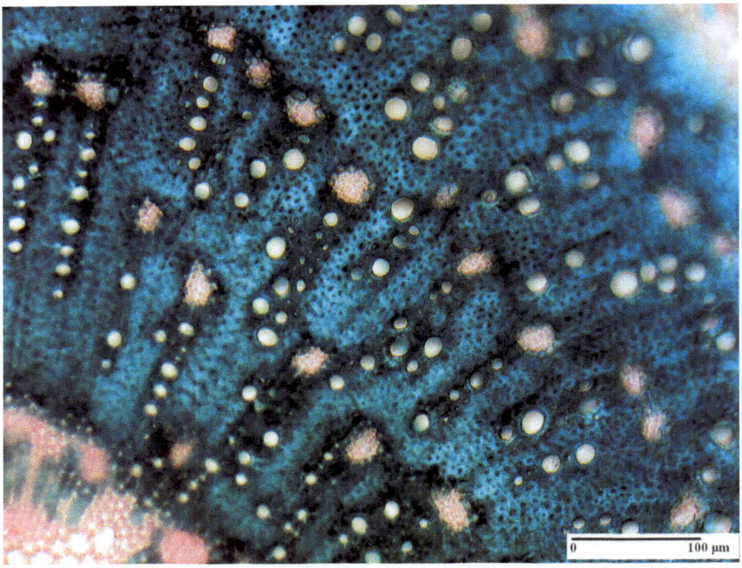

Fig. 7.27 Successive cambia in *Petrosimonia oppositifolia* (root, RO)

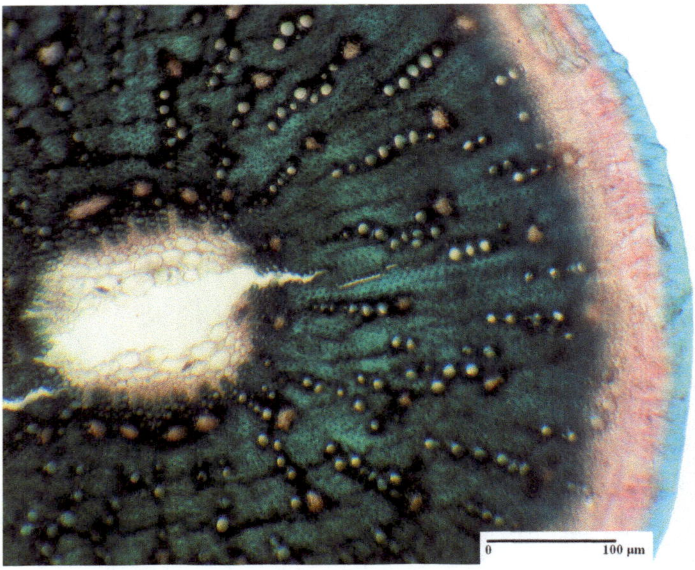

Fig. 7.28 Successive cambia in *Petrosimonia oppositifolia* (root, RO)

The cross sections made even on the basis of the organ (Fig. 7.43d) show the same structure as for the root, resulting from the activity of several successive cambia so that the number of phloem and xylem concentric rings is higher (three to four).

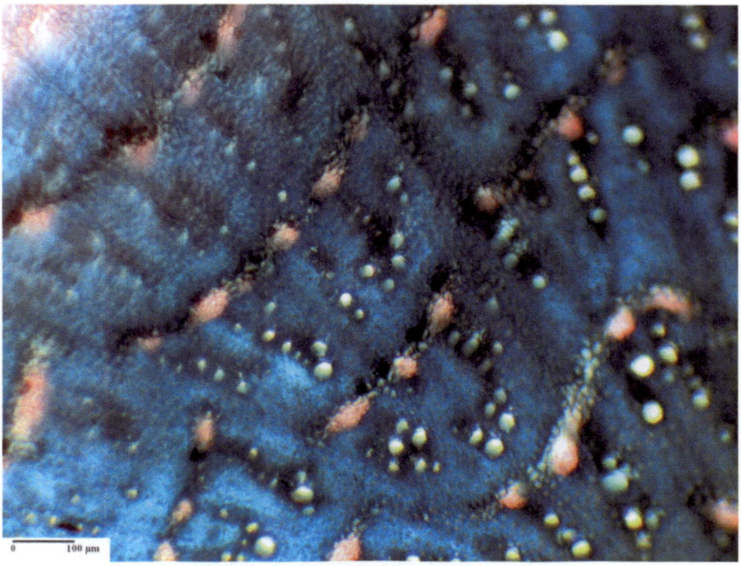

Fig. 7.29 Successive cambia in *P. triandra* (root, RO)

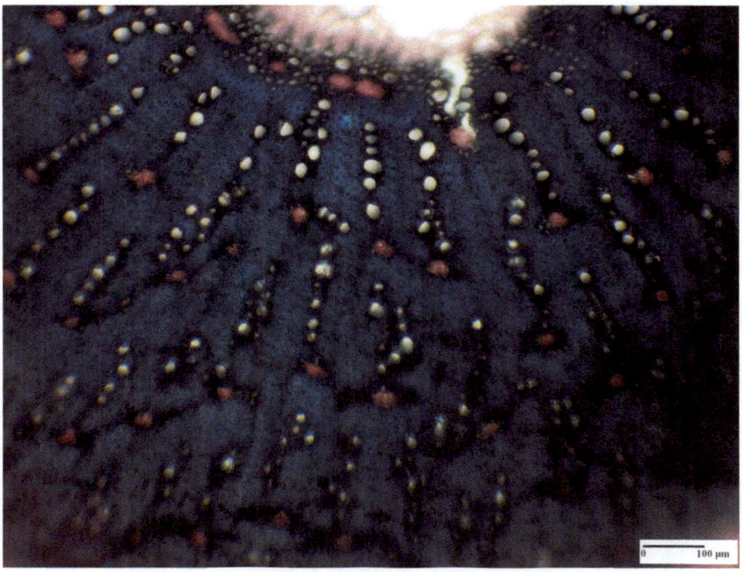

Fig. 7.30 Successive cambia in *P. triandra* (root, RO)

In *Atriplex littoralis* (Figs. 7.42) (Grigore and Toma 2007), in the lower level of root, the central cylinder presents a number of three to four rings of vascular tissues resulting from the activity of the successive cambia, each of them having inward

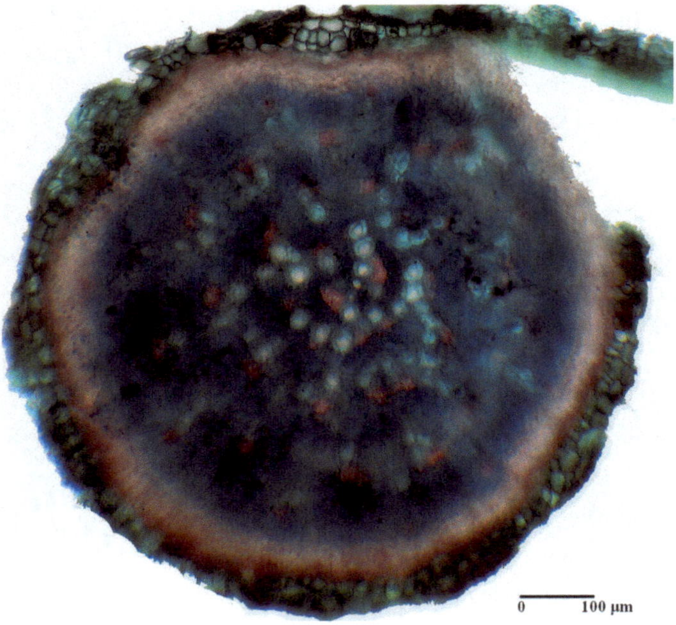

Fig. 7.31 Successive cambia in *Salicornia europaea* (root, RO)

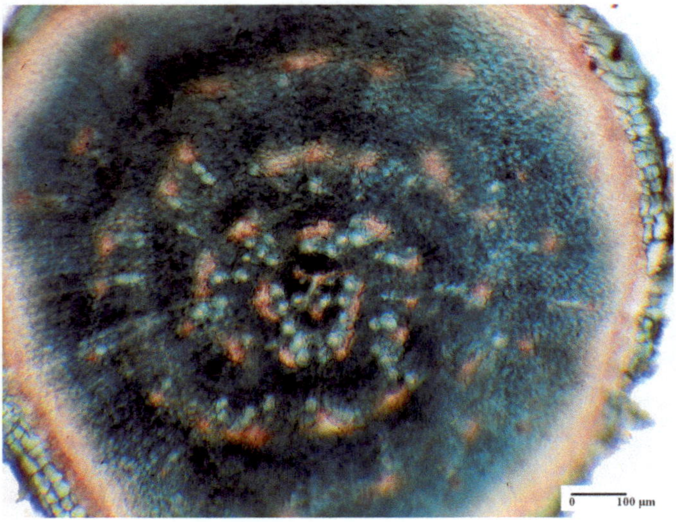

Fig. 7.32 Successive cambia in *Salicornia europaea* (root, RO)

xylem (much thicker) and outward phloem (much thinner). In each ring, the xylem is fully sclerified and lignified, with few irregularly dispersed vessels. The xylem fibers, which predominate in the xylem, have intensely thickened and moderately lignified wall. The phloem consists of sieved tubes, companion cells, and phloem

Fig. 7.33 Successive cambia in *S. ramosissima* (root, ESP)

parenchyma cells. The last ring of vascular tissues is still emerging, being uneven as the thickness on the circumference of the root.

In the middle level, the general structure remains the same, with the same number of concentric vascular rings. Xylem vessels have a smaller diameter axial area but have more intensely lignified walls. Libriform has fibers with extremely strong thickened walls but only partially lignified.

In the upper level, the root is thicker, having in its particular structure 5(6) concentric rings of vascular tissues; nonetheless, the phloem does not form continuous rings, but discontinuous ones, as isles are completely surrounded by the xylem; the latter is sclerified and intensely lignified, with the libriform predominating in it.

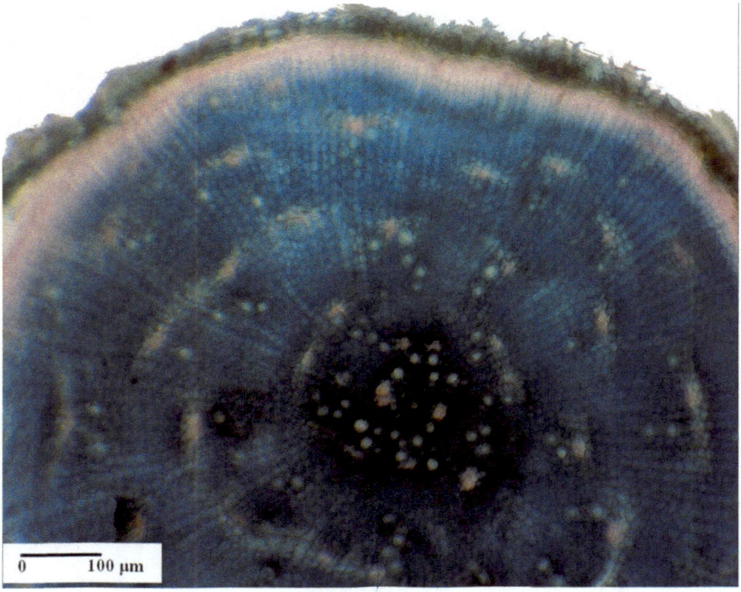

Fig. 7.34 Successive cambia in *Suaeda maritima* (root, RO)

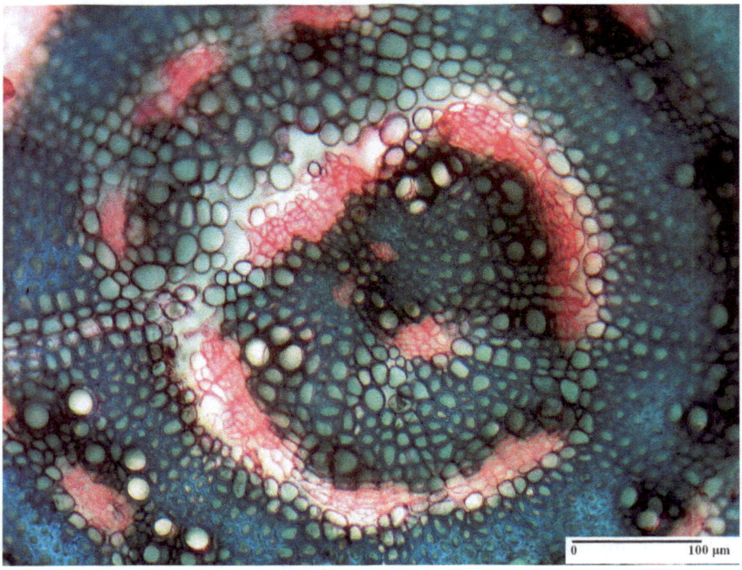

Fig. 7.35 Successive cambia in *Suaeda spicata* (root, ESP)

Fig. 7.36 Successive cambia in *Suaeda spicata* (root, ESP)

Fig. 7.37 Successive cambia in *Suaeda splendens* (root, ESP)

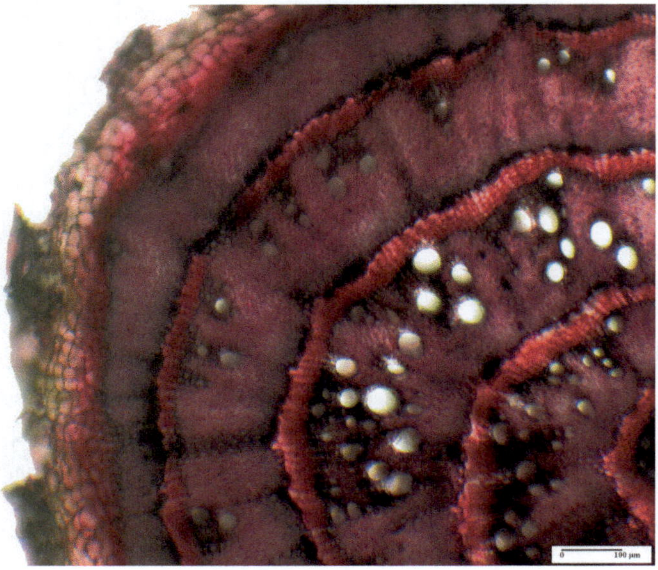

Fig. 7.38 Successive cambia in *Salsola kali* (root, ESP)

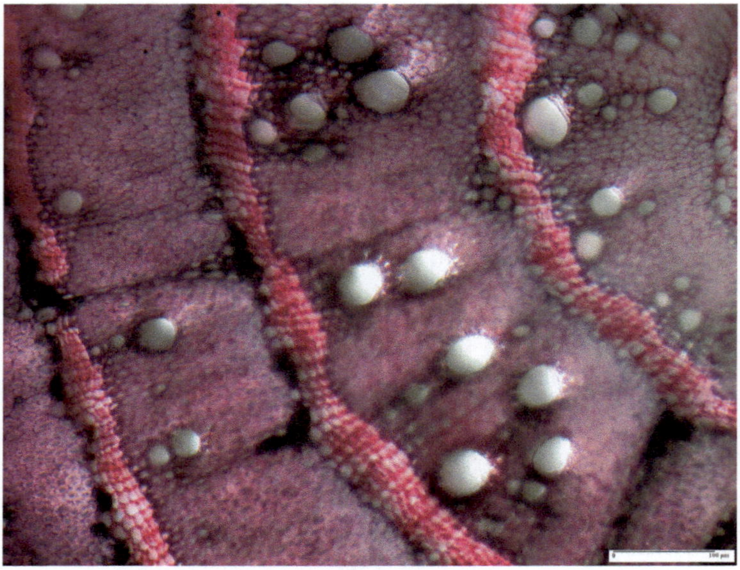

Fig. 7.39 Successive cambia in *Salsola kali* (root, ESP)

In the upper level of the stem, the structure is normal, a fact underlined by Metcalfe and Chalk (1972), who state that successive cambia do occur in species with a thick stem and are not noticeable at the top of the stem or in the species with thin stems. The central cylinder contains a large number (14–16) of vascular

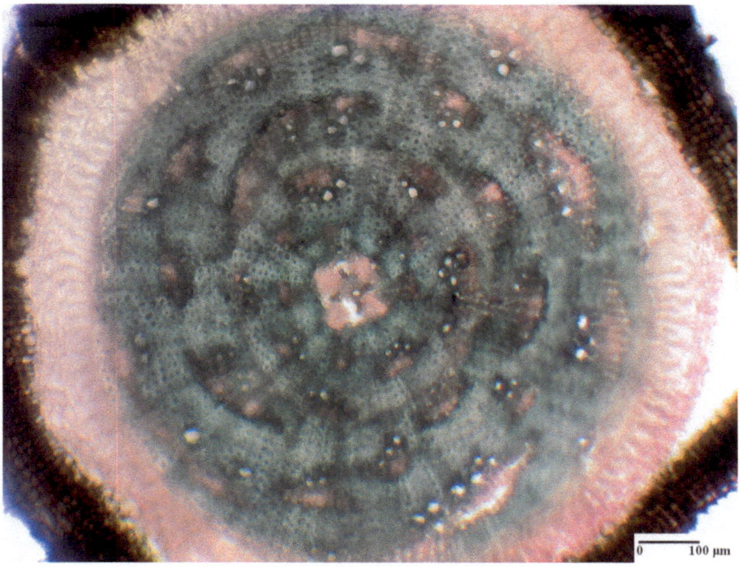

Fig. 7.40 Successive cambia in *Sarcocornia fruticosa* (root, ESP)

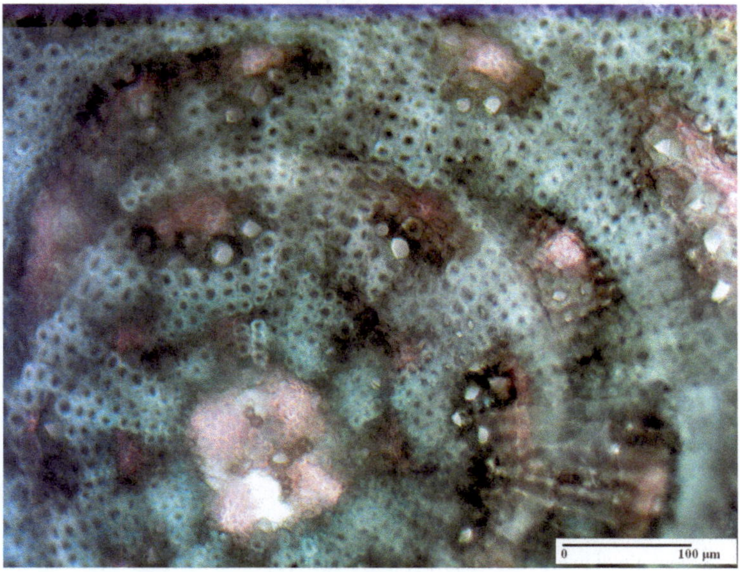

Fig. 7.41 Successive cambia in *Sarcocornia fruticosa* (root, ESP)

bundles, of open-collateral type and different sizes; the largest bundles protrude deeply into the medulla.

In the central cylinder, in addition to the vascular bundles analyzed at the previous level, there is a thick libriform ring, in which several bundles (blades) of

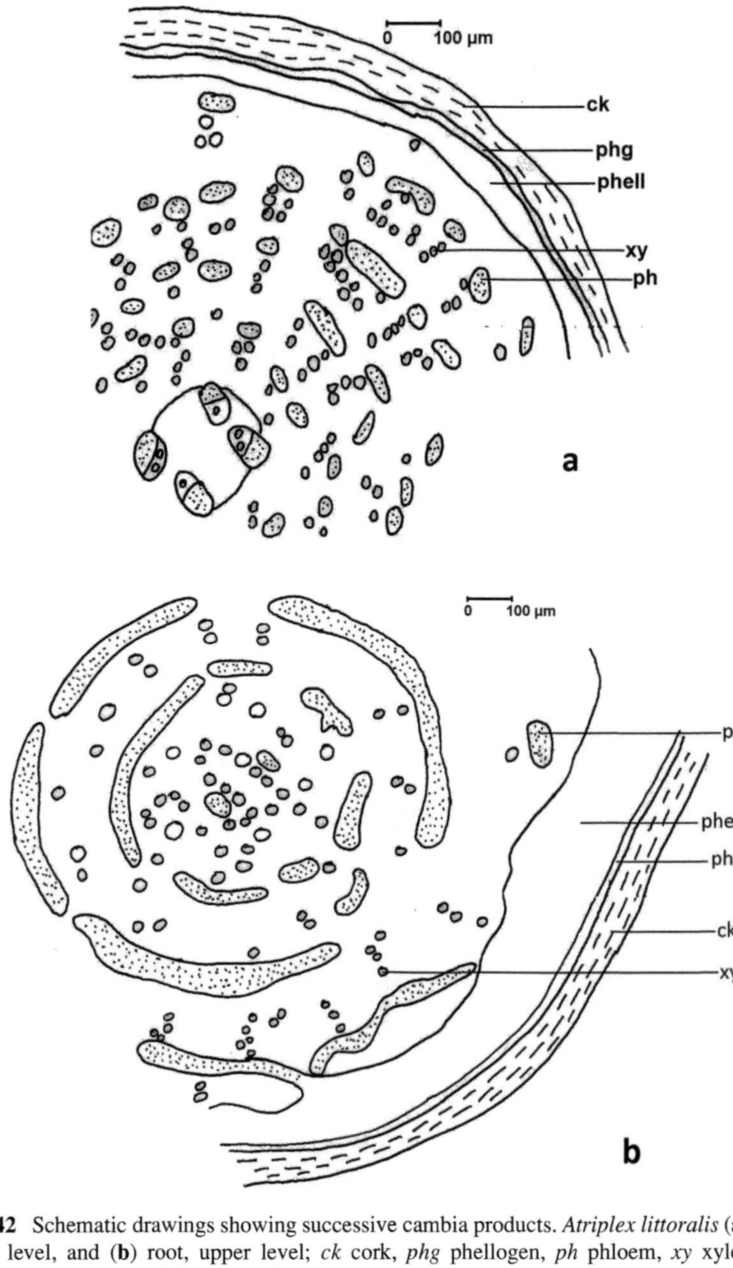

Fig. 7.42 Schematic drawings showing successive cambia products. *Atriplex littoralis* (**a**) root, at middle level, and (**b**) root, upper level; *ck* cork, *phg* phellogen, *ph* phloem, *xy* xylem, *phell* phelloderm) (Grigore and Toma 2010)

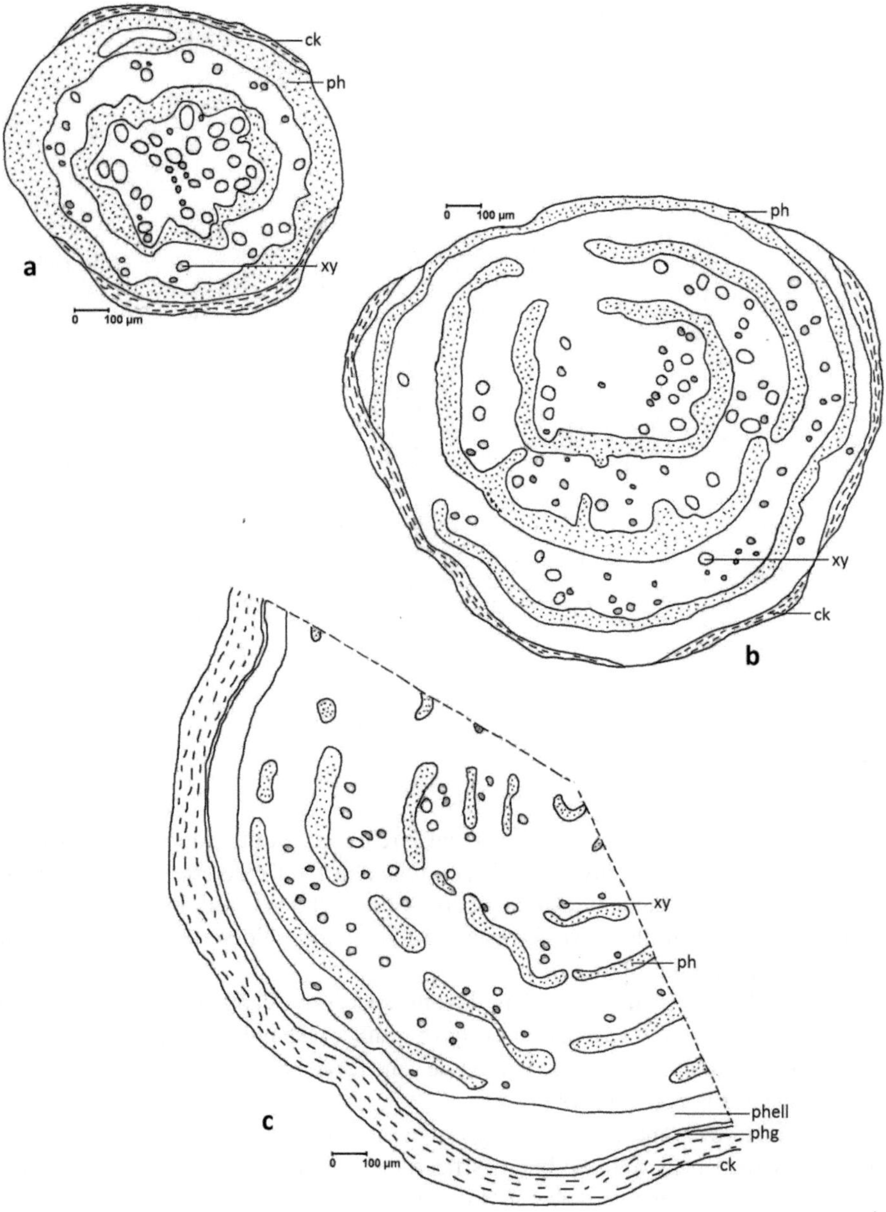

Fig. 7.43 Schematic drawings showing successive cambia products. *Atriplex prostrata* (**a**) root, at lower level; (**b**) middle level; and (**c**) upper level; *d* stem, at the base; *ck* cork, *phg* phellogen, *ph* phloem, *xy* xylem, *phell* phelloderm) (Grigore and Toma 2010). (**d**) Schematic drawings showing successive cambia products. *Atriplex prostrata*

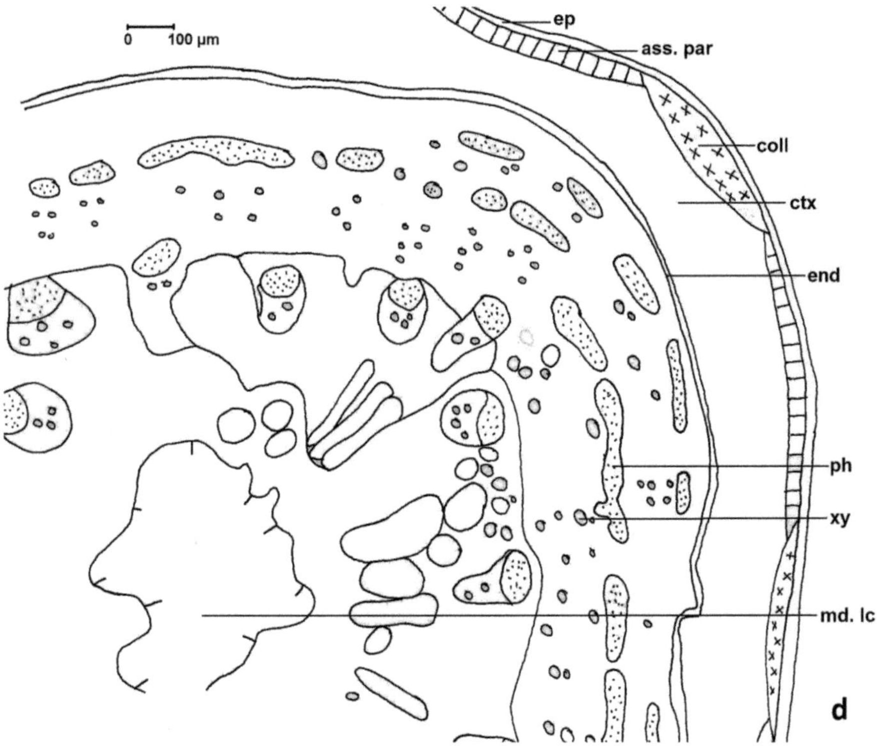

Fig. 7.43 (continued)

xylem vessels and isles of phloem elements are located at the exterior. The phloem isles have visibly collenchymatic elements, and at their exterior, one may notice cords (caps) of sclerenchymatous fibers with thick, yet cellulosed walls. The additional cambium that gave form to the xylem and phloem already mentioned is continuous and multilayered.

Toward the base of the organ, based on the successive cambia's activity, three to four concentric rings of vascular tissue have resulted, the phloem appearing as rings completely surrounded by the xylem tissue. The fascicular type structure noticed in previous levels is not so visible here.

The libriform is more intensely sclerified and lignified, the xylem vessels are irregularly dispersed, and the bundles from the initial primary structure are fully deepened in the parenchymatic-cellulosic medulla.

In *Halimione verrucifera* (Figs. 7.46 and 7.47) at the root level, the central cylinder has the typical structure of the *Chenopodiaceae* family. This structure is mostly due to the activity of the successive cambia.

The root axis is occupied by a massive xylem body completely sclerified and lignified, in which three radially strings of primary xylem vessels are arranged on a single line; based on these strings of primary xylem, it can be considered that the central cylinder from the primary structure is of triarch type.

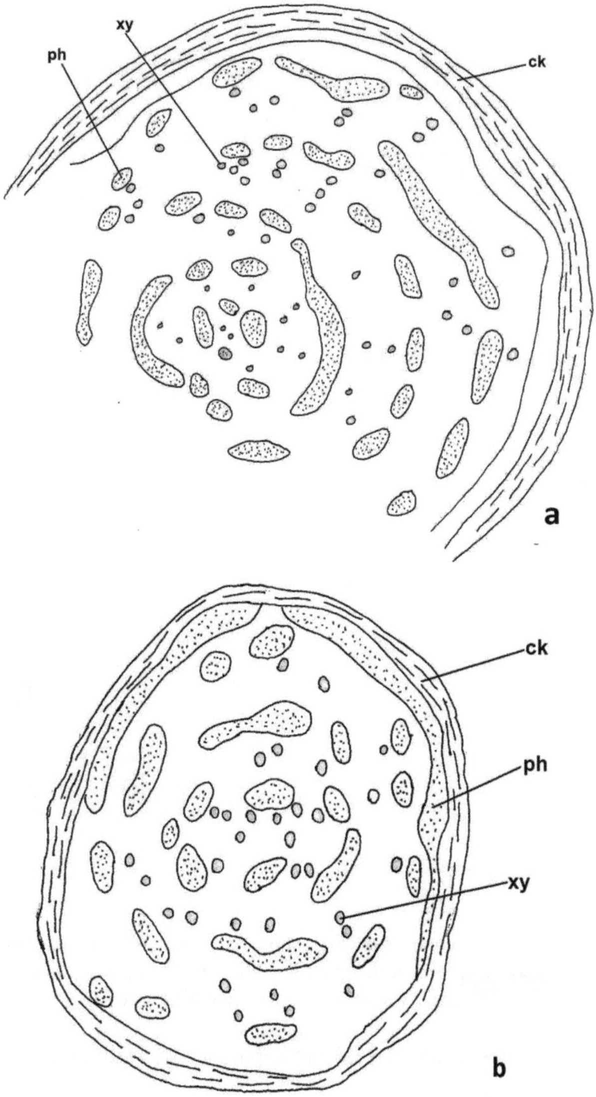

Fig. 7.44 Schematic drawings showing successive cambia products—*A. tatarica* (**a**) root, middle level, and (**b**) upper level; *ep* epidermis, *coll* collenchyma, *end* endodermis, *md lc* medullary lacuna, *ck* cork, *phg* phellogen, *ph* phloem, *xy* xylem, *phell* phelloderm) (Grigore and Toma 2010)

This compact and central xylem massive is surrounded by three phloem bundles which alternate with three radial blades of primary xylem and are separated by cellulosic parenchyma elements.

After that, two sinuous xylem rings can be noticed, as a result of the activity of the additional cambia, separated by areas of phloem completely cellulosic, and resulting from the activity of the same cambia.

The xylem rings are intensely sclerified and lignified, with a predominance of libriform and vessels of a different diameter; often, these vessels are grouped together, and next to them, at the periphery, there can be noticed phloem isles.

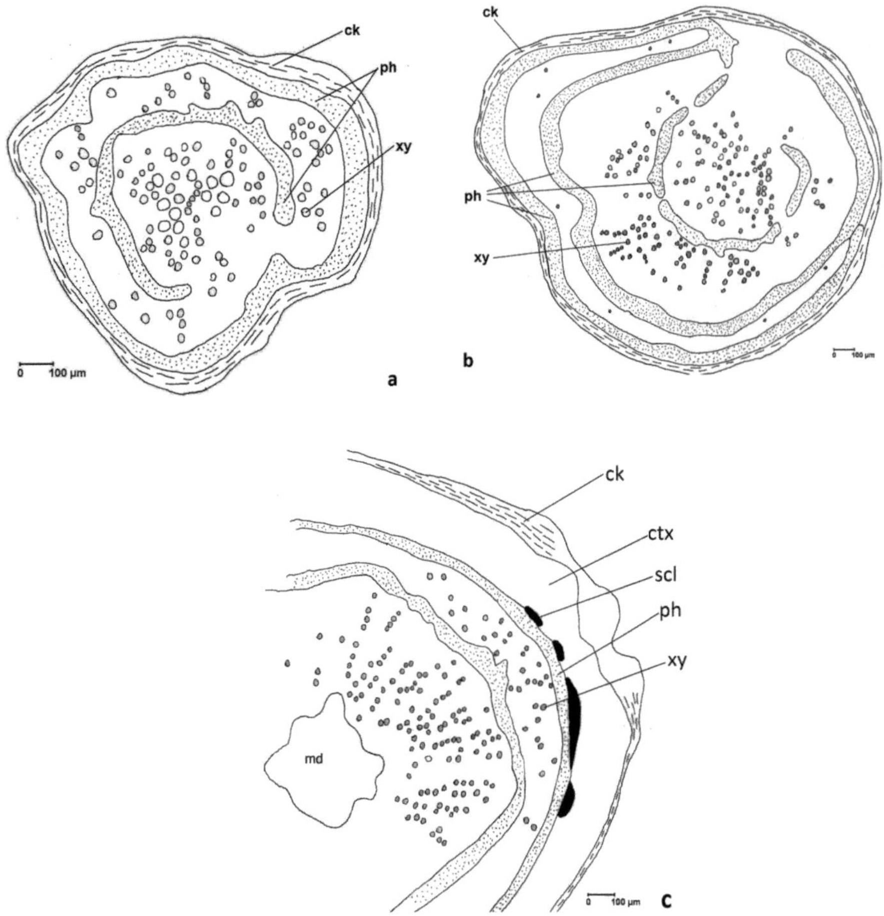

Fig. 7.45 Schematic drawings showing successive cambia products. *Camphorosma annua* (**a**) root, lower level; (**b**) root, upper level; and (**c**) stem, at the base; *md* medullary lacuna, *ck* cork, *ctx* cortex, *ph* phloem, *xy* xylem) (Grigore and Toma 2010)

These isles (eight to ten) are separated by radial blades consisting of sclerified and lignified elements.

The development of the second (external) xylem ring is still in process; therefore, it contains few vessels, and at their exterior, the cambium has produced only few phloem elements.

Toward the middle level, the general structure is the same except that the second discontinuous phloem ring (of isle-type) is already formed now and is surrounded at the exterior by the xylem of the third ring.

As in the previous level, the central part of the central cylinder has a slight trilobite shape overlapped to the three phloem bundles.

In the stem (Fig. 7.47), the central cylinder contains more (eight) vascular bundles of different sizes, a great part of the xylem and phloem being of primary origin; in the vicinity of the phloem, the vessels are separated by few libriform elements.

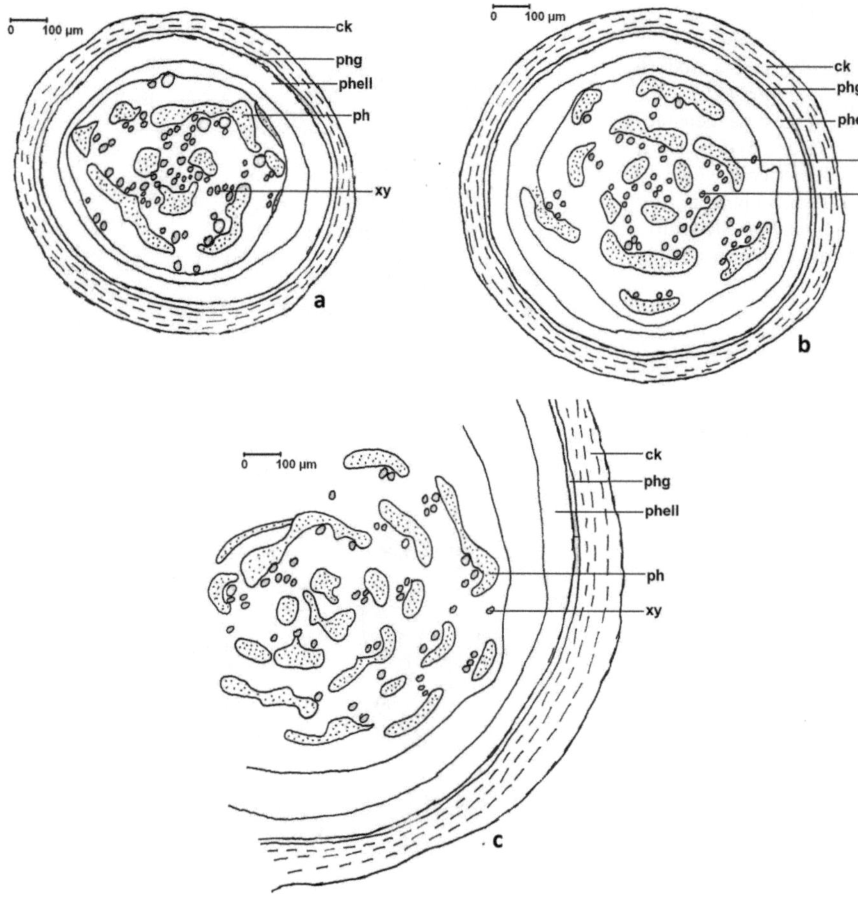

Fig. 7.46 Schematic drawings showing successive cambia products. *Halimione verrucifera* (**a**) root, lower level; (**b**) root, middle level; and (**c**) root, upper level; *ck* cork, *phg* phellogen, *ph* phloem, *xy* xylem, *phell* phelloderm) (Grigore and Toma 2010)

At the bundles' periphery and between them, there is a sclerenchyma tissue which forms a thick ring with cells having intensely thickened and lignified walls; thus, the phloem of the vascular bundles appears as isles completely surrounded by sclerified and lignified tissues.

At the periphery of the sclerenchyma ring, the first additional cambium (of pericyclic origin), which formed from place to place few xylem vessels inward and several phloem elements outward, can be noticed, the latter together with the cambium achieving almost a continuous ring. At the middle level, the central cylinder is thicker than the one from the previously analyzed level; it contains an intensely sclerified and lignified ring of fundamental parenchyma and a medulla with eight vascular bundles protruding from the primary structure.

Within the sclerified and lignified ring, one can notice here and there xylem vessels with an irregular disposal but sometimes with discontinuous radial strings, each one of these strings having at the periphery a phloem tissue isle.

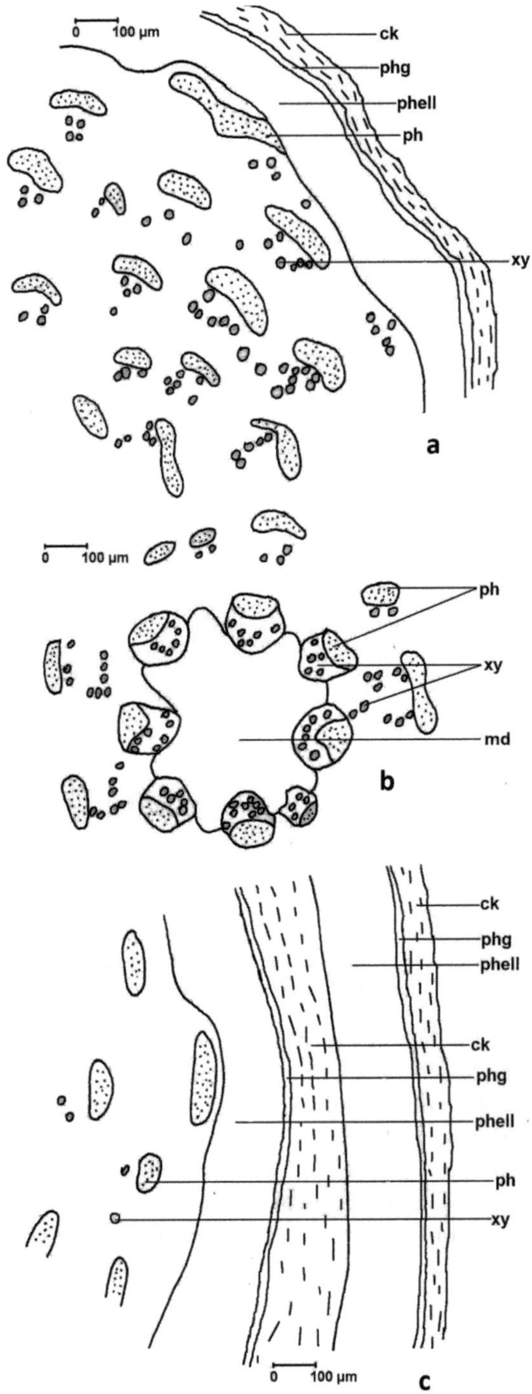

Fig. 7.47 Schematic drawings showing successive cambia products. *Halimione verrucifera* (**a**) stem, at the base, general view; (**b** and **c**) details; *md* medulla, *ck* cork, *phg* phellogen, *ph* phloem, *xy* xylem, *phell* phelloderm) (Grigore and Toma 2010)

The first string of xylem vessels together with the corresponding phloem isle (cap) forms a thick ring of "bundles" deeply embedded into the fundamental libriform mass.

The second ring of vascular bundle is much thinner, having phloem isles (caps) surrounded by an intensely lignified sclerenchyma tissue.

Here and there, on the internal part of the primary cortex, one may notice some solitary sclerenchyma cells located at the periphery of the still developing last ring of successive cambium and latter formed phloem elements.

In what concerns the central cylinder, this is much thicker than in the previously analyzed levels, containing a number of four to five cellulosic ring isles of phloem; these correspond—together with the xylem from the internal cortex—to as many collateral-type vascular bundles rings embedded into the fundamental mass of an intensely sclerified and lignified libriform.

In *Petrosimonia triandra* (Fig. 7.49) (Grigore and Toma 2007), in the general structure of the root, at its lower level, six to seven concentric rings of vascular tissues stand out, resulting from the activity of a corresponding number of additional cambia. It can be noticed that there are much libriform, a smaller number of xylem vessels irregularly dispersed, and numerous tangential thin stripes of phloem tissue, which marks the location and number of vascular bundles arisen based on the additional cambia activity.

All mechanical elements of sclerenchyma (libriform) have a very thick wall and mostly intensely lignified.

At the middle level, in the central part of the root the diarch-type primary structure is still distinguishable; after it follows a relatively homogeneous structure, represented by several concentric areas of vascular bundles, among which there are sclerified and lignified rays.

The xylem vessels from the primary structure have a small diameter and intensely lignified walls; the vessels from the bundles resulting from the successive cambia activity have a much larger diameter and a very thick wall, yet poorly lignified, with libriform irregularly dispersed between them.

The vascular bundles form thin tangential layers interrupted here and there by narrow strings of mechanical cells with intensely thickened and moderately lignified walls, like those of libriform from the xylem structure.

At the sectioned level, four concentric rings (circles) of vascular tissues are noticed; these result from the activity of four successive cambia.

In the stem, at the upper level, the central cylinder contains a few (five to seven) vascular bundles, of different sizes, separated by parenchymatic-cellulosic medullary rays.

All vascular bundles have phloem consisting of sieved tubes, companion cells, and xylem formed of radial strings of vessels, separated by cellulosic parenchyma cells. Therefore, the structure is typically primary.

At the periphery of the vascular bundles has already appeared the first additional cambium, of pericyclic origin starting to function in two opposite directions, giving xylem inward and phloem outward; the tracheo-genesis process is still ongoing, some xylem vessels having very thin, cellulosic walls.

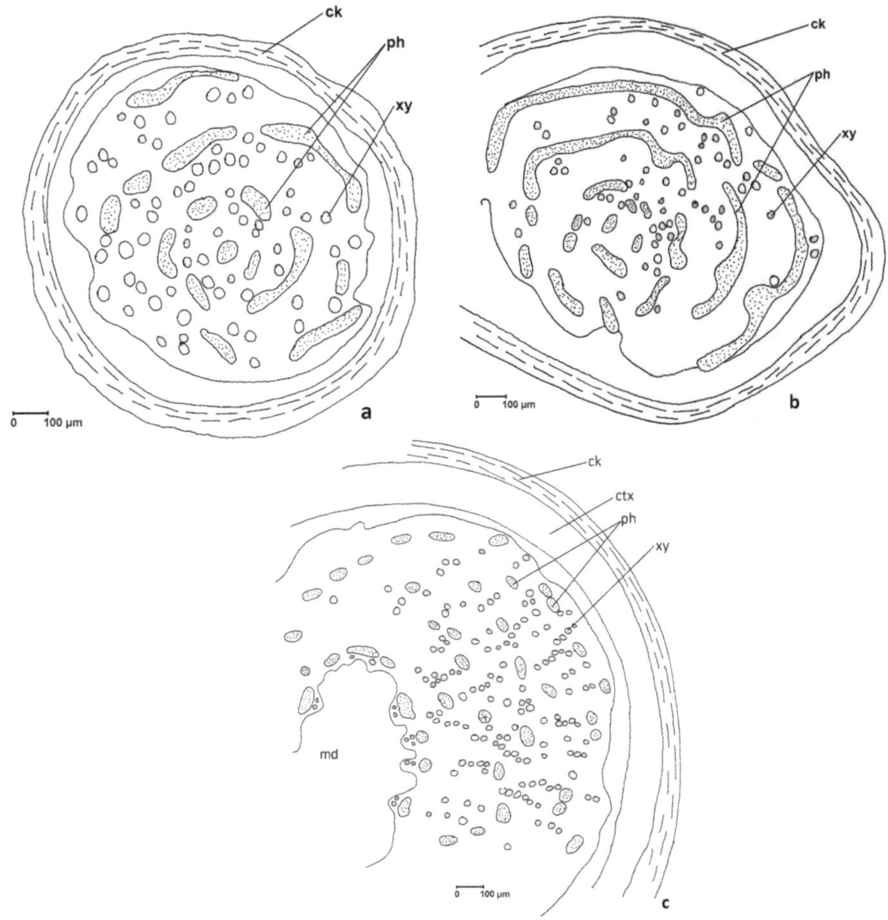

Fig. 7.48 Schematic drawings showing successive cambia products. *Petrosimonia oppositifolia* (**a**) root, lower level; (**b**) root, middle level; and (**c**) root, upper level; *md* medulla, *ctx* cortex, *ck* cork, *ph* phloem, *xy* xylem) (Grigore and Toma 2010)

It must be stressed upon that this first additional cambium forms a continuous, multilayered ring, which will produce vascular bundles, and here and there, medullary rays.

In the middle level, the tertiary structure, resulting from the successive cambium activity, characterized by a large quantity of libriform, intensely sclerified, but slightly lignified, where xylem vessels are irregularly dispersed. At the periphery of this thick tertiary xylem ring, a thin ring of phloem elements is noticed.

Toward the lower level of the stem, the structure remains the same, except that the central cylinder is much thicker, containing a large amount of libriform, relatively few xylem vessels, irregularly dispersed or forming discontinuous radial strings, and very few small isles of phloem tissue.

In *Petrosimonia oppositifolia* (Fig. 7.48), in the root, the secondary structure at this level is completed by that derived from the activity of the successive cambia

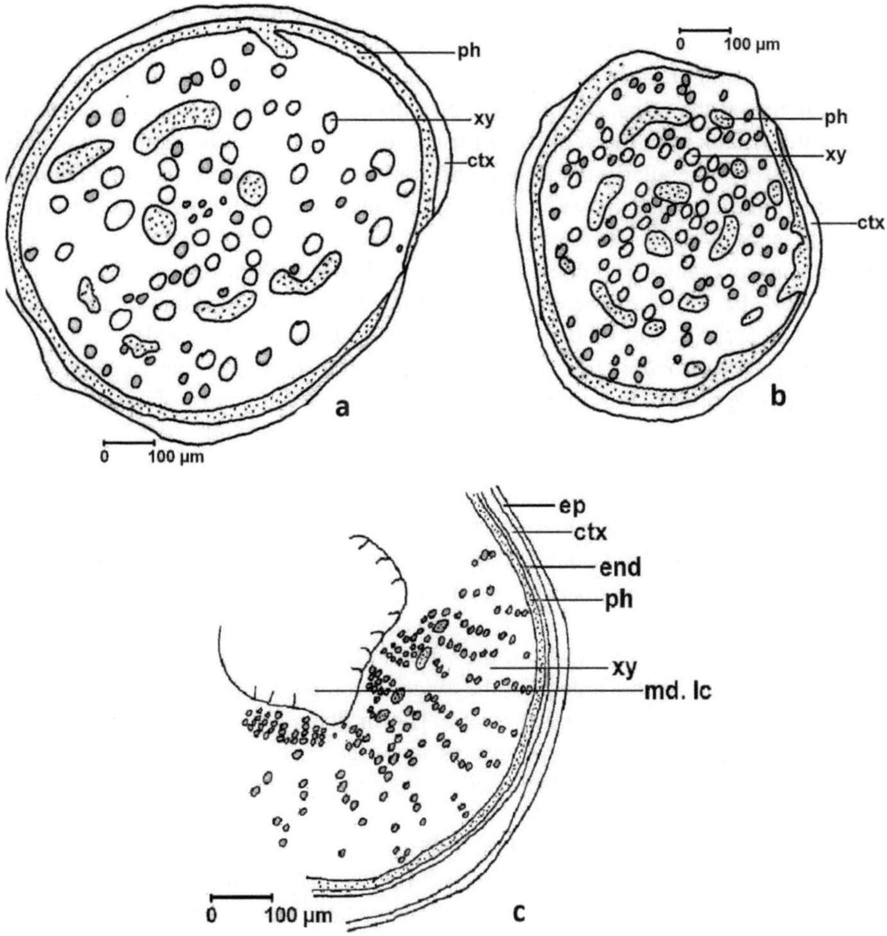

Fig. 7.49 Schematic drawings showing successive cambia products. *Petrosimonia triandra* (**a**) root, lower level; (**b**) root, middle level; and (**c**) stem, at the base; *md lc* medullary lacuna, *ep* epidermis, *end* endodermis, *ctx* cortex, *ck* cork, *ph* phloem, *xy* xylem) (Grigore and Toma 2010)

(generally two cambia), the vascular tissues forming incomplete rings and phloem and xylem arches.

In the center, one can notice the compact secondary xylem that has on both sides two isles of secondary phloem. Around this normal secondary structure, there are several areas of xylem with irregularly disposed of vessels, separated by a large quantity of intensely sclerified and lignified libriform. The root is thicker at this level, the general structure remaining the same; there are several (three to four) concentric areas with xylem and phloem. The latter is thinner, is discontinuous, and forms wide-open arches or isles.

In *Atriplex tatarica* (Fig. 7.44), in the lower level of the root, the secondary structure is completed by tissues issued from the activity of the successive cambia (two at this level) producing a larger quantity of xylem inward and a smaller quantity of phloem outward.

In *Salicornia europaea*, the successive cambia occur both in the stem and in the non-articulated stem. In the root (Fig. 7.50), the central cylinder is very thick, but it especially results from the activity of the successive cambia. They generate mainly xylem vessels (irregularly dispersed) and libriform cells, and in the fundamental, sclerified, and lignified mass. In the non-articulated stem, the structure resulting from the activity of the successive cambia is represented by a very thick libriform ring, where several small phloem islands are embedded. Their internal side has few xylem vessels, whose diameter does not differ too much from one of the xylem fibers; they are different by the slightly thinner and less lignified wall.

In *Suaeda maritima*, in the root, the activity of the additional cambia generates a number of five rings of vascular tissues, each of them having most of it occupied by the libriform, with few vessels irregularly dispersed in it, and numerous phloem islands, separated by cellulosic or poorly lignified parenchyma; these islands are not equidistant and they do not have the same size either.

Based on a more careful analysis, several radial strings of xylem vessels at the level of the phloem islands can be distinguished; these are separated by a very big quantity of libriform; the fibers of the latter have extremely thickened and moderately lignified walls.

In the upper part of the stem, all the vascular bundles have an initial secondary structure, which is visible at least at the xylem level where, among the vessels with extremely thick walls, there are several libriform fibers too.

None of the *Chenopodiaceae* species investigated by us or by other authors presented xylem vessels with thickened, but poorly lignified walls, like the ones of the libriform fibers.

In *Camphorosma annua* root (Fig. 7.45), from the activity of both additional cambia result in two arches, almost closed by intensely lignified libriform, in the consistency of which we can distinguish irregularly dispersed and of different diameter vessels. These arches are slightly strangulated here and there, giving the impression of a very close number of xylem bundles, separated by parenchymatic-lignified rays.

In the stem (Fig. 7.45c) of this species, successive cambia were evidenced only in the basal part; at this level of the stem, an additional cambium has been formed. The result of its activity is a relatively thick xylem ring and another one, thinner, formed of phloem. In the xylem ring, vessels are rare, and at the periphery of the phloem ring, several very thin sclerenchymatous fiber cords with moderately thickened and poorly lignified walls can be noticed.

Interestingly, in *Bassia hirsuta,* in the structure of vegetative organs, no additional cambia have been evidenced.

Visible massive lignification in the central cylinder of roots in the species affected by successive cambia could be induced by excessive soil salinity (Bickenbach 1932). Kozlowski (1997) stated that salinity increases the quantity of fiber. We must take into consideration the fact that in the context of multiple adaptations of plants to salinity, each species responds through certain metabolic

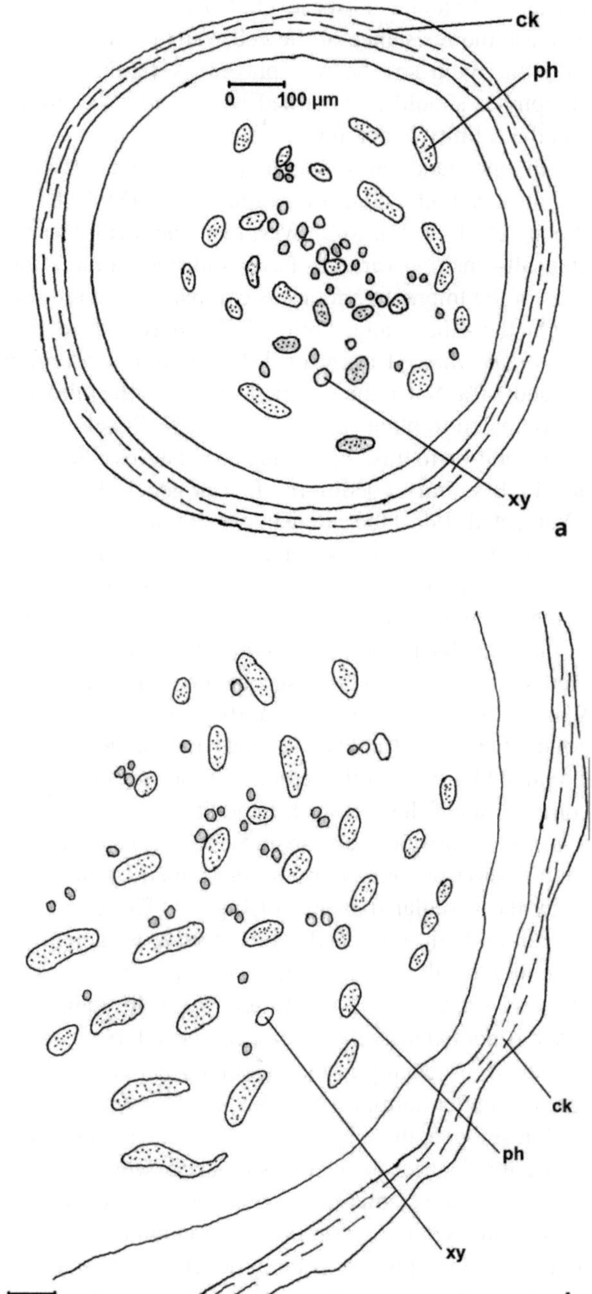

Fig. 7.50 Schematic drawings showing successive cambia products. *Salicornia europaea* (**a**) root, middle level, and (**b**) root, upper level, *ck* cork, *ph* phloem, *xy* xylem) (Grigore and Toma 2010)

changes. It should not be excluded for lignin to be such a response, in order to increase the resistance of the cell walls to the high osmotic pressure from the soil solution. Even so, the role played by lignin in providing resistance to salt for halophytes should be regarded with caution because the connection between this and the salinity is not always relevant.

On one hand, in some varieties of rice (resistant, respectively sensitive to salinity), a high value of salinity (40 mM NaCl) increases the lignin content (Seung et al. 2004). Moreover, histochemical observations on wheat salt-tolerant and salt-sensitive varieties have confirmed a more intense lignification in the root cells of the tolerant varieties as compared with the sensitive ones (Jbir et al. 2001).

On the other hand, high salinity may reduce lignin content in internodes of *Suaeda maritima* (Hagege et al. 1988). In *Atriplex prostrata* grown in solutions with 0.5 and 1.0% NaCl, the lignified surface in the third and fourth internodes decreases as compared to plants grown in 0.0% NaCl solution (Wang et al. 1997).

In addition to this, interesting correlations can be made between lignin, extensin, and high salinity conditions. It was generally noticed that in *Atriplex prostrata* (Wang et al. 1997) the extensin content (a glycoprotein contained in the cell wall) is increased in plants grown under salinity. Thus, in the first and second internodes, plants grown in solutions of 2.0% NaCl have registered the highest level of extensin in the vascular tissue, cortex, epidermis, and medulla. The extensin level increased in the vascular tissue as the salinity did so too in the third internode; it has also increased in the vascular tissue and medulla of the fourth internode. The findings of Wang et al. (1997) showed that there was a greater quantity of soluble extensin in young internodes (first and second) than in the older ones (third and fourth). This is explained by the fact that extensin molecules tend to bind to each other or to other components of the cell wall. In addition to this, extensin is closely correlated with lignin's deposition and can provide mechanical support to cells subjected to a compression or torsion stress determined by the increase of the vascular surface or by the vascular transport of liquids (Tiré et al. 1994; Li and Showalter 1996). Therefore, both lignin and extensin serve to increase cell wall resistance. Lignin is generally localized in xylem, whereas extensin is found mostly in the phloem (Ye and Varner 1991; Showalter 1993). In young internodes, where the vascular tissue is not extensively developed, the soluble extensin content is increased by the salt stress, especially in the medullary parenchyma, which has a small quantity of extensin under normal conditions. In older internodes, where xylem and phloem are well developed, there is no increase in the medulla but in the vascular tissue. Thus, the lignified surface (xylem) is reduced, whereas the phloem area has increased in saline conditions; this may be correlated with the fact that certain organic substances are synthesized in order to maintain osmotic potential under saline conditions so that the phloem is more developed to translocate these organic substances. Consequently, extensin becomes abundant in the stem, but the lignin content decreases. These results might suggest that extensin can replace lignin to increase cell wall resistance under salt stress. Moreover, it is known that lignin is an important component of the secondary cell wall and extensin is a regular component of the primary cell wall. These observations could be correlated with the fact that plants under salt stress may remain at the juvenile stages of development due to delay in growth and thereby the "proportion" of the primary cell wall is increased.

As a general observation, we can state, at least based on the analysis of halophyte species included in our study, that lignification has a higher proportion at the root level than at the stem one, aspect already mentioned above. In our opinion, this is not a coincidence, as it can be placed in the context of the root's position and role in the plant's life. Excluding maybe marine halophyte species subjected to salt spray, it can be stated that the root is the key organ most exposed to salinity. In this sense, it is logical to support the presence at the root level of general mechanisms (structures) able to control and manage the saltwater intrusion in the plant's organs and possibly its retention at this level.

All these details support the root's importance and the role it plays in plants exposed to high salinity conditions. As well as the other authors, our opinion is that lignin plays a major role in ensuring cell wall resistance for those cells that must withstand very high osmotic pressure. Therefore, these realities make from the root the key interface between salinity at the level of rhizosphere and the plant as a whole. As stated, not by accident the stem of many halophytes species is unaffected by successive cambia phenomenon, at least not entirely, so the lignin's proportion is much lower (Grigore and Toma 2006, 2007).

Nonetheless, it is not easy to find a convenient explanation regarding the ecological or ecophysiological significance of this phenomenon. Halophytes are an ecological heterogeneous group; their habitats vary greatly, so it is difficult to establish accurate correlations. Rather each of the species may provide sources of interpretation.

Carlquist (2007) discusses the implications of the phenomenon in the storage and recovery of photosynthates and water. He believes that the alternation between vascular complexes (vascular increments) and parenchyma, caused by successive cambia, may provide an "ideal" histological plan for water storage and recovery of photosynthates and water.

In the sense of this idea, we have formulated since 2006 a hypothesis regarding the ecological significance of successive cambia phenomenon in halophytes (Grigore and Toma 2006). Now, it must be emphasized again that the presence of the successive cambia phenomenon in halophytes could be related to environmental adjusting factors. It is well known that there are several mechanisms regulating the salt content. One of them is the salt dilution by the growth of the organ (Greenway and Thomas 1965). Another mechanism is related to retention of salt in roots and stems (Black 1956; Eshel and Waisel 1965; Jacoby 1964, 1965) as well as retransportation of salts inside the roots and their removal into the environment (Willert 1968; Cooil et al. 1965). All this could be related to an increased internal surface, if we consider only the high capacity of retention and "storage" of the saltwater in root and stem. On the other hand, the cork outward the root could also delay water absorption. Therefore, salts penetrate slowly in roots, but once arrived there, they would be dispersed in this increased surface. Literally, the water distribution to the rest of the plant's organs seems to be "delayed." Increasing this surface would inevitably mean a dispersion area for salts, which are also diluted, thus these being ultimately less harmful to the plant. Undoubtedly, the number and diameter of xylem vessels may play a role in this mechanism. Therefore, successive cambia phenomenon can be considered—based on its effects (numerous vessels, high "internal" surface) as a beneficial "compromise" for the plant between the growth limitation necessity (controlled by the abscisic acid ABA) and the necessity of imposing dilution strategies of salts in the plant's organ, except perhaps the higher regions of stem. The stem's apex as a growth region would thus be protected from the harmful effects of salt because it is known that young tissues are more sensitive to salts, as well as flowering, considered an extremely important stage in the plant's life (Waisel 1972).

It has been considered that, during the course of evolution, different groups of plants have undergone various modifications, which may be biochemical, morphological, or structural. These modifications helped the plants to adapt to particular climatic or ecological conditions. Among these structural modifications, the patterns of secondary thickening include formation of successive cambia, rayless xylem, and pedomorphosis and the formation of included phloem or of internal phloem (Rajput et al. 2008). Stems and roots with successive cambia have great adaptive potential. The relative amounts of parenchyma, fibers, vessels, and sieve tubes can easily be reallocated by this ontogenetic system so as to provide more mechanical strength, more flexibility, or more storage capacity (Elbar 2015).

Elbar (2015) found that the increments of stem diameter of *Sesuvium verrucosum* caused by successive cambia activity produce numerous functional vascular strands scattered throughout the old stem. Thus, a much greater area of the studied stem is probably available for conduction by secondary phloem and secondary xylem than in a dicotyledon with a single cambium. So, the prolonged conductive activity in these vascular increments is increased. This is an adapted feature of the halophyte *S. verrucosum* which grows in saline habitat and subjected to water stress.

He shows that in stem of *S. verrucosum*, the fibers are often organized as sheaths around the individual vessels or intervened the clusters of vessels; perhaps, this is a mechanism that helps in the protection of water columns from embolism.

References

Artschwager EF (1920) On the anatomy of *Chenopodium album* L. Am J Bot 7(6):252–260

Artschwager E (1926) Anatomy of the vegetative organs of the sugar beet. J Agric Res 33:143–176

Bickenbach K (1932) Zur Anatomie und Physiologie einiger Strand und Dünenpflanzen. Beitrage zum Halophytenproblem. Beitr Biol Pflanz 15:334–370

Black RF (1956) Effect of NaCl in water cultures on the ion uptake and growth of *Atriplex hastata*. Aust J Biol Sci 9:65–80

Bonnier G, Du Sablon L (1905) Cours de Botanique. Phanérogames. Librairie Générale de l' Enseignement, Paris

Carlquist S (1975) Wood anatomy of Onagraceae, with notes on alternative modes of photosynthate movement in dicotyledonous woods. Ann Mo Bot Gard 62:386–424

Carlquist S (1996) Wood, bark, and stem anatomy of Gnetales: a summary. Int J Plant Sci 157 (6 suppl):558–576

Carlquist S (2001) Comparative wood anatomy, 2nd edn. Springer, Berlin

Carlquist S (2003) Wood and stem anatomy of woody Amaranthaceae s.s.: ecology, systematics and the problems of defining rays in dicotyledons. Bot J Linn Soc 143:1–19

Carlquist S (2007) Successive cambia revisited: ontogeny, histology, diversity, and functional significance. J Torrey Bot Soc 134(2):301–332

Cockrell RA (1941) A comparative study of the wood of several south American species of *Strychnos*. Am J Bot 28:32–41

Cooil BJ, de la Fuente RK, de la Pena RS (1965) Absorption and transport of sodium and potassium in squash. Plant Physiol 40:625–632

De Bary A (1877) Vergleichende Anatomie der Vegetationsrgane der phanerogamen und farne. In: Hofmeister W (ed) Handbuch der physiologischen Botanik, vol 3. Wilhelm Engelmann, Leipzig

Droysen K (1877) Beiträge zur Anatomie und Entwicklungsgeschichte der Zuckerrübe. Halle a. S. (Inaug. Diss.)

Elbar OHA (2015) Development of the successive cambia in *Sesuvium verrucosum* Raf (*Aizoaceae*). Ann Agric Sci 60(2):203–208

Esau K, Cheadle VI (1969) Secondary growth in *Bougainvillea*. Ann Bot 33:807–819

Eshel Y, Waisel Y (1965) The salt relations of *Prosopis farcta* (Banks et Sol.) Eig Isr J Bot 14:50–51

Fahn A, Zimmermann MH (1982) Development of the successive cambia in *Atriplex halimus* (Chenopodiaceae). Bot Gaz 143(3):353–357

Fron G (1899) Recherches anatomiques sur la racine et la tige des Chénopodiacées. Ann Sc Nat 8-ème sér Bot 9:157–240

Gernet CAV (1859) Notizen ueber den Bau des Holzkoerpers einiger Chenopodiaceen. Bull Soc Imp Nat Mosc 32:164–188

Gheorhgieff S (1887) Beitrag zur vergleichenden Anatomie der Chenopodiaceen. Bot Centralbl ser 3 31:23–57, 53–57, 113–116, 151–154, 214–218, 251–255

Greenway H, Thomas DA (1965) Plant response to saline substrates. V. Chloride regulation in the individual organs of *Hordeum vulgare* during treatment with sodium chloride. Aust J Biol Sci 18:505–524

Greguss P (1968) Xylotomy of the living cycads. Academiai Kiado, Budapest

Grew N (1682) The anatomy of plants. With idea of a philosophical history of plants and several other lectures, read before the royal society. Printed by W. Rawlins for the Author, London

Grigore M-N (2008) Introducere în Halofitologie. Elemente de anatomie integrativă. Edit. Pim, Iaşi

Grigore M-N (2012) Romanian salt tolerant plants. Taxonomy and ecology. Edit. Tehnopress, Iasi

Grigore M-N, Toma C (2005) Contributions to the knowledge of anatomical structure of some halophytes I. Stud Cerc Şt biol Univ Bacău 10:125–128

Grigore M-N, Toma C (2006) Evidencing the successive cambia phenomenon on some halophylous representatives among *Chenopodiaceae* and its possible ecological-adaptive implications. Stud Com Complexul Muzeal St Nat "Ion Borcea" 21:87–93

Grigore M-N, Toma C (2007) Histo-anatomical strategies of *Chenopodiaceae* halophytes: adaptive, ecological and evolutionary implications. WSEAS Trans Biol Biomed 12(4):204–218

Grigore M-N, Toma C (2008) Ecological anatomy of halophyte species from the *Chenopodiaceae* family. In: Advanced topics on mathematical biology and ecology. Proceedings of the 4th WSEAS International Conference on Mathematical Biology and Ecology–MABE '08, Acapulco, Mexico, 25–27 Jan, pp 62–67

Grigore M-N, Toma C (2010) Halofitele. Aspecte de anatomie ecologică. Edit. Univ. "Al. I. Cuza", Iaşi

Grigore M-N, Toma C, Zamfirache M-M, Ivănescu L (2012) A survey of anatomical adaptations in Romanian halophytes. Towards an ecological interpretation. Fresenius Environ Bull 21 (11b):3370–3375
Grigore M-N, Toma C, Zamfirache M-M, Ivănescu L, Daraban I (2013) Anatomical and ecological observations in succulent (articulated) halophytes from *Chenopodiaceae*. Lucr Şt (Horticultură), USAMV "Ion Ionescu de la Brad" Iaşi 56(2):19–24
Grigore M-N, Ivănescu L, Toma C (2014) Halophytes. An integrative anatomical study. Springer, Cham
Hagege D, Kevers C, Boucaud J, Gaspar T (1988) Activités peroxydasiques, production d'éthylène, lignification et limitation de croissance chez *Suaeda maritima* cultivé en l'absence de NaCl. Plant Physiol Biochem 26:609–614
Hayward HE (1938) The structure of economic plants. Macmillan, New York
Hérail J (1885) Recherches sur l'anatomie comparée de la tige des Dicotylédones. Ann Sci Nat sér 7 Bot 2:203–314
Jacoby B (1964) Function of bean roots and stems in sodium retention. Plant Physiol 39:445–449
Jacoby B (1965) Sodium retention in excised bean stems. Physiol Plant 18:730–779
Jbir N, Chaibi W, Ammar S, Jemmali A, Ayadi A (2001) Effet du NaCl sur la croissance et la lignification des racines de deux espèces de blé différant par leur sensibilité au sel (abstract). Comp Rend Acad Sci ser III Sci de la Vie 324(9):863–868
Kirchoff BK, Fahn A (1984) Initiation and structure of the secondary vascular system in *Phytolacca dioica* (Phytolaccaceae). Can J Bot 62:2580–2586
Kozlowski TT (1997) Response of woody plants to flooding and salinity. Physiol Monograph 1:1–29
Li S, Showalter AM (1996) Immunolocalization of extension and potato tuber lectin in carrot, tomato, and potato. Physiol Plant 97:708–718
Mennega A (1980) Anatomy of the secondary xylem. In: Leeuwenberg AJM (ed) Angiospermae: Ordnung Gentianales fam. Logtaniaceae, Die natiirlichen Pflanzenfamilien, vol 28b(l), 2nd edn. Ducnker and Hum-blot, Berlin, pp 112–161
Metcalfe CR, Chalk L (1972) Anatomy of the dicotyledons, vol 2. Clarendon, Oxford
Mikesell JE, Popham RH (1976) Ontogeny and correlative relationship of the peimary thickening in four-o'clock plants (Nyctaginaceae) maintained under long and short photoperiods. Am J Bot 63:427–437
Morot L (1885) Recherches sur le péricycle ou couche périphérique dy cylindre central chez les Phanérogames. Ann Sci Nat 6-ème sér Bot 20:217–309
Pfeiffer H (1926) Das abnorme Dickenwachstum. In: Handbuch der Pflanzenanatomie, vol 9(2). Gebrüder Bontraeger, Berlin, pp 1–272
Prillieux M (1877) Anatomie comparée de la tigelle et du pivot de la Betterave pendant la germination. Bull Soc Bot France 24:239–244
Rajput KS, Rao KS (1999) Structural and developmental studies on cambial variant in *Pupalia lappacea (Amaranthaceae)*. Ann Bot Fenn 36:137–141
Rajput KS, Patil VS, Shah DG (2008) Formation of successive cambia and stem anatomy of Sesuvium sesuvioides (Aizoaceae). Bot J Linn Soc 158:548–555
Rao KS, Rajput KS (2003) Cambial variants in the roots of *Glinus lotoides* L. and *G. oppositifolius* (L.) A. DC. (*Molluginaceae*). Acta Bot Hungar 45(1–2):183–191
Regnault M (1860) Recherches sur les affinités de structure des tiges des partes du groupe des Cyclospermées. Ann Sci Nat Bot 4-ème sér 14:73–166
Sanio C (1863) Einige Bemerkungen über den Gerbstoff und seine Vebreitung bei den Holzpflanzen. Bot Zeit 3:17–23
Schenck H (1893) Beiträge zur Biologie und Anatomie der Lianen. Biol Mitheil Trop 5:1–271
Şerbănescu-Jitariu G, Toma C (1980) Morfologia şi anatomia plantelor. Ed. Did. şi Ped, Bucureşti
Showalter AM (1993) Structure and function of plant cell wall proteins. Plant Cell 5:9–23
Stevenson DW, Popham RA (1973) Ontogeny of the primary thickening meristem in seedlings of *Bougainvillea spectabilis*. Am J Bot 60:1–9

Terrazas T (1991) Origin and activity of successive cambia in *Cycas* (Cycadales). Am J Bot 78 (10):1335–1344

van Tieghem PH (1870–1871) Recherches sur la symètrie de structure des plantes vasculaires. Ann Sci Nat Bot 5-ème sér 13:5–314

Tiré C, De Rycke M, De Loose D, Inzé D, Van Montagu M, Engler G (1994) Extensin gene expression is induced by mechanical stimuli leading to local cell wall strenghtening in *Nicotiana plumbaginifolia*. Planta 195:175–181

Toma C, Niţă M, Zavaleche V (1991) Research of ecological, compared and ontogenetic anatomy upon some infraunits of *Salsola kali* L. An Şt Univ "Al. I. Cuza" Iaşi s. II a (Biol.) 37:5–21

Van Vliet GJCM (1979) Wood anatomy of the *Combretaceae*. Blumea 25:141–223

Volkens G (1893) *Chenopodiaceae*. In: Engler A, Prantl K (eds) Die Natürlichen Pflanzenfamilien, 3 (1a), Leipzig, W. Engelmann, pp 36–91

Waisel Y (1972) Biology of halophytes. Academic, New York

Wang L-W, Showalter AM, Ungar A (1997) Effect of salinity on growth, ion content, and cell wall chemistry in *Atriplex prostrata* (*Chenopodiaceae*). Am J Bot 84(9):1247–1255

Weiss E (1883) Das markständige Gefässbündelsystem eininger Dikotyledonen in seiner Beziehung zu den Blattspuren. Bot Centralbl 3 ser 15:280–295. 318–327, 358–367, 390–397, 401–415

Seung GW, Kim JS, Kim JH, Baek M, Yang D, Lee MC, Chung BY (2004) Effects of salinity on lignin and hydroxycinnamic acid contents in rice. Korean Journ Crop Sci 49(5):368–372

Wiessner J (1867) Einleitung in die technische Mikroskopie nebst mikroskopisch-technischen Untersuchungen. W. Braumüller, Wien

Willert DJV (1968) Tagesschwankungen des Ionengehaltes in *Salicornia europaea* in Abhängigkeit vom Sandort und von der Überflutung. Ber Deut Bot Ges 81:442–449

Ye ZH, Varner JE (1991) Tissue-specific expression of cell wall protein in developing soybean tissues. Plant Cell 3:23–37

Chapter 8
Bulliform Cells

As in the case of successive cambia, bulliform cells are not treated here as a distinct and well-defined adaptation of halophytes, but rather as a feature found in several halophyte species; since this feature and ecological spectra of halophytes where they occur are connected and the role of bulliform cells is still a problematic issue, we dedicate them a separate chapter.

The role of bulliform cells in halophytes' adaptations to environmental conditions has been largely discussed by Grigore and Toma (2011), when studying several Romanian salt-tolerant plants: *Juncus gerardii, Bolboschoenus maritimus, Carex distans, C. vulpina, Agrostis stolonifera, Alopecurus arundinaceus,* and *Puccinellia distans.*

Bulliform cells are structural features found especially on species included in *Poaceae* (Duval-Jouve 1875; Holm 1891, 1892, 1895; Mateu Andres 1991; Zhang and Clark 2000; Arriaga 2000; Peterson 2000; Tipping and Murray 2000; Khan 2002; Kirkham 2005; Gibson 2009), *Cyperaceae* (Duval-Jouve 1871; Beal 1886; Mazel 1891; Metcalfe 1971), and *Juncaceae* (Duval-Jouve 1871). Bulliform cells, sometimes named in various ways by different authors over time, are large, regularly with thin-walled cell. Despite that they were recognized for many years, their ecological significance in plant adaptation to salinity remains unclear.

It is well known that many leaves are capable of rolling up in dry, unfavorable conditions and reopening again under conditions when there is no water stress and have special, thin-walled water-containing cells that enable them to make these movements (Cutler et al. 2007). These are the bulliform or motor cells, which under conditions of water deficit lose turgor and thus constrict in upon themselves, causing lamina to fold or roll inward edge to edge (Dickison 2000).

Although bulliform cells also occur on non-halophytic monocotyledons species—suggesting probably an origin in a common ancestor—their presence on species exposed on both drought and salinity conditions requires a detailed and prudent analysis in accordance with environmental factors.

Mateu Andres (1991) found bulliform cells in several plants from coastal Mediterranean salt marshes: *Parapholis filiformis* (Fig. 8.1), *Puccinellia festuciformis* (Fig. 8.2), *Spartina versicolor* (Fig. 8.3), and *Aeluropus littoralis* (Fig. 8.4).

M.-N. Grigore, C. Toma, *Anatomical Adaptations of Halophytes*,
DOI 10.1007/978-3-319-66480-4_8

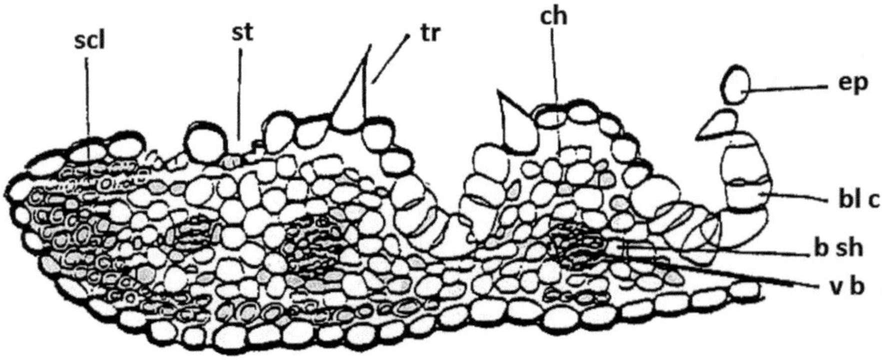

Fig. 8.1 Cross section through lamina of *Parapholis filiformis* (*ep* epidermis; *bl c* bulliform cells, *b sh* bundle sheath, *v b* vascular bundle, *st* stomata, *scl* sclerenchyma, *tr* trichome, *ch* chlorenchyma) (Mateu Andres 1991)

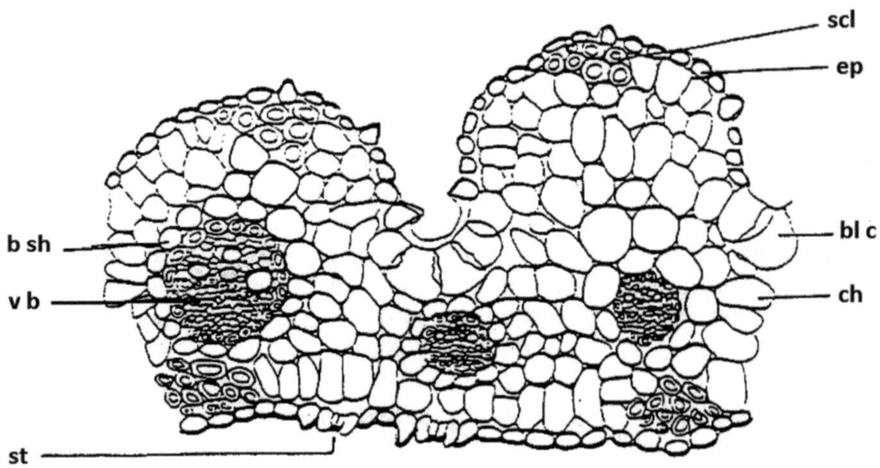

Fig. 8.2 Cross section through lamina of *Puccinellia festuciformis* (*ep* epidermis, *bl c* bulliform cells, *b sh* bundle sheath, *v b* vascular bundle, *st* stomata, *scl* sclerenchyma, *ch* chlorenchyma) (Mateu Andres 1991)

Toma et al. (1987) evidenced bulliform cells at the level of epidermis in the lamina of *Puccinellia distans* (Fig. 8.5), vegetating in areas differently affected by salinization.

Haberlandt (1914) included the *Poaceae* bulliform cells among motor, hygroscopic system of plants; the author stated that the leaves of many xerophilous grasses become folded, or curl up, when they are insufficiently supplied with water, in order to avoid excessive transpiration. Other botanists have also assigned a similar function of these cells, closely related to xerophytic environment value. Beal (1886) called also these cells "blister" cells; according to him, when dry, these cells contract and aid in closing the leaf in two or three ways. When moist the leaf

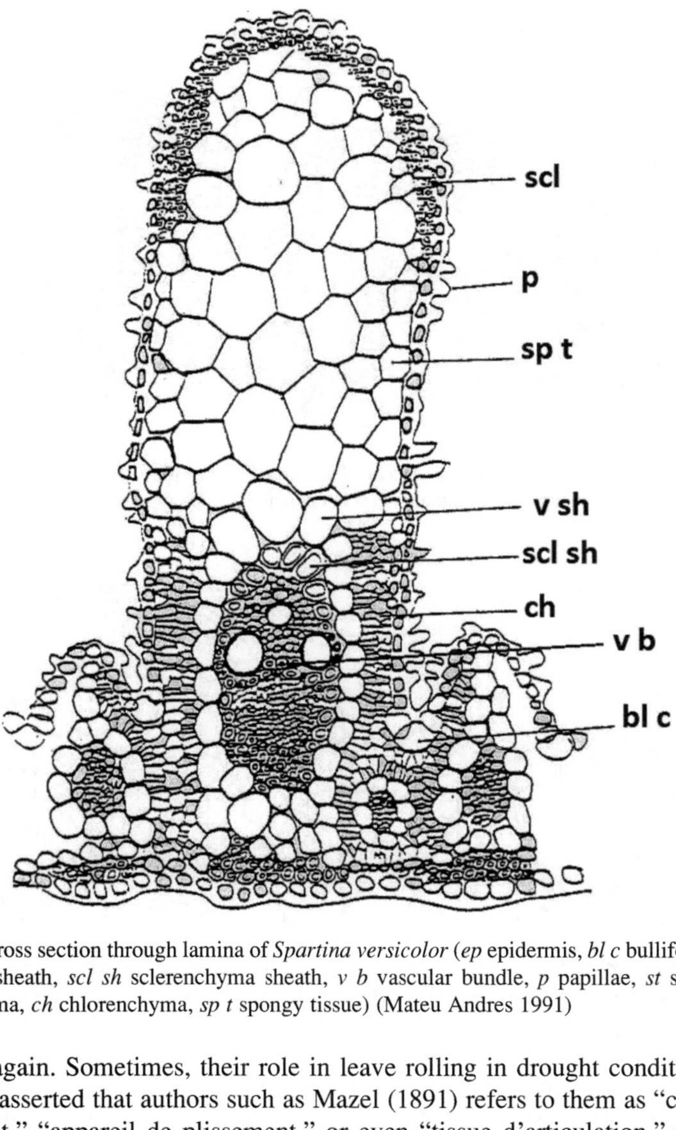

Fig. 8.3 Cross section through lamina of *Spartina versicolor* (*ep* epidermis, *bl c* bulliform cells, *b sh* bundle sheath, *scl sh* sclerenchyma sheath, *v b* vascular bundle, *p* papillae, *st* stomata, *scl* sclerenchyma, *ch* chlorenchyma, *sp t* spongy tissue) (Mateu Andres 1991)

expands again. Sometimes, their role in leave rolling in drought conditions is so intensely asserted that authors such as Mazel (1891) refers to them as "cellules de plissement," "appareil de plissement," or even "tissue d'articulation," related to their role among some *Carex* species.

Brongniart (1860) is among the first botanist who observed movements of leaves on a *Poaceae* species. Nothing is mentioned about bulliform cells, but it is interesting that the author was able to distinguish between this kind of movement and typical, motor movement of dicotyledon species.

Duval-Jouve (1875) has gone more deeply concerning the presence and the role of bulliform cells on *Poaceae*; moreover, he even proposed a classification system of grasses, according to the number and disposition of bulliform cells. This French botanist discerned exactly that the rolling movement of grass leaves in drought

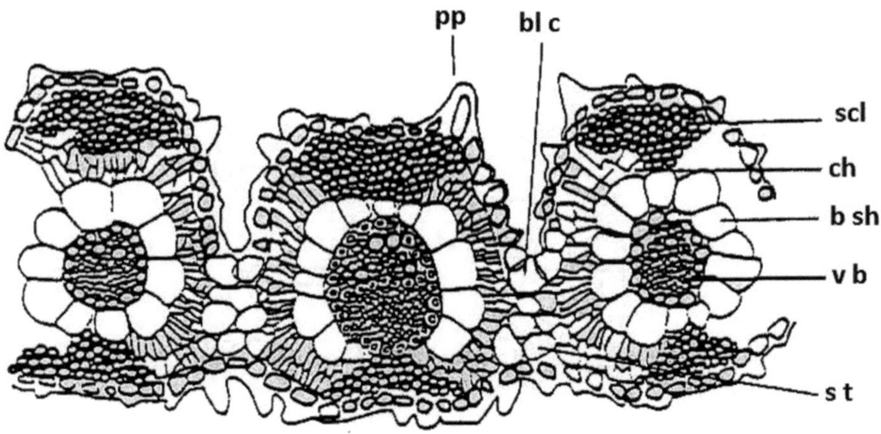

Fig. 8.4 Cross section through lamina of *Aeluropus littoralis* (*bl c* bulliform cells, *b sh* bundle sheath, *v b* vascular bundle, *pp* papillae, *st* stomata, *scl* sclerenchyma, *ch* chlorenchyma) (Mateu Andres 1991)

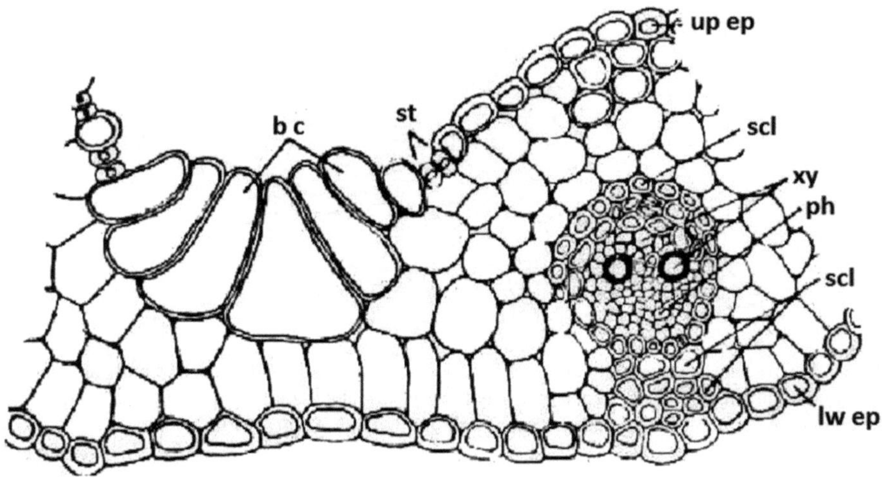

Fig. 8.5 Cross section through the lamina of *Puccinellia distans* (*b c* bulliform cells, *lw ep* lower epidermis, *up ep* upper epidermis, *ph* phloem, *scl* sclerenchyma, *xy* xylem, *st* stomata) (Toma et al. 1987)

conditions is different from that expressed on dicotyledons. The movement induced by bulliform cells is very slow and is involved in the diminution of leaves' transpiration surface.

Kearney (1900) identified bulliform cells on some grasses, observing that the margins of leaves become more or less involute, when the supply of water is small, becoming flat when moisture is plentiful.

Britton (1903) opined that the presence of bulliform cells, considered by him "water cells" on grasses and sedges represents one of the most interesting and striking examples of special adaptations to xerophytic conditions. These cells are found at the bottom of the grooves of the upper surface of the leaf. The stomata are situated along the slopes of the groove and when the bulliform cells give up their water the grooves close up, thus preventing to a large extent the further escape of water through the stomata.

Warming (1909) considered that leave rolling of *Poaceae* and *Cyperaceae* species represents a manner in which the transpiring surface is reduced. In these movements, a part is played by bulliform cells (called by Warming "hinge cells") lying in furrows on the upper face of the leaves. These cells are deeper than the other epidermal cells and their cellulose walls are easily folded as the leaf curls.

Fahn and Cutler (1992) stated that bulliform cells of grasses are a xeromorphic adaptation. Moreover, the bulliform cells were found to be more developed in desert ecotypes than in mesophytic ecotypes of some plants investigated (Waisel 1963), clearly suggesting that these cells are involved in plant adaptive response to water and salt stress.

For other authors, these cells were considered as water storage (Prat 1948; Eleftheriou and Noistakis 1978; Vecchia et al. 1998) and can participate in the young leaf expansion. Their implication in leaf rolling and/or folding of mature leaves was discussed by some researchers (Shields 1951; Jane and Chiang 1991). According to Esau (1965), during excessive water loss, the bulliform cells, together with or without colorless cells, became flaccid and enabled leaf either to fold or to roll. Clayton and Renvoize (quoted by Alvarez et al. 2008) opined that bulliform cells favored the light entrance in the mesophyll cells. In some species, bulliform cells were not actively or specifically related to unfolding and hygroscopic leaf movement, since they accumulated large amounts of silicon and their outermost walls might thicken and cutinize, becoming stiff (Ellis 1976).

It was shown that in water stress conditions, the activity of these cells becomes more intense. For instance, *Loudetiopsis chrysothrix* and *Tristachya leiostachya* showed leaf rolling of mature and young leaves during water stress (Alvarez et al. 2008). According to Moulia (1994), the leaf rolling is a xeromorphic characteristic and has adaptive value, reducing light interception and transpiration and protecting the leaf from dehydration and overheating. This would be a mechanism to minimize light exposition and water transpiration, thus keeping the stomata in a microclimate with higher humidity, preventing drought conditions (Clarke 1986; Silva et al. 2001).

Other species exposed to water stress show, among different adaptations, bulliform cells, such as *Carex ligerica* (Toma and Dumitru 1973), *Zea mays* (Ristic and Cass 1991), common bean (Silva et al. 1999), and tomato (Sam et al. 2000).

Nawazish et al. (2006) showed that on a species collected from xeric and saline habitat, *Cenchrus ciliaris*, the bulliform cells were well developed in severe drought; it was assumed that these cells are very crucial under moisture limited environments as these are responsible for leaf curling and ultimately checking water loss through leaf surface (Albernethy et al. 1998; Alvarez et al. 2003).

But Ellis (1976) suggested caution in assigning bulliform cells a role in leaf movement; Shields (1951) described that the subepidermal sclerenchyma and other elements of mesophyll rather than bulliform cells contributed to involution in some xeric grasses.

Our ecological short notes in the field sustain the abovementioned observations (Grigore 2012; Grigore and Toma 2014; Grigore et al. 2014). The halophytic species investigated by us are mainly hygrophilous, some of them being salt marsh species. The temporary characters of soil moisture and atmosphere humidity induce anyway the necessity of some xeromorphic adaptations, as a response to both water stress and salt stress. It was already stated that salt stress has a high dehydration component.

As far as we are concerned, another discussion looks like a challenge to us. Most of the existing interpretations about these bulliform cells converge toward their xeric feature, toward the connection between the water stress and the function of bulliform cells. Nevertheless, our interpretations, the ecological characterization made by other authors in relation to the species that we have investigated, and especially the observations that we have noted in the field all converge toward another direction. All the investigated halophyte species (to which we also add *Puccinellia distans*, *Juncus gerardii*, and *Bolboschoenus maritimus*, where bulliform cells are present) are plants of wet habitats, therefore hygro-halophilous. Therefore, does this look like a contradiction? We believe it does not. Actually, interpretations must be made in an integrative manner, taking into account the multitude of the environment factors, their intensity, and, especially, the permanence or intermittence characteristic of their action on plants.

Some foreign authors (Sculthorpe 1967; Font-Quer 1970, quoted by Arriaga and Jacobs 2006) described as "amphibious" those species that may live in dry soils flooded during a period of the year. Amphibious species provide, in a structural and an ecological sense, a gradual transition between truly terrestrial and truly aquatic species. The plants we have investigated, and which we refer to, could be included in the same category: they are hygro-halophyte species, but the wet feature of their habitat is relative and it is not constant. We have noticed these species growing also in dry conditions, through water evaporation, because of prolonged drought periods. This is the only way we can explain the presence of these xeric characteristics in plants construed as hygrophilous. In fact, if one takes this definition into account, they are amphibious halophytes (Grigore and Toma 2010). Some of the enlisted species also have air-storing lacunae, which is a typical adaptation for hygrophytes.

We believe that this is an aspect omitted by many authors, but which is essential in understanding the adaptations of certain plants under the influence of several factors, not of only one factor, which never has constant, static values.

Not accidentally, Grigore and Toma (2010) delineated these amphibious halophytes within a complex of halophytes' classification, based on the anatomical features found in halophytes and their ecological requirements (Fig. 8.6).

In another study (Grigore and Toma 2011), a correlation between the presence of bulliform cells in several investigated halophytes (amphibious) and their ecological

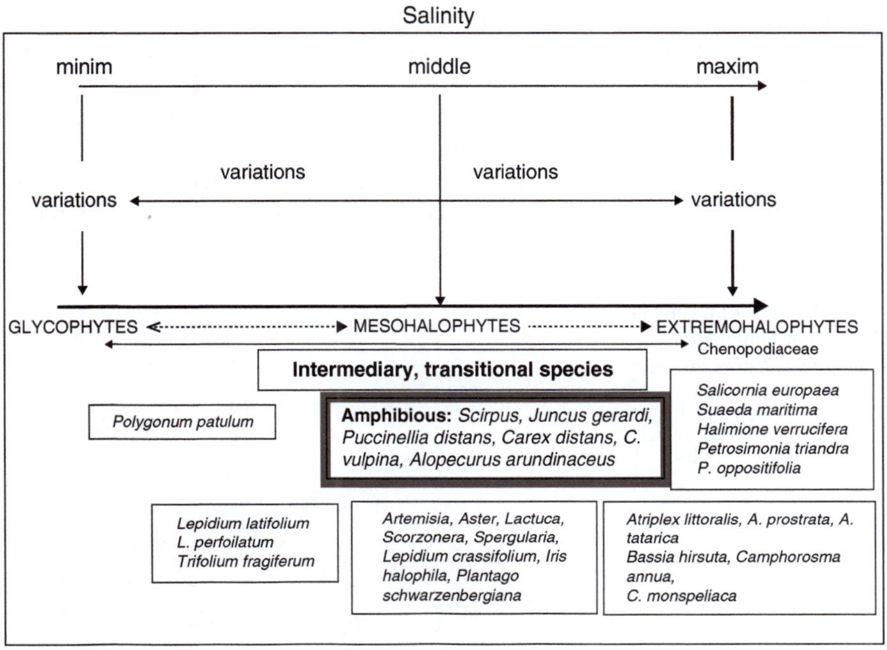

Fig. 8.6 Halophytes classification and the place of amphibious halophytes (Grigore and Toma 2010)

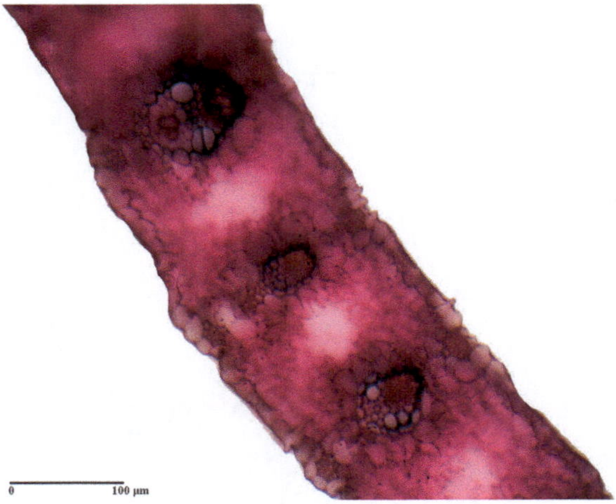

Fig. 8.7 Bulliform cells in Romanian "amphibious" halophytes: *Bolboschoenus maritimus*

spectra has been proposed. Thus, *Bolboschoenus maritimus* (Fig. 8.7) is a neohalophyte species (Bucur et al. 1961), preferential one (Ţopa 1954), being included in the second category by Prodan (1939); *Carex distans* (Fig. 8.8) is a preferential halophyte (Ţopa 1954), meso-hygrophyte, facultative halophyte

(Ciocârlan 2000); *Carex vulpina* (Figs. 8.9 and 8.10) was characterized as a neohalophyte (Bucur et al. 1961); *Juncus gerardii* (Figs. 8.11 and 8.12) was described as a euhalophyte (Bucur et al. 1960), preferential halophyte (Ţopa 1954), meso-hygro-halophyte (Ciocârlan 2000), included in the first category by Prodan (1939). *Agrostis stolonifera* (Fig. 8.13) is a neohalophyte (Bucur et al. 1961), supporting halophyte (Ţopa 1954) and introduced in the first category by Prodan (1939); *Alopecurus arundinaceus* (Figs. 8.14 and 8.15) is a preferential

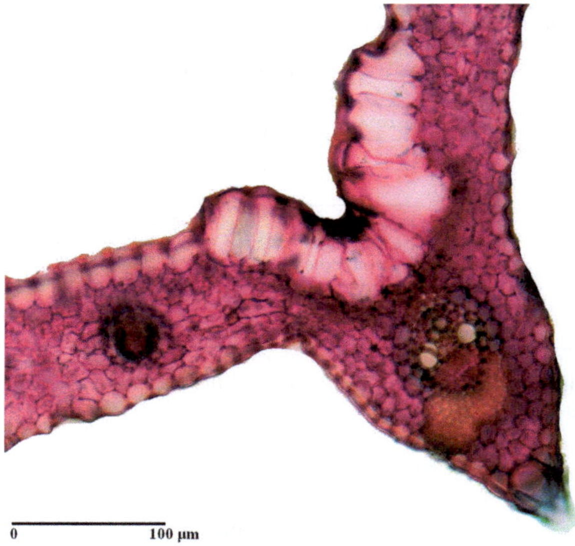

Fig. 8.8 Bulliform cells in Romanian "amphibious" halophytes: *Carex distans*

Fig. 8.9 Bulliform cells in Romanian "amphibious" halophytes: *Carex vulpina*

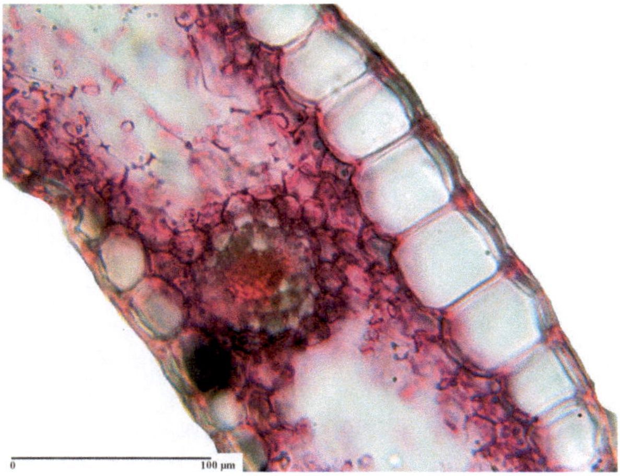

Fig. 8.10 Bulliform cells in Romanian "amphibious" halophytes: *Carex vulpina*

Fig. 8.11 Bulliform cells in Romanian "amphibious" halophytes: *Juncus gerardii*

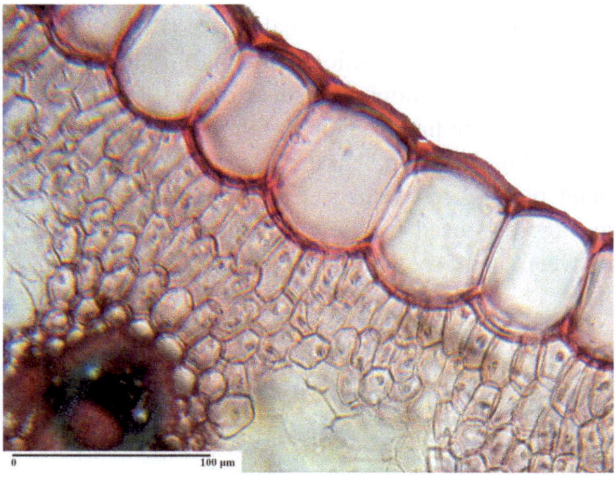

Fig. 8.12 Bulliform cells in Romanian "amphibious" halophytes: *Juncus gerardii*

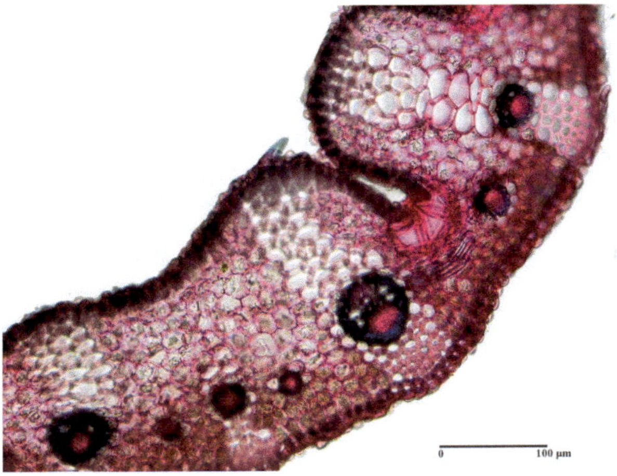

Fig. 8.13 Bulliform cells in Romanian "amphibious" halophytes: *Agrostis stolonifera*

Fig. 8.14 Bulliform cells in Romanian "amphibious" halophytes: *Alopecurus arundinaceus*

halophyte (Ţopa 1954), neohalophyte (Bucur et al. 1961), and meso-hygrophyte, facultative halophyte (Ciocârlan 2000). *Puccinellia distans* (Figs. 8.16 and 8.17) is considered a euhalophyte (Bucur et al. 1960), preferential halophyte (Ţopa 1954), included in the first category by Prodan (1939). It is obvious, considering the above exposed information, that these taxa have been differently characterized by various plant biology researchers. For further explanations and comments regarding the terminology used by these botanists, different classifications, and equivalencies between them, see Grigore's works (2008, 2012).

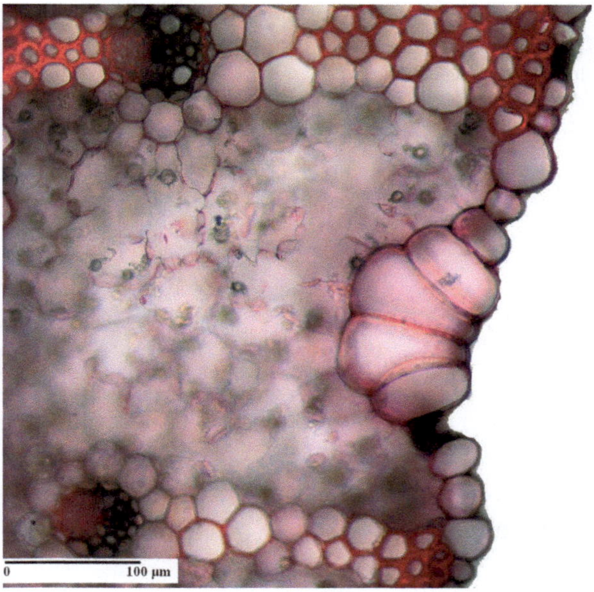

Fig. 8.15 Bulliform cells in Romanian "amphibious" halophytes: *Alopecurus arundinaceus*

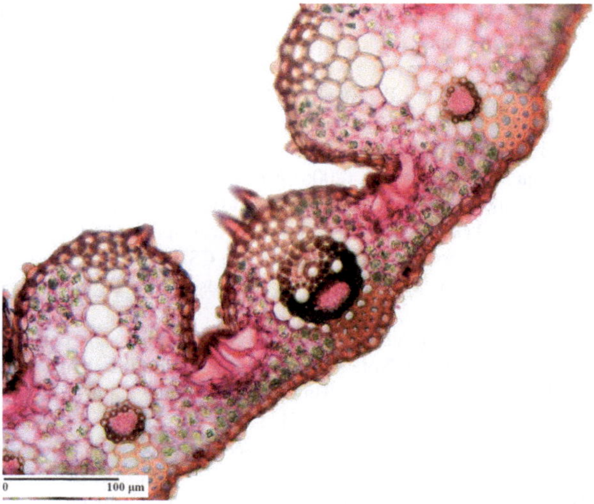

Fig. 8.16 Bulliform cells in Romanian "amphibious" halophytes: *Puccinellia distans*

Obviously, the problems raised in these paragraphs are much more complicated. Which may be even more fascinating are the evolutionary correlations, since it is well known that there are theories suggesting that different angiosperm groups have "retained" different levels of plasticity, which are manifested under the form of characteristic abilities of adaptation to some aquatic environments, for instance.

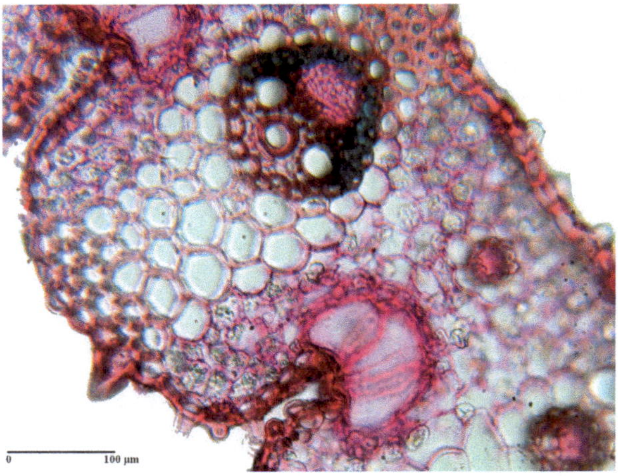

Fig. 8.17 Bulliform cells in Romanian "amphibious" halophytes: *Puccinellia distans*

References

Albernethy GA, Fountain DW, Mcmanus MT (1998) Observations on the leaf anatomy of *Festuca novae-zelandiae* and biochemical responses to a water deficit. New Zeal J Bot 36(1):113–123

Alvarez JM, Rocha JF, Machado SR (2003) Ultrastructural aspects of bulliform cells in two Cerrado Grass species. In: Proc of the XIX Congr Brazil soc microscopy and microanalysis (abstract).

Alvarez JM, Rocha JF, Machado SR (2008) Bulliform cells in *Loudetiopsis chrysothrix* (Ness) Conert and *Tristachya leiostachya* Nees (Poaceae): structure in relation to function. Braz Arch Biol Technol 51(1):113–119

Arriaga MO, Jacobs W L (2006) An anatomo-ecological experiment in *Austrostipa aristiglumis*, a lowland Stipoid species. Telopea, 11(2):161–170

Arriaga MO (2000) Austral South American species of *Eriochloa*. In: Jacobs SWL, Everett J (eds) Grassees. systematics and evolution. CSIRO Publishing, Collingwood, pp 141–148

Beal WB (1886) The bulliform or hygroscopic cells of grasses and sedges compared. Bot Gaz 2:321–326

Britton WE (1903) Vegetation of the North Haven sand plains. Bull Torr Bot Club 30:571–620

Brongniart A (1860) Note sur le sommeil des feuilles dans une plante de la famille de Graminées, le *Strephium guianense*. Bull Soc Bot France **7**:470–472

Bucur N, Dobrescu C, Turcu G, Lixandru G, Teşu C (1960) Contribuţii la studiul halofiliei plantelor din păşuni şi fâneţe de sărătură din Depresiunea Jijia-Bahlui (partea a II-a). Stud şi Cerc (Biol şi Şt Agricole) Acad R.P.Române, filiala Iaşi 11(2):333–347

Bucur N, Dobrescu C, Turcu G, Lixandru G, Teşu C (1961) Contribuţii la studiul halofiliei plantelor din păşuni şi fâneţe de sărătură din Depresiunea Jijia-Bahlui (partea a III-a). Stud şi Cerc (Biol şi Şt Agricole) Acad R.P.R., filiala Iaşi 12(1):169–190

Ciocârlan V (2000) Flora ilustrată a României. Edit. Ceres, Bucureşti

Clarke JM (1986) Effect of leaf rolling on leaf water loss in *Triticum* spp. Can J Plant Sci 66:885–891

Cutler DF, Botha T, Stevenson DW (2007) Plant anatomy: an applied approach. Blackwell, Australia

Dickison WC (2000) Integrative plant anatomy. Harcourt Academic, San Diego

Duval-Jouve J (1871) Sur quelques tissus de *Joncées*, de *Cyperacées* et de *Graminées*. Bull Soc Bot France 18:231–239

Duval-Jouve J (1875) Histotaxie des feuilles de Graminées. Ann Sci Nat 6 ser Bot 1:294–371

Eleftheriou EP, Noistakis B (1978) A comparative study on the leaf anatomy of the grasses *Andropogon ischaemum* and *Chrysopogon gryllus*. Phyton 19:27–36

Ellis RP (1976) A procedure for standardizing comparative leaf anatomy in the Poaceae. I. The leaf-blade as viewed in transverse section. Bothalia 12:65–109

Esau K (1965) Plant anatomy, 2nd edn. Wiley, New York

Fahn A, Cutler DF (1992) Xerophytes. Handbuch der Pflanzenanatomie (band XIII, teil 3). Gebrüder Borntraeger, Berlin, Stuttgart

Font-Quer P (1970) Diccionario de Botánica. Labor: Barcelona

Gibson DJ (2009) Grasses and grassland ecology. Oxford University Press, Oxford

Grigore M-N (2008) Halofitotaxonomia. Lista plantelor de sărătură din România. Pim, Iaşi

Grigore M-N (2012) Romanian salt tolerant plants. Taxonomy and ecology. Tehnopress, Iasi

Grigore M-N, Toma C (2010) A proposal for a new halophytes classification, based on integrative anatomy observations, 2010. Muz Olteniei Craiova Stud şi Com Şt Nat 26(1):45–50

Grigore M-N, Toma C (2011) Ecological implications of bulliform cells on halophytes, in salt and water stress natural conditions. Studia Universitatis 'Vasile Goldiş', Ser. Şt. Vieţii, 21(4):785-792

Grigore M-N, Toma C (2014) Integrative ecological notes on halophytes from "Valea Ilenei" (Iaşi) nature reserve. Memoirs Scientific Sect Romanian Acad 37:19–36

Grigore M-N, Ivănescu L, Toma C (2014) Halophytes. An integrative anatomical study. Springer, Cham, Heidelberg

Haberlandt G (1914) Physiological plant anatomy. Macmillan, London

Holm T (1891) A study of some anatomical characters of North American Gramineae. Bot Gaz 16:166–171. 219–225, 275–281

Holm T (1892) A study of some anatomical characters of North American Gramineae. Bot Gaz 17:358–362

Holm T (1895) A study of some anatomical characters of North American Gramineae. Bot Gaz 20:362–365

Jane WN, Chiang SHT (1991) Morphology and development of bulliform cells in *Arundo formosana* Hack. Taiwania 36:85–97

Kearney TH (1900) The plant covering of Ocracoke Island: a study in the ecology of the North Carolina strand vegetation. Contr U.S. Nat Herb 5:261–319

Khan A (2002) Plant anatomy and physiology. Kalpaz, Delhi

Kirkham MB (2005) Principles of soil and plant water relations. Elsevier Academic, London

Mateu AI (1991) Leaf anatomy of plants from coastal Mediterranean salt-marshes. Monocotyledons. Candollea 46(2):345–358

Mazel A (1891) Études d'anatomie comparée sur les organes de végétation dans le genre *Carex*. Thèse. Gèneve, Imprimerie Soullier, Rue de la Cité, 19, Université de Gèneve

Metcalfe CR (1971) Anatomy of the monocotyledons. V. Cyperaceae. Clarendon, Oxford

Moulia B (1994) Biomechanics of leaf rolling. Biomimetics 2:267–281

Nawazish S, Hameed M, Naurin S (2006) Leaf anatomical adaptations of *Cenchrus ciliaris* L. from the Salt Range, Pakistan against drought stress. Pak J Bot 38(5):1723–1730

Peterson PM (2000) Systematics of the *Muhlenbergiinae* (*Poaceae: Eragrostidae*). In: Jacobs SWL, Everett J (eds) Grasses. Systematics and evolution. CSIRO Publishing, Collingwood, pp 195–212

Prat K (1948) General features of the epidermis in *Zea mays*. Ann Missouri Bot Garden 35:341–351

Prodan I (1939) Flora pentru detrminarea şi descrierea plantelor ce cresc în România. II. (ediţia a II-a). Cartea Românească, Cluj-Napoca

Ristic Z, Cass DD (1991) Leaf anatomy of *Zea mays* L. in response to water shortage and high temperature: a comparison of drought-resistant and drought-sensitive lines. Bot Gaz 152 (2):173–185

Sam O, Jerez E, Dell'Amico J, Ruiz-Sanchez MC (2000) Water stress induced changes in anatomy of tomato leaf epidermis. Biol Plant 43(2):275–277

Sculthorpe CD (1967) The Biology of Aquatic Vascular Plants. Edward Arnold: London

Shields LM (1951) The involution mechanism in leaves of certain xeric grasses. Phytomorphology 1:225–241

Silva H, Martinez JP, Baginsky C, Pinto M (1999) Effect of water deficit on the leaf anatomy of six cultivars of the common bean, *Phaseolus vulgaris*. Rev Chil Hist Natural 72(2):219–235

Silva S, Soares AM, Oliveira LEM, Magalhaes PC (2001) Respostas fisiologicas de gramineas promissoras para revegetacao ciliar de reservatorios hidreletricos, submetidas a deficiencia hidrica. Ciencia Agrotecnica 25:124–133

Tipping C, Murray DR (2000) Effects of elevated atmospheric [CO_2] in *Panicum* species of different photosynthetic modes (*Poaceae: Panicoideae*). In: Jacobs SWL, Everett J (eds) Grasses. Systematics and evolution. CSIRO Publishing, Collingwood, pp 259–266

Toma C, Dumitru E (1973) Contribuţii la studiul histo-anatomic al organelor vegetative de la *Carex ligerica* J. Gay. Stud şi Com Muz Şt Nat Suceava 3:5–18

Toma C, Moţiu T, Niţă M (1987) Structura organelor vegetative de *Puccinellia distans* (L.) Park. în funcţie de gradul de sărăturare a solului. Culegere de Stud şi artic de Biol, Univ. "Al.I.Cuza" Iaşi (Grăd Bot) (Lucrările seminarului ştiinţific "Valorificarea resurselor vegetale ale României") 1:117–126

Ţopa E (1954) Vegetaţia terenurilor sărate din R.P.R. Natura 6(1):57–76

Vecchia FD, Asmar TF, Calamassi R, Rascio N, Vazzana C (1998) Morphological and ultrastructural aspects of dehydration and rehydration in leaves of *Sporobolus stapfianus*. Plant Growth Reg 24:219–228

Zhang W, Clark LG (2000) Phylogeny and classification of the *Bambusoideae* (*Poaceae*). In: Jacobs SWL, Everett J (eds) Grasses. Systematics and evolution. CSIRO Publishing, Collingwood, pp 35–42

Waisel Y (1963) Ecotypic differentiation in the flora of Israel. III. Anatomical studies of some ecotype pairs. Bull Res Counc Israel, Sec D11:183–190

Warming E (1909) *Oecology of plants: an introduction to the study of plant-communities*. Clarendon, Oxford

MIX
Papier aus verantwortungsvollen Quellen
Paper from responsible sources
FSC® C105338

If you have any concerns about our products,
you can contact us on
ProductSafety@springernature.com

In case Publisher is established outside the EU,
the EU authorized representative is:
Springer Nature Customer Service Center GmbH
Europaplatz 3, 69115 Heidelberg, Germany

Printed by Libri Plureos GmbH
in Hamburg, Germany